Lecture Notes in Physics

Edited by H. Araki, Kyoto, J. Ehlers, München, K. Hepp, Zürich
R. L. Jaffe, Cambridge, MA, R. Kippenhahn, München, D. Ruelle, Bures-sur-Yvette
H. A. Weidenmüller, Heidelberg, J. Wess, Karlsruhe and J. Zittartz, Köln
Managing Editor: W. Beiglböck

379

J.D. Hennig   W. Lücke   J. Tolar   (Eds.)

Differential Geometry, Group Representations, and Quantization

Springer-Verlag Berlin Heidelberg GmbH

Editors

Jörg-Dieter Hennig
Wolfgang Lücke
Arnold Sommerfeld Institute for Mathematical Physics
Technical University Clausthal
Leibnizstraße 10, D-3392 Clausthal, FRG

Jiří Tolar
Faculty of Nuclear Science and Physical Engineering
Czech Technical University
Břehová 7, CS-115 19 Prague, Czechoslovakia

This book was processed by the authors using the T_EX macro package

ISBN 978-3-662-13870-0   ISBN 978-3-540-46473-0 (eBook)
DOI 10.1007/978-3-540-46473-0

Preface

This book is dedicated to Prof. Dr. Heinz-Dietrich Doebner on the occasion of his 60th birthday. Professor Doebner was born in Berlin on 11 May, 1931. He studied physics at the Free University of Berlin, and there he received his doctorate in 1962 under Prof. G. Ludwig. He completed his habilitation in 1965 at the University of Marburg, and in 1969 he became Professor of Theoretical Physics and the Director of the Institute for Theoretical Physics at the Technical University of Clausthal.

His many additional accomplishments include his chairmanship of the committee to establish the University of Osnabrück between 1971 and 1974 (he served as its first rector in 1974) and his chairmanship since 1977 of the University Advisory Board for the government of Lower Saxony. He became Dean of the Physics Department of the Technical University of Clausthal in 1977, and since 1981 he has also been chairman of the Faculty of Mathematics and Natural Sciences.

The scientific research of Professor Doebner has contributed to several fields of mathematical physics, especially to differential geometric techniques, quantization methods, and representations of groups and algebras. His current interests include diffeomorphism groups, quantum groups, and the foundations of quantum mechanics.

Professor Doebner has actively promoted scientific exchanges with scientists from various countries – mediated by the International Centre for Theoretical Physics, Trieste, or by the Humboldt Foundation and the German Academic Exchange Agency (DAAD). He has played a major role in sustaining and organizing international conference series, notably the series on Differential Geometric Methods in Mathematical Physics starting in 1971. His annual Summer Workshops at the Arnold Sommerfeld Institute in Clausthal have brought together leading researchers in special areas of mathematical physics from throughout the world.

In 1989 several colleagues promoted the idea of dedicating a special volume to Professor Doebner. This idea attracted very strong support. It has been our privilege to collect a set of contributions from his friends that we consider to be fairly representative of his special research interests. Besides essays of a general nature which review important developments in the application of group theoretic and differential geometric methods to classical and quantum physics, we have included several additional short contributions presenting new results.

We hope that this book provides additional stimulation for Professor Doebner's scientific activities, and we wish him all the best for the years to come.

Clausthal   J.D. Hennig
January 1991   W. Lücke
                                                                         J. Tolar

Acknowledgements

We are indebted to numerous colleagues from various countries for encouragement, and we regret that we could not invite them all to contribute to this volume. We are particularly grateful to Prof. A. Bohm for his efficient communication with the publisher, and to the authors, all of whom agreed with pleasure to contribute, and who prepared their contributions with great care. Last but not least we thank Springer-Verlag, especially Professor Beiglböck, for substantial support and for providing the special package of TEX macros used for this book.

Contents

VIII

I

Differential Geometric Techniques in Physics

Global Differential Geometric Methods in Elasticity and Hydrodynamics

E. Binz

Dept. of Mathematics and Computer Science
University of Mannheim, A5
D-6800 Mannheim 1, F.R.G.

Abstract: A formalism in the framework of global analysis and largely based on translational symmetry is used to relate in a one-to-one correspondence force densities of elasticity and constitutive laws. The customary setting of elasticity developed e.g. in [17] is included in our approach. Moreover, a structural viscosity coefficient can be naturally introduced and a dynamical setting based on d'Alembert's principle yields a Navier-Stokes type of equation.

1 Introduction

The purpose of this review article is to show a formalism in the framework of global analysis, which relates in a one-to-one correspondence two notions in the theory of elasticity and hydrodynamics. This correspondence is given by a Neumann problem converting force densities into constitutive maps which in turn characterize constitutive laws. This method is largely based on translational symmetry. As we will see below the customary setting of elasticity developed e.g. in [17] is included in our approach. Moreover, this formalism offers a natural way to introduce a structural viscosity coefficient.

To describe in short what we mean by a constitutive law, we begin by looking at a moving deformable bounded body in $\mathbb{R}^n$. The material should constitute a deformable medium. The medium forming the boundary may differ from the one forming the inside of the body. We make the geometric assumption that at any time the body is diffeomorphic to a compact, connected, oriented and smooth manifold of fixed dimension with (oriented) boundary. The boundary needs not to be connected. These assumptions allow us to think of a standard body M. Consequently a configuration is a smooth embedding from M into $\mathbb{R}^n$. The configuration space is hence the collection $E(M, \mathbb{R}^n)$ of all smooth embeddings of M into $\mathbb{R}^n$. This set equipped with Whitney's C^∞-topology is a Fréchet manifold (cf. [8]). A smooth motion of the body in $\mathbb{R}^n$, therefore, is described by a smooth curve in $E(M, \mathbb{R}^n)$. (The calculus on Fréchet manifolds adopted in the sequel is the one presented in [8], which in our setting coincides with the one developed in [9].)

The physical qualities of the deforming medium enter certainly the work $F(J)(L)$ needed to deform (infinitesimally) the material at any configuration $J \in E(M, \mathbb{R}^n)$ in any direction L. The directions are tangent vectors to $E(M, \mathbb{R}^n)$ and thus are nothing else but maps in $C^\infty(M, \mathbb{R}^n)$ and vice versa.

In the following we take F, which is assumed to be a smooth one-form on $E(M, \mathbb{R}^n)$, as the basic entity of our notion of a constitutive law. In specifying the notion of a constitutive law somewhat further, we require that the constitutive properties should not be affected by the particular location of the body in $\mathbb{R}^n$, hence F has to be invariant under the operation of the translation group $\mathbb{R}^n$ of the real vector space $\mathbb{R}^n$. In addition to this we require that $F(J)(L) = 0$, for any constant map L and any $J \in E(M, \mathbb{R}^n)$, a condition which will be interpreted a few lines below.

The forms F, which have these two properties, can be regarded as one-forms on $\{dJ | J \in E(M, \mathbb{R}^n)\}$, where dJ is the differential of J. This reduced configuration space is a Fréchet manifold as well and is denoted by $E(M, \mathbb{R}^n)/\mathbb{R}^n$. It can naturally be identified via the differential operator d with the space of all those configurations, which have a fixed center of mass. A smooth one-form on $E(M, \mathbb{R}^n)/\mathbb{R}^n$ will be denoted by $F_{\mathbb{R}^n}$. It describes the work if the center of mass is fixed. We thus deal with one-forms of the type $F = d^* F_{\mathbb{R}^n}$. The space $E(M, \mathbb{R}^n)/\mathbb{R}^n$ admits a natural metric $\mathcal{G}$ of an L_2-type, which is $SO(n)$-invariant and closely related to the classical Dirichlet integral.

A one-form F being of this type is called a constitutive law, provided $F_{\mathbb{R}^n}$ admits an integral representation. It turns out that any constitutive law F is determined by some smooth map $\mathcal{H} \in C^\infty(E(M, \mathbb{R}^n)/\mathbb{R}^n, C^\infty(M, \mathbb{R}^n))$, called a constitutive map.

Hence in our setting we characterize the medium as far as the internal physical properties are encodable in the function $\mathcal{H}$. This constitutive function $\mathcal{H}$ determines at any $dJ \in E(M, \mathbb{R}^n)/\mathbb{R}^n$ two smooth force densities $\Phi(dJ)$ and $\varphi(dJ)$ linked to $\mathcal{H}(dJ)$ by the following system of equations:

$$\Delta(J)\mathcal{H}(dJ) = \Phi(dJ) \quad \text{and} \quad d\mathcal{H}(dJ)(\mathbf{n}) = \varphi(dJ) \,.$$

Here $\Delta(J)$ is the Laplacian determined by the pull back under J of a fixed scalar product on $\mathbb{R}^n$ and $\mathbf{n}$ is the positively oriented unit normal of ∂M in M. The integrability condition necessary to solve this Neumann problem of which the force densities are given and the function $\mathcal{H}$ is the unknown, is equivalent with the requirement that

$$F(J)(z) = 0 \quad \forall J \in E(M, \mathbb{R}^n) \text{ and } \forall z \in \mathbb{R}^n \,.$$

This condition can be interpreted by saying that the resulting forces acting upon the fixed center of mass vanish. The constitutive map $\mathcal{H}$ determines a stress tensor

T given by

$$T(J)(X,Y) :=< \mathrm{d}\mathcal{H}(J)X, \mathrm{d}JY > \qquad \forall J \in E(M,N).$$

X, Y vary among all smooth vector fields on M. Vice versa any stress tensor yields a constitutive map via the force densities mentioned above, if $\dim M = n$.

Since F is also affected by the material forming the boundary, we treat in an analogous way the boundary material and exhibit in analogy to $\mathcal{H}$ a characteristic constitutive map $\mathfrak{h}$ (which in turn determines its own force density along ∂M) which differs from $\mathrm{d}\mathcal{H}(\mathrm{d}J)(\mathbf{n})$. This observation allows us to decode the influence of the whole body on the physical quality of the boundary material. What we have described so far is presented in the first six sections.

In Sect. 7 we show that $\mathcal{H}$ and $\mathfrak{h}$ are structured in the sense that in both of them the work needed to deform volume, area and shape of the body and the boundary, respectively, is naturally encoded. This observation is illustrated on two bubble models in Sect. 8. It will be apparent that if the qualities of the material depend on the shape, then the bubble (if it exists at all !) has to be an immersed torus. In Sect. 9 we touch the influence of the action of the rotation group on the configuration space.

The remaining sections deal with a dynamics for boundary-less bodies which yield a Navier-Stokes type of equations. This dynamics is based on d'Alembert's principle, where the constitutive law is given by the virtual work and thus determines the deviation from the free motions. The key to this application of the formalism developed earlier is to generalize the notion of a constitutive law in such a way that it depends not only on configurations but also on velocities.

In the first three appendices some technical tools are developed. The last appendix deals with a natural metric on the space of embeddings. This metric is based on a mass density. Its geodesics describe the collections of free motions of the pointlike material particles of the body in the ambient space.

2 Configuration and Phase Space

Let us think of a deformable material body moving and deforming in the Euclidean space $\mathbb{R}^n$. We make the geometric assumption that at any time the body maintains the shape of a m-dimensional, compact, connected, oriented and smooth manifold with (oriented) boundary. We assume $m \leq n$. The boundary shall not necessarily be connected. The physical qualities of the medium forming the boundary may differ from the ones forming the inside of the body.

Hence a *configuration* is a smooth embedding $J : M \longrightarrow \mathbb{R}^n$ and the *space of configurations* is $E(M, \mathbb{R}^n)$, the collection of all smooth embeddings of M into $\mathbb{R}^n$, endowed with the C^∞-topology. It is thus a Fréchet manifold (cf. [8] or [14]).

Clearly each $J \in E(M, \mathbb{R}^n)$ induces a smooth embedding $J|\partial M : \partial M \longrightarrow \mathbb{R}^n$ of the boundary i.e. a configuration of the boundary ∂M of the body. Let us denote the collection of all smooth embeddings of ∂M into $\mathbb{R}^n$ by $E(\partial M, \mathbb{R}^n)$. The latter space endowed with the C^∞-topology is a Fréchet manifold, too. In fact it is a principal bundle with the diffeomorphism group of M as structure group (cf. [6]).

The *phase space* of the body is

$$TE(M, \mathbb{R}^n) = E(M, \mathbb{R}^n) \times C^\infty(M, \mathbb{R}^n).$$

Here $C^\infty(M, \mathbb{R}^n)$ is the Fréchet space consisting of the collection of all smooth maps from M to $\mathbb{R}^n$, endowed with the C^∞-topology. It contains $E(M, \mathbb{R}^n)$ as an open subset. Proceeding for ∂M as for M we obtain $E(\partial M, \mathbb{R}^n)$ as an open subset of the Fréchet space $C^\infty(\partial M, \mathbb{R}^n)$ (cf. [14]). Its tangent bundle is obviously trivial, too, i.e. the *phase space* of the boundary is $E(\partial M, \mathbb{R}^n) \times C^\infty(\partial M, \mathbb{R}^n)$.

In the sequel of these notes we write O_∂ instead of $\{J|\partial M | J \in E(M, \mathbb{R}^n)\}$. The map assigning to any $J \in E(M, \mathbb{R}^n)$ its restriction $J|\partial M$ is called R.

On the configuration space of the body we have two natural actions, namely

$$t : E(M, \mathbb{R}^n) \times \mathbb{R}^n \longrightarrow E(M, \mathbb{R}^n)$$

and

$$s : SO(n) \times E(M, \mathbb{R}^n) \longrightarrow E(M, \mathbb{R}^n)$$

assigning to each $J \in E(M, \mathbb{R}^n)$ and each $z \in \mathbb{R}^n$ the embedding $J + z$ and $g \circ J$ for each $g \in SO(n)$, respectively. These actions reflect the *translational* and the *rotational* symmetry on $E(M, \mathbb{R}^n)$ respectively. t and s extend obviously to $C^\infty(M, \mathbb{R}^n)$. The groups $\mathbb{R}^n$ and $SO(n)$ both act accordingly on $E(\partial M, \mathbb{R}^n)$. These actions restrict to O_∂ and obviously both also extend to $C^\infty(\partial M, \mathbb{R}^n)$.

The orbit spaces of the respective actions of the translation group $\mathbb{R}^n$ are denoted by $C^\infty(M, \mathbb{R}^n)/\mathbb{R}^n$, $C^\infty(\partial M, \mathbb{R}^n)/\mathbb{R}^n$, $E(M, \mathbb{R}^n)/\mathbb{R}^n$, $E(\partial M, \mathbb{R}^n)/\mathbb{R}^n$ and $O_\partial/\mathbb{R}^n$.

The nature of these spaces is easily understood if we introduce for any map $L \in C^\infty(M, \mathbb{R}^n)$ the differential dL which is locally given by the Fréchet derivative. The tangent map TL of L is, therefore, (L, dL). The respective notion of $l \in C^\infty(\partial M, \mathbb{R}^n)$ is introduced accordingly. Hence the orbit spaces mentioned above are nothing else but spaces of differentials of the elements of those spaces, on which $\mathbb{R}^n$ acts.

For our later investigations we observe that M and ∂M inherit via respective embeddings into $\mathbb{R}^n$ some basic geometric structures described in the first appendix. In particular each $J \in E(M, \mathbb{R}^n)$ and each $j \in E(\partial M, \mathbb{R}^n)$ yield, by pulling back the scalar product $<,>$ of $\mathbb{R}^n$ to M, the two *Riemannian metrics* $m(J)$ and $m(j)$ on M and ∂M respectively. In turn each J and each j also define in a unique way *Riemannian volume* elements $\mu(J)$ on M and $\mu(j)$ on ∂M respectively. In case $j := J|\partial M$ these are related to each other by $i_\mathbf{n}\mu(J) = \mu(j)$ with $\mathbf{n}$ the positively *oriented unit normal* along $\partial M \subset M$. Clearly, this unit normal $\mathbf{n}$ depends on J!

3 The Notion of Work and Fixing the Center of Mass

We will characterize the type of the material constituting the body M in so far as it affects the work caused by an infinitesimal distortion of M (cf. [13], [12],[3],[7]). This idea is formalized by giving a smooth one-form on $E(M,\mathbb{R}^n)$, i.e. a smooth map

$$F : E(M,\mathbb{R}^n) \times C^\infty(M,\mathbb{R}^n) \longrightarrow \mathbb{R},$$

which varies linearly in the second argument. We interpret $F(J)(L)$ as the *work* done if M is distorted by $L \in C^\infty(M,\mathbb{R}^n)$ at the configuration $J \in E(M,\mathbb{R}^n)$. We call the medium described by F a *smoothly deformable medium*.

In order to describe only internal qualities of the medium we expose F to the translational symmetry and require that

$$F(J + z) = F(J), \qquad \forall J \in E(M,\mathbb{R}^n), \quad \forall z \in \mathbb{R}^n. \tag{1}$$

This means that the work caused by (only internal) physical processes does not depend on the particular location of $J(M)$ within $\mathbb{R}^n$. Moreover, we require that a constant distortion by any $z \in \mathbb{R}^n$ causes no work. It is a considerably weaker condition than to assume that the internal force densities only depend on the metric relations between the particles of the body (cf. [21], where problems of isometric deformations are studied). Formally expressed we impose the further restriction:

$$F(J)(z) = 0, \qquad \forall J \in E(M,\mathbb{R}^n), \quad \forall z \in \mathbb{R}^n \tag{2}$$

on F. This restriction is very fundamental in our development. We will interpret it further below. To implement the possibility of extracting force densities from our basic notion of work, we need a little more structure associated with our forms satisfying (1) and (2). We will do this in the next section. But first we investigate the notion of work more closely:

To do so let us denote the collection of all smooth $\mathbb{R}^q$-valued one-forms on a manifold Q (finite or infinite dimensional!) by $A^1(Q,\mathbb{R}^q)$. From Sect. 1 it is clear that any $F \in A^1(E(M,\mathbb{R}^n),\mathbb{R})$ satisfying (1) and (2) is of the form

$$F(J)(L) = F_{\mathbb{R}^n}(dJ)(dL), \qquad \forall J \in E(M,\mathbb{R}^n) \quad \text{and} \quad \forall L \in C^\infty(M,\mathbb{R}^n), \tag{3}$$

where $F_{\mathbb{R}^n} \in A^1(E(M,\mathbb{R}^n)/\mathbb{R}^n,\mathbb{R})$. Instead of (3) we write $F = d^*F_{\mathbb{R}^n}$.

By introducing the *center of mass* for any $J \in E(M,\mathbb{R}^n)$ we can interpret the forms in $A^1(E(M,\mathbb{R}^n)/\mathbb{R}^n,\mathbb{R})$ as follows: Let us choose a map $o \in C^\infty(E(M,\mathbb{R}^n),\mathbb{R})$, for which the mass $\mathbf{m}$ defined by

$$\mathbf{m}(J) := \int_M o(J)\mu(J), \qquad \forall J \in E(M, \mathbb{R}^n),$$

is positive. The *center of mass* $z_0(J)$ is given by the equation

$$\mathbf{m}(J) \cdot z_0(J) = \int_M o(J) \cdot J\mu(J)$$

or equivalently

$$\mathbf{m}(J) < z_0(J), z >= \int_M o(J) < J, z > \mu(J), \tag{4}$$

for all $J \in E(M, \mathbb{R}^n)$ and all $z \in \mathbb{R}^n$. Let $E_0(M, \mathbb{R}^n)$ be the collection of all $J \in E(M, \mathbb{R}^n)$, for which the center of mass $z_0(J)$ is fixed, e.g. $z_0(J) = 0$. Then we realize $E(M, \mathbb{R}^n)/\mathbb{R}^n$ as a space of configurations via the map

$$\mathrm{d} : E_0(M, \mathbb{R}^n) \longrightarrow E(M, \mathbb{R}^n)/\mathbb{R}^n,$$

sending any J in the domain into $\mathrm{d}J$. It is a diffeomorphism. Moreover, we observe

$$E_0(M, \mathbb{R}^n) \oplus \mathbb{R}^n = E(M, \mathbb{R}^n)$$

(if $z_0(J) = 0$, then this splitting is orthogonal with respect to (4)) and hence that

$$TE(M, \mathbb{R}^n) = TE_0(M, \mathbb{R}^n) \oplus T\mathbb{R}^n.$$

Having the meaning of $E_0(M, \mathbb{R}^n)$ in mind, equation (3) tells us that $F_{\mathbb{R}^n}$ can be interpreted as the work in case the center of mass is kept fixed.

4 The Notion of a Constitutive Law and the Dirichlet Integral

The purpose of this section is to define the notion of a constitutive law. However, the additional structure mentioned above relies on an integral representation of a one-form $F_{\mathbb{R}^n}$ on $E(M, \mathbb{R}^n)/\mathbb{R}^n$. To prepare this notion we need to represent $\mathbb{R}^n$-valued one-forms relative to the differential of embeddings. To do so let $\gamma \in A^1(M, \mathbb{R}^n)$ and $J \in E(M, \mathbb{R}^n)$ be given and let us consider the two-tensor $T(\gamma, \mathrm{d}J)$ determined by $\langle \gamma X, \mathrm{d}JY \rangle$ for all $X, Y \in \Gamma TM$. This-two tensor yields a unique strong bundle map $A(\gamma, \mathrm{d}J)$ of TM defined by

$$T(\gamma, \mathrm{d}J)(X, Y) = m(J)(A(\gamma, \mathrm{d}J)X, Y), \qquad \forall X, Y \in \Gamma TM \tag{5}$$

From this equation we read off :

$$\gamma X = \mathrm{d}J A(\gamma, \mathrm{d}J)X + (\gamma X)^\perp, \qquad \forall X \in \Gamma TM$$

with $(\gamma X)^\perp$ being pointwise orthogonal to $\mathrm{d}JTM$ (which appears if $\dim M < n$).

Let us rewrite $\bigcup_{p \in M} T_{J(p)} \mathbb{R}^n$ by $T\mathbb{R}^n | J(M)$ which is the restriction of the tangent bundle $T\mathbb{R}^n$ to $J(M)$. Hence we find a bundle map

$$c(\gamma, \mathrm{d}J) : T\mathbb{R}^n | J(M) \longrightarrow T\mathbb{R}^n | J(M)$$

mapping the vector space $\mathrm{d}J T_p M$ into the normal space of $\mathrm{d}J T_p M$ and vice versa for each $p \in M$. Without loss of generality we may assume that $c(\gamma, \mathrm{d}J)(p) : \mathbb{R}^n \longrightarrow \mathbb{R}^n$ is skew symmetric with respect to $< , >$ for each $p \in M$. We therefore arrive at our desired unique representation:

$$\gamma X = c(\gamma, \mathrm{d}J)\mathrm{d}J X + \mathrm{d}J A(\gamma, \mathrm{d}J)X.$$

For any two one-forms $\gamma_1, \gamma_2 \in A^1(M, \mathbb{R}^n)$ along an embedding $J \in E(M, \mathbb{R}^n)$ we define the *dot product* of γ_1 and γ_2 relative to J by

$$\gamma_1 \cdot \gamma_2 := -\frac{1}{2}\mathrm{Tr}\, c(\gamma_1, \mathrm{d}J)c(\gamma_2, \mathrm{d}J) + \mathrm{Tr}\, A(\gamma_1, \mathrm{d}J) \cdot \tilde{A}(\gamma_2, \mathrm{d}J), . \qquad (6)$$

Here $\tilde{A}(\gamma_2, \mathrm{d}J)$ is the adjoint of $A(\gamma_1, \mathrm{d}J)$ formed fiber-wise with respect to $m(J)$. Associated with this product is a type of a scalar product $\mathcal{G}(J)$ on $A^1(M, \mathbb{R}^n)$ defined by

$$\mathcal{G}(J)(\gamma_1, \gamma_2) := \int_M \gamma_1 \cdot \gamma_2, \mu(J), . \qquad (7)$$

We call the right hand side of (7) the *Dirichlet-integral* (cf. [5]), it is SO(n) invariant.

We equip $A^1(M, \mathbb{R}^n)$ with the C^∞-topology (cf. [8]). The real number $\mathcal{G}(\mathrm{d}J)(\gamma_1, \gamma_2)$ depends smoothly on all its variables J, γ_1 and γ_2. Moreover, $\mathcal{G}$ is a quadratic structure on the trivial bundle $E(M, \mathbb{R}^n)/\mathbb{R}^n \times A^1(M, \mathbb{R}^n)$ and hence yields a metric on $E(M, \mathbb{R}^n)/\mathbb{R}^n$, denoted by $\mathcal{G}$, too.

We say that $F_{\mathbb{R}^n}$, a one-form on $E(M, \mathbb{R}^n)/\mathbb{R}^n$, admits an *integral representation* if there exists a smooth map

$$\alpha : E(M, \mathbb{R}^n) \longrightarrow A^1(M, \mathbb{R}^n),$$

called the *kernel* of $F_{\mathbb{R}^n}$, such that

$$F_{\mathbb{R}^n}(\mathrm{d}J)(\mathrm{d}L) = \int_M \alpha(\mathrm{d}J) \cdot \mathrm{d}L\mu(J) = \mathcal{G}(\mathrm{d}J)(\alpha(J), \mathrm{d}L)$$

holds true for any choice of $\mathrm{d}J \in E(M, \mathbb{R}^n)/\mathbb{R}^n$ and $\mathrm{d}L \in C^\infty(M, \mathbb{R}^n)/\mathbb{R}^n$.

Definition 1 A *constitutive law* F is a smooth one-form on $E(M, \mathbb{R}^n)$ with the following two properties

   i.) $F = \mathrm{d}^* F_{\mathbb{R}^n}$

   ii.) $F_{\mathbb{R}^n}$ admits an integral representation (with kernel α, say).

The kernel of a constitutive law is not unique at all. The following theorem provides us with a natural splitting of the kernel (and later allows us to extract from it a unique kernel of a specific kind). The proof can be found in [4].

Theorem 2 *Let $\gamma \in A^1(M, \mathbb{R}^n)$ and $J \in E(M, \mathbb{R}^n)$. There exists a uniquely determined differential $d\mathcal{H} \in C^\infty(M, \mathbb{R}^n)/\mathbb{R}^n$ called the exact part of γ and a uniquely determined $\beta \in A^1(M, \mathbb{R}^n)$ such that*

$$\gamma = d\mathcal{H} + \beta, \tag{8}$$

where the exact part of β vanishes. Both $d\mathcal{H}$ and β depend smoothly on J. If $\mathcal{H}(p_0)$ for some $p_0 \in M$ is kept constant in J, then also $\mathcal{H}$ varies smoothly in J.

The effect of the splitting of kernels of constitutive laws is described in the following theorem, proved in [4] (cf. also [5]) :

Theorem 3 *Let F be any constitutive law. There exists a smooth map $\mathcal{H} \in C^\infty(E(M, \mathbb{R}^n)/\mathbb{R}^n, C^\infty(M, \mathbb{R}^n))$, such that for any $J \in E(M, \mathbb{R}^n)$ and any $L \in C^\infty(M, \mathbb{R}^n)$*

$$F(J)(L) = \int_M d\mathcal{H}(dJ) \cdot dL\mu(J) = \mathcal{G}(dJ)(d\mathcal{H}(dJ), dL). \tag{9}$$

The kernel $d\mathcal{H} : E(M, \mathbb{R}^n)/\mathbb{R}^n \longrightarrow C^\infty(M, \mathbb{R}^n)/\mathbb{R}^n \subset A^1(M, \mathbb{R}^n)$ is uniquely determined by F, the map $\mathcal{H}$ can be chosen such that $(id, \mathcal{H}) \in \Gamma(TE_0(M, \mathbb{R}^n))$, the latter choice is unique, too.

Since the constitutive law F is determined by $\mathcal{H}$, we call this map a *constitutive map*. With any constitutive law the smooth two-tensor in (5) associated with $\gamma = d\mathcal{H}(dJ)$ is referred to as *stress tensor* $T(J)$ at the configuration J.

5 The Relation Between the Customary Notion of the Stress Tensor and the Constitutive Map

To see that the setting in [17] is included in the treatment presented here, we assume that dim $M = n$ and the work $F(J)(L)$ depends on the metric $m(J)$ and its derivative $Dm(J)(L)$ at J in the direction of L rather than J and L, themselves. (We may work at a fixed J.) Moreover let us suppose that F admits for all its variables an integral representation of the form

$$F(J)(L) = \int_M T(J) \cdot Dm(J)(L)\mu(J),$$

with a two-tensor $T(J)$ as kernel. The dot-product in the integrand is defined by representing both $T(J)$ and $Dm(J)(L)$, respectively as strong bundle maps $K_1(J)$ and $K_2(J)(L)$ of TM via the metric $m(J)$ and then proceeding as in (6), i.e.

$$T(J) \cdot Dm(J)(L) := \text{Tr } K_1(J) \cdot K_2(J)(L).$$

Using (8) and (9) the verification of the following is straightforward :

$$F(J)(L) = \int_M \text{Tr } K_1(J) \cdot K_2(J)(L)\mu(J) = \int_M \text{Tr } K_1(J) \cdot \tilde{A}(\mathrm{d}L, \mathrm{d}J)\mu(J)$$

$$= \int_M \alpha(\mathrm{d}J) \cdot \mathrm{d}L\mu(J) = \int_M \mathrm{d}\mathcal{H}(\mathrm{d}J) \cdot \mathrm{d}L\mu(J)$$

(10)

with $\alpha(\mathrm{d}J) := \mathrm{d}JK_1(J)$, holding for all the variables of F. In case J is an equilibrium condition the tensors $\mathcal{T}(J)$ and $\frac{1}{2}\mathrm{D}m(J)(L)$ correspond to the *stress tensor* and to the *deformation tensor*, respectively. The calculation in (10) shows, moreover, that to each smooth stress tensor assignment in the setting of [17] there is a constitutive map. (Nevertheless the two-tensors $T(J)$ and $\mathcal{T}(J)$ may still differ !) Reading (10) backwards shows that it suffices to work with $\mathcal{T}(J)$ instead of $\mathcal{H}(J)$. This difference however is obsolete with respect to the resulting force densities (cf. remark to Theorem 4).

6 Force Densities Associated with Constitutive Laws

The purpose of this section is to present how to associate with any constitutive law at any configuration some well defined force densities, one acting upon the whole body, and another one acting upon the boundary only. Vice versa any given pair of force densities satisfying an integrability condition will be obtained via a suitable constitutive law.

Throughout this section F is a constitutive law with a kernel α. By the previous theorem we may assume that $\alpha(E(M, \mathbb{R}^n)/\mathbb{R}^n) \subset C^\infty(M, \mathbb{R}^n)/\mathbb{R}^n$. The following theorem shows how to associate the force densities mentioned above to any constitutive map (cf. Append. 2). The existence of a solution of the Neumann problem can found in [15].

Theorem 4 *Every constitutive law $F \in A^1(E(M, \mathbb{R}^n), \mathbb{R})$ admits a smooth constitutive map*

$$\mathcal{H} : E(M, \mathbb{R}^n)/\mathbb{R}^n \longrightarrow C^\infty(M, \mathbb{R}^n),$$

such that F can be expressed as

$$F(J)(L) = \int_M <\Delta(J)\mathcal{H}(\mathrm{d}J), L> \mu(J) + \int_{\partial M} <\mathrm{d}\mathcal{H}(\mathrm{d}J)(\mathbf{n}), L> i_\mathbf{n}\mu(J),$$

for each $J \in E(M, \mathbb{R}^n)$ and each $L \in C^\infty(M, \mathbb{R}^n)$. For all $J \in E(M, \mathbb{R}^n)$ the map $\mathcal{H}$ defines the force densities Φ and φ respectively by

$$\Phi(\mathrm{d}J) := \Delta(J)\mathcal{H}(\mathrm{d}J)$$

(11)

and

$$\varphi(\mathrm{d}J) := \mathrm{d}\mathcal{H}(\mathrm{d}J)(\mathbf{n}),$$

(12)

which satisfy, due to the fundamental properties of F, the equation

$$0 = \int_M \Phi(\mathrm{d}J)\mu(J) + \int_{\partial M} \varphi(\mathrm{d}J)i_n\mu(J). \tag{13}$$

*Given vice versa two smooth maps $\Phi \in C^\infty(E(M,\mathbb{R}^n)/\mathbb{R}^n, C^\infty(M,\mathbb{R}^n))$ and $\varphi \in C^\infty(E(M,\mathbb{R}^n)/\mathbb{R}^n, C^\infty(\partial M,\mathbb{R}^n))$, for which the equation (13) holds as an **integrability condition**, then there exists also a smooth map $\mathcal{H} \in C^\infty(E(M,\mathbb{R}^n)/\mathbb{R}^n, C^\infty(M,\mathbb{R}^n))$ satisfying (11) and (12) for which $\mathcal{H}(\mathrm{d}J)$ is uniquely determined up to a constant for each $J \in E(M,\mathbb{R}^n)$. Moreover $\mathcal{H}$ is a constitutive map for the constitutive law F given by the force densities via the formula*

$$F(J)(L) = \int_M < \Phi(\mathrm{d}J), L > \mu(J) + \int_{\partial M} < \varphi(\mathrm{d}J), L > i_n\mu(J), \tag{14}$$

holding for all $J \in E(M,\mathbb{R}^n)$ and for all $L \in C^\infty(M,\mathbb{R}^n)$. In case $\partial M = \emptyset$ then (12) and also the second terms on the right hand sides of both (13) and (14) disappear.

Remark. This theorem shows to us that our notion of constitutive laws (based on translational invariance) is equivalent with the $\mathbb{R}^n$-valued solution of Neumann problems formulated on M and hence is equivalent with a pair of force densities satisfying the integrability condition (13), which tells us that the resulting force density to the center of mass vanishes. (13) is obviously the analog of (2). Moreover, by Theorem 3 a general kernel α and its exact part both determine the same force densities!

7 The Interplay Between Constitutive Laws of Boundary and Body

The deformable media forming the inside of the body and the boundary respectively may differ and each separate material hence has to be described on one hand by different constitutive laws, as we have done in the previous sections. On the other hand both materials together form one body and should be describable by only one constitutive law holding for the whole body. Since constitutive laws behave additively, the comparison between the two procedures allows us to decode the influence of the whole body to the constitutive properties of the boundary material.

Let F be the constitutive law of the deformable medium forming the whole body. According to Theorem 4 the one-form F is determined by a smooth constitutive map $\mathcal{H}$. It affects the quality of the boundary material: $\mathcal{H}$ yields according to Theorem 4 force densities $\Phi \in C^\infty(E(M,\mathbb{R}^n)/\mathbb{R}^n, C^\infty(M,\mathbb{R}^n))$ and $\varphi \in C^\infty(E(M,\mathbb{R}^n)/\mathbb{R}^n, C^\infty(\partial M,\mathbb{R}^n))$.

The force density acting on ∂M, is defined by

$$\varphi(\mathrm{d}J) = \mathrm{d}\mathcal{H}(\mathrm{d}J)(\mathbf{n}), \qquad \forall \mathrm{d}J \in E(M, \mathbb{R}^n)/\mathbb{R}^n.$$

Having the integrability condition (13) for $\Delta(J|\partial M)$ in mind, we split this force density φ into

$$\varphi(\mathrm{d}J) = \varphi_{\mathbb{R}^n}(\mathrm{d}J) + \psi(\mathrm{d}J), \qquad \forall \mathrm{d}J \in E(M, \mathbb{R}^n)/\mathbb{R}^n,$$

where $\varphi_{\mathbb{R}^n}(\mathrm{d}J)$ is characterized for each $\mathrm{d}J \in E(M, \mathbb{R}^n)/\mathbb{R}^n$ by the equation

$$\int_{\partial M} \varphi_{\mathbb{R}^n}(\mathrm{d}J) i_{\mathbf{n}} \mu(J) = 0. \tag{15}$$

The map $\psi \in C^\infty(E(M, \mathbb{R}^n)/\mathbb{R}^n, \mathbb{R}^n)$ is L_2-orthogonal to $\varphi_{\mathbb{R}^n}$. According to Theorem 4 the condition (15) allows us to choose some map

$$\hbar \in C^\infty(E(M, \mathbb{R}^n)/\mathbb{R}^n, C^\infty(\partial M, \mathbb{R}^n)),$$

such that for all $\mathrm{d}J \in E(M, \mathbb{R}^n)/\mathbb{R}^n$ the equation

$$\varphi_{\mathbb{R}^n}(\mathrm{d}J) = \Delta(J|\partial M)\hbar(\mathrm{d}J)$$

holds true. With these data we easily verify the following :

Theorem 5 *Any smoothly deformable medium is characterized by a constitutive map $\mathcal{H} \in C^\infty(E(M, \mathbb{R}^n)/\mathbb{R}^n, C^\infty(M, \mathbb{R}^n))$, that determines itself two smooth maps $\hbar \in C^\infty(E(M, \mathbb{R}^n)/\mathbb{R}^n, C^\infty(\partial M, \mathbb{R}^n))$ and $\psi \in C^\infty(E(M, \mathbb{R}^n), \mathbb{R}^n)$, which are linked to $\mathcal{H}$ by the boundary condition*

$$\mathrm{d}\mathcal{H}(\mathrm{d}J)(\mathbf{n}) = \Delta(J|\partial M)\hbar(\mathrm{d}J) + \psi(\mathrm{d}J) \tag{16}$$

for each $J \in E(M, \mathbb{R}^n)$. The differential $\mathrm{d}\hbar$ and $\psi \in C^\infty(E(M, \mathbb{R}^n)/\mathbb{R}^n, \mathbb{R}^n)$ are unique. For each $J \in E(M, \mathbb{R}^n)$ the map $\mathcal{H}$ satisfies

$$0 = \int_M \Delta(J)\mathcal{H}(\mathrm{d}J)\mu(J) + \int_{\partial M} \psi(\mathrm{d}J) i_{\mathbf{n}} \mu(J).$$

The constitutive law on $E(M, \mathbb{R}^n)$ describing the constitutive properties of the materials forming the body together with its boundary is determined by $\mathcal{H}$, which yields $\hbar$ and ψ and thus is given via the formula

$$F(J)(L) = \int_M <\Delta(J)\mathcal{H}(\mathrm{d}J), L> \mu(J)$$

$$+ \int_{\partial M} <\Delta(J|\partial M)\hbar(J|\partial M) + \psi(\mathrm{d}J), L> i_{\mathbf{n}} \mu(J)^{,,}$$

valid for all variables of F. The work of any distortion $l \in C^\infty(\partial M, \mathbb{R}^n)$ of the deformable material forming the boundary, regarded as being detached from the body, is for any $J \in E(M, \mathbb{R}^n)$ given by the constitutive law

$$F_{\partial M}(\mathrm{d}J)(l) = \int_{\partial M} <\Delta(J|\partial M)\hbar_\partial(\mathrm{d}J), l> i_{\mathbf{n}} \mu(J),$$

for some constitutive map $\hbar_\partial \in C^\infty(O_\partial/\mathbb{R}^n, C^\infty(\partial M, \mathbb{R}^n))$. Hence the difference $\hbar - \hbar_\partial$ and ψ describe how the constitutive properties of the material forming the boundary of the body are affected by the fact that this material is incorporated into the material forming the whole body. (Any $\hbar_\partial(\mathrm{d}J)$ can be harmonically extended to all of M).

8 A General Decomposition of Constitutive Laws

In this section we will exhibit decompositions of $\mathrm{d}\mathcal{H}$ and $\mathrm{d}\hbar$ as induced by (16). Both splittings are based on the examples in Append. 3. We need the maps $\mathcal{V}, \mathcal{A}$ and N, describing for each configuration the volume, the area and the shape, respectively (cf. (iii) in Append. 3). In particular we will show that $\mathrm{D}\mathcal{V}$ and $R^*\mathrm{D}\mathcal{A}$ (cf. Sect. 2 for R) multiplied with appropriate $\mathbb{R}$-valued functions are part of any constitutive law F defined on $E(M, \mathbb{R}^n)$. Again let $j := J|\partial M$.

To obtain the desired decomposition of a given constitutive law we broaden our scope a little and, first of all, introduce the Hilbert space A_j consisting of all maps $\gamma_1, \gamma_2 : T\partial M \longrightarrow \mathbb{R}$ linear on the fibers of $T\partial M$ for which the right hand side of

$$G(\mathrm{d}j)(\gamma_1, \gamma_2) := \int_{\partial M} \gamma_1 \cdot \gamma_2 \mu(j)$$

exists. Clearly $\mathrm{d}\hbar_\mathbf{n}(\mathrm{d}J)$ (defined in Append. 3), $\mathrm{d}j$ and $\mathrm{d}N(j)$ all belong to A_j and are generically linearly independent. In the special case of $j(\partial M)$ being a $(n-1)$-sphere in $\mathbb{R}^n$ however, $N(j), j$ and $\hbar_\mathbf{n}(\mathrm{d}J)$ are not linearly independent. The set O_3 of all $J \in E(M, \mathbb{R}^n)$ for which these three differentials are linearly independent forms a dense open set in $E(M, \mathbb{R}^n)$.

For each $J \in O_3$ we split the differential of $\hbar(\mathrm{d}J)$ into components along the span of the three mentioned differentials and a component perpendicular to it. The next step is to define maps $(\hbar_\mathbf{n}(\mathrm{d}J))_M, j_M$ and $N(j)_M$ associated with $\hbar_\mathbf{n}(\mathrm{d}J), j$ and $N(j)$, respectively. This is done by solving the following Visik problem (cf. [15]) : Let $f \in C^\infty(\partial M, \mathbb{R}^n)$ be given. We define $f_M(J) \in C^\infty(M, \mathbb{R}^n)$ by

$$\Delta(J)f_M(J) = 0 \quad rmand \quad \mathrm{d}f_M(J)(\mathbf{n}) - \Delta(j)f = 0$$

for any $J \in E(M, \mathbb{R}^n)$. In particular $\hbar_\mathbf{n}$ satisfies

$$\mathrm{d}(\hbar_\mathbf{n}(\mathrm{d}J))_M = \mathrm{d}J, \qquad \forall \in E(M, \mathbb{R}^n).$$

f_M depends smoothly on J. The above mentioned decomposition of $\mathrm{d}\mathcal{H}$ is then presented in the following theorem partly based on Theorem 5 :

Theorem 6 *Let F be a constitutive law on $E(M, \mathbb{R}^n)$. Then any of its constitutive maps $\mathcal{H} \in C^\infty(E(M, \mathbb{R}^n)/\mathbb{R}^n, C^\infty(M, \mathbb{R}^n))$ determines uniquely three smooth functions*

$$a_1, a_2, a_3 : O_3/\mathbb{R}^n \subset E(M,\mathbb{R}^n)/\mathbb{R}^n \longrightarrow \mathbb{R} \qquad (17)$$

and the smooth maps

$$\hbar, \hbar_2 : O_3/\mathbb{R}^n \subset E(M,\mathbb{R}^n)/\mathbb{R}^n \longrightarrow C^\infty(\partial M, \mathbb{R}^n),$$

$$\psi : O_3 \subset E(M,\mathbb{R}^n) \longrightarrow \mathbb{R}^n,$$

linked to $\mathcal{H}$ by

$$d\mathcal{H}(dJ)(\mathbf{n}) = \Delta(j)\hbar(dJ) + \psi(dJ),$$

such that the following splitting holds for any $J \in O_3 \subset E(M,\mathbb{R}^n)$

$$d\hbar(dJ) = a_1(dJ) \cdot d\hbar_\mathbf{n}(dJ) + a_2(dJ)dj + a_3(dJ) \cdot dN(j) + d\hbar_2(dJ) \qquad (18)$$

with $j := J/\partial M$. The differential $d\hbar_2(dJ)$ is with respect to $G(dj)$ orthogonal to the span of $d\hbar_\mathbf{n}(dJ), dj$ and $dN(j)$. The differential $d\mathcal{H}(dJ)$ decomposes for each $J \in O_3$ accordingly into

$$d\mathcal{H}(dJ) = a_1(dJ) \cdot dJ + a_2(dJ) \cdot dj_M + a_3(dJ) \cdot dN_M(dJ) + d\mathcal{H}_2(dJ), \qquad (19)$$

where $d\mathcal{H}_2(dJ)$ is such that (19) holds. The splittings (18) and (19) are valid for the respective maps if the representatives of the differentials are chosen from $E_0(M, \mathbb{R}^n)$.

9 Two Simple Bubble Models

Let us illustrate the meaning of the coefficients a_1 and a_2 in (17) by a simple bubble model. (dI and di as arguments will be omitted.) Let $\dim M = 3$ and let us think of M as a bubble and of ∂M as the *middle surface* of a very thin shell bounding the bubble. Moreover we assume $a_3 = 0$ and $\mathcal{H}_2 = 0$ as well and suppose that there is an equilibrium configuration $I \in E(M, \mathbb{R}^n)$, i.e. to say we suppose that $F(I) = 0$. Since ∂M represents two bounding surfaces, (18) turns into

$$0 = a_1 \cdot d\hbar_\mathbf{n} + 2 \cdot a_2 di.$$

Using the respective formulas in Append. 3, we find

$$a_1 + 2 \cdot a_2 \cdot H = 0,$$

with H the mean curvature of $i(M) \subset \mathbb{R}^3$. Thus we observe that H is a constant map on M and deduce from the classical bubble models, therefore, that a_1 and a_2 can be interpreted as an internal pressure and as a capillarity, respectively.

If the quality of the deformable medium depends in addition on the shape, i.e. if $a_3 \neq 0$ (but still $\mathcal{H}_2 = 0$) at the equilibrium configuration I (which might be an immersion only), we find due to (18) and (A5) the following equation

$$0 = a_1 + 2a_2 \cdot H + 2a_3 \cdot H^2 - 2a_3\kappa, \qquad (20)$$

with κ the Gaussian curvature. By integrating (20) and by using the theorem of Gauss-Bonnet we deduce

$$0 = a_1 + 2a_2 \cdot H + 2a_3 \cdot H^2 - 4\pi a_3 \frac{\chi(\partial M)}{A}, \qquad (21)$$

with A the area of ∂M. Equation (21) in turn yields

$$\chi(\partial M) = \frac{1}{2\pi} \cdot \kappa \cdot A.$$

One now easily shows that H is constant and that in this case ∂M cannot be a sphere. If I exists at all, $i(\partial M) \subset \mathbb{R}^3$ has to be an immersed torus as seen by a theorem of Efimov (cf. [2]).

10 The Rotational Symmetry

From the two symmetries described in Sect. 1 we have used so far the translational symmetry only. It is clear, however, that internal physical processes are also invariant under the rotational symmetry. In this section we will show what additional property any constitutive map inherits from this symmetry.

Invariance under $SO(n)$ of a given constitutive law F with $\mathcal{H}$ as a constitutive map yields

$$F(g \circ J)(g \circ L) = F(J)(L)$$

for all the variables of F and for any $g \in SO(n)$. Thus the constitutive map satisfies

$$\int_M d\mathcal{H}(g \circ dJ) \cdot d(g \circ L)\mu(g \circ J) = \int_M d\mathcal{H}(dJ) \cdot dL\mu(J)$$

and hence

$$\int_M d(g^{-1} \circ \mathcal{H}(g \circ dJ) - \mathcal{H}(dJ)) \cdot dL\mu(J) = 0,$$

for all $g \in SO(n)$, for all $J \in E(M, \mathbb{R}^n)$ and for all $L \in C^\infty(M, \mathbb{R}^n)$. From the last equation and from the general procedure of representing differentials via embeddings we read off :

Proposition 7 *Given a constitutive law F with constitutive map*

$$\mathcal{H} \in C^\infty(E(M, \mathbb{R}^n)/\mathbb{R}^n, C^\infty(M, \mathbb{R}^n)),$$

then F is invariant under $SO(n)$ iff $\mathcal{H}$ is equivariant, i.e. iff

$$g^{-1} \circ d\mathcal{H}(g \circ dJ) = d\mathcal{H}(dJ), \qquad \forall G \in SO(n). \qquad (22)$$

The infinitesimal version of (22) is

$$d D\mathcal{H}(dJ)(c) = c d\mathcal{H}(dJ), \qquad \forall c \in so(n).$$

The validity of (22) implies in particular the following identity:

$$A(\mathrm{d}\mathcal{H}(g \circ \mathrm{d}J), g \circ \mathrm{d}J) = A(\mathrm{d}\mathcal{H}(\mathrm{d}J), \mathrm{d}J)$$

holding for all $g \in \mathrm{SO}(n)$ and for all $J \in E(M, \mathbb{R}^n)$. The stress tensor $T(J)$ determined by $\mathcal{H}$ is hence invariant under $\mathrm{SO}(n)$ for any $J \in E(M, \mathbb{R}^n)$.

If F factors to $T_J E(M, \mathbb{R}^n)/so(n) \cdot J$, and thus $F(J)(cJ) = 0$ for all $c \in so(n)$ then

$$\int_M < c\phi(\mathrm{d}J), J > \mu(J) + \int_{\partial M} < c\varphi(\mathrm{d}J), J >_{i_n} \mu(J) = 0$$

holds. This is the analog to (2). If dim $M = 3$, then $E(M, \mathbb{R}^n)/\mathrm{SO}(n)$ plays a fundamental role in continuum mechanics (cf. [19]).

11 A Generalization of the Notion of a Constitutive Law and the Introduction of a Dynamics

In this section we introduce a dynamics in case $\partial M = \emptyset$. All constructions could be performed for any codimension of the body in $\mathbb{R}^n$. However, due to simplicity we require dim $M = n - 1$. Here we follow closely [7]. First of all we extend the notion of a constitutive law. In doing so we consider special one-forms on $TE(M, \mathbb{R}^n)$, namely those of the type

$$\hat{F}_{\mathbb{R}^n} : T^2 \left(E(M, \mathbb{R}^n)/\mathbb{R}^n \right) \longrightarrow \mathbb{R},$$

satisfying

$$\hat{F}_{\mathbb{R}^n}(\mathrm{d}j, \mathrm{d}k)(\mathrm{d}l, \mathrm{d}s) = \hat{F}_{\mathbb{R}^n}(\mathrm{d}j, \mathrm{d}k)(\mathrm{d}l, 0),$$

for any $\mathrm{d}j \in E(M, \mathbb{R}^n)/\mathbb{R}^n$ and for any choice of $\mathrm{d}k, \mathrm{d}l, \mathrm{d}s \in C^\infty(M, \mathbb{R}^n)/\mathbb{R}^n$.

A constitutive law (depending not only on the configurations but also on the velocities) is then a one-form F on $TE(M, \mathbb{R}^n)$ satisfying for all $\mathrm{d}j \in E(M, \mathbb{R}^n)/\mathbb{R}^n$ and all $\mathrm{d}k, \mathrm{d}l \in C^\infty(M, \mathbb{R}^n)/\mathbb{R}^n$

$$F(j, k)(l) = \hat{F}_{\mathbb{R}^n}(\mathrm{d}j, \mathrm{d}k)(\mathrm{d}l, 0),$$

and admitting an integral representation. The constitutive map $\mathcal{H}$ depends hence on configurations and velocities!

Next we will define motions subjected to a constitutive law by using d'Alembert's principle:

A *motion* on $E(M, \mathbb{R}^n)$ in our model is described by a smooth curve of embeddings

$$\sigma : (-\lambda, \lambda) \longrightarrow E(M, \mathbb{R}^n),$$

for some positive real λ. This motion is subjected to the constitutive law F, if

$$F(\dot{\sigma}(t))(\sigma(t), h) = \mathcal{B}(\sigma(t))(\ddot{\sigma}(t), h), \qquad \forall h \in T_{\sigma(t)} E(M, \mathbb{R}^n),$$

is satisfied for all $t \in (-\lambda, \lambda)$ with $\mathcal{B}$ as in Append. 4. Hence

$$\int_M < \Phi(\sigma(t), \dot{\sigma}(t)), h > \mu(\sigma(t)) = \int_M \rho(\sigma(t)) < \ddot{\sigma}(t), h > \mu(\sigma(t))$$

has to hold for h and for all t, mentioned above (cf. [13],[20],[22]). Here Φ is the force density determined by F. This implies Newton's third law of motion on the density level:

$$\rho(\sigma(t))\ddot{\sigma}(t) = \Phi(\sigma(t), \dot{\sigma}(t)) = \Delta((\sigma(t)) \mathcal{H} (\mathrm{d}\sigma(t), \mathrm{d}\dot{\sigma}(t)), \quad \forall t \in (-\lambda, \lambda). \quad (23)$$

Clearly σ parameterizes a straight line segment, i.e. a *geodesic* of $\mathcal{B}$, iff $\Phi(\sigma(t), \dot{\sigma}(t)) = 0$ for all $t \in (-\lambda, \lambda)$.

Let σ be a motion satisfying (23). We rewrite $\sigma(t)$ according to the principal bundle structure of $E(M, \mathbb{R}^n)$ (cf. [6]), by proceeding as follows: At first we note that $\dot{\sigma}(t)$ admits in $\mathbb{R}^n$ the pointwise splitting

$$\dot{\sigma}(t) = \mathrm{d}\sigma(t)Z(t) + \dot{\sigma}(t)^{\perp},$$

and easily verify

$$\begin{aligned}
\mathrm{d}\dot{\sigma}(t)Z(t) &= \mathrm{d}(\mathrm{d}\sigma(t)Z(t))Z(t) + \mathrm{d}\dot{\sigma}(t)^{\perp}Z(t) \\
&= \sigma(t)\nabla(\sigma(t))_{Z(t)}Z(t) + S(\sigma(t))(Z(t), Z(t)) \\
&\quad + \mathrm{d}\sigma(t)W(\sigma(t))Z(t) + (\mathrm{d}\dot{\sigma}(t)^{\perp}Z(t))^{\perp}
\end{aligned}$$

where "$\perp$" denotes the pointwise formed component normal to $\sigma(t)(M)$. Moreover we have for each $t \in (-\lambda, \lambda)$ and for some well-defined functions $\epsilon(\sigma(t), \dot{\sigma}(t)) \in C^{\infty}(M, \mathbb{R})$ and $K_F(\sigma(t), \dot{\sigma}(t)) \in C^{\infty}(M, \mathbb{R})$

$$\dot{\sigma}(t)^{\perp} = \epsilon(\sigma(t), \dot{\sigma}(t))N(\sigma(t))$$

and

$$\Phi(\sigma(t), \dot{\sigma}(t))^{\perp} = K_F(\sigma(t), \dot{\sigma}(t))N(\sigma(t)),$$

respectively. Using the formula

$$\begin{aligned}
\dot{N}(\sigma(t)) &= \mathrm{D}N(\sigma(t))(\dot{\sigma}(t)) \\
&= \mathrm{d}\sigma(t)(W(\sigma(t))Z(t) - \mathrm{grad}_{\sigma(t)}\epsilon(\sigma(t), \dot{\sigma}(t))),
\end{aligned}$$

we find the following system for a motion on $E(M, \mathbb{R}^n)$ subjected to a constitutive law:

$$\begin{aligned}
\nabla(\sigma(t))_{Z(t)}Z(t) &+ \dot{Z}(t) + 2\epsilon(\sigma(t), \dot{\sigma}(t))W(\sigma(t))Z(t) \\
&- \epsilon(\sigma(t), \dot{\sigma}(t))\mathrm{grad}_{\sigma(t)}\epsilon(\sigma(t), \dot{\sigma}(t)) \\
&= \rho^{-1}(\sigma(t))Y_F(\sigma(t), \dot{\sigma}(t)),
\end{aligned} \qquad (24)$$

and

$$\dot{\epsilon}(\sigma(t), \dot{\sigma}(t)) = \rho^{-1}(\sigma(t))K_F(\sigma(t), \dot{\sigma}(t))$$
$$+ \mathbf{f}(\sigma(t))(Z(t), Z(t)) - \mathrm{d}\epsilon(\sigma(t), \dot{\sigma}(t))Z(t),$$

with $Y_F(\sigma(t), \dot{\sigma}(t)) \in \Gamma T M$ such that $\mathrm{d}J Y_F(\sigma(t), \dot{\sigma}(t)) = \Phi(\sigma(t), \dot{\sigma}(t))^\mathsf{T}$ for all $t \in (-\lambda, \lambda)$. Here $\mathbf{f}(\sigma(t))$ denotes the second fundamental form of $\sigma(t)$ (cf. Append. 1) and " $^\mathsf{T}$" denotes the pointwise formed component tangential to $\sigma(t)(M)$.

Remark. Equation (24) describes a straight line segment in $E(M, \mathbb{R}^n)$ iff both $Y_F(\sigma(t), \dot{\sigma}(t)) = 0$ and $K_F(\sigma(t), \dot{\sigma}(t)) = 0$ hold for all $t \in (-\lambda, \lambda)$. In this case the equations (24) read as

$$\nabla(\sigma(t))_{Z(t)} Z(t) + \dot{Z}(t) + 2\epsilon(\sigma(t), \dot{\sigma}(t))W(\sigma(t))Z(t)$$
$$-\epsilon(\sigma(t), \dot{\sigma}(t))\mathrm{grad}_{\sigma(t)}\epsilon(\sigma(t), \dot{\sigma}(t)) = 0,$$
$$\dot{\epsilon}(\sigma(t), \dot{\sigma}(t)) = \mathbf{f}(\sigma(t))(Z(t), Z(t)) - \mathrm{d}\epsilon(\sigma(t), \dot{\sigma}(t))Z(t).$$

Otherwise (24) describes the deviation from geodesics.

12 A Navier-Stokes Type of Equation

Again M, F, and $\mathcal{H}$ are as in the previous section. The introduction of a structural *viscosity* coefficient relies on a mathematical observation : If we write any $k \in C^\infty(M, \mathbb{R}^n)$ in the form

$$k = \mathrm{d}j X(k, j) + \theta(k, j) \cdot N(j), \tag{25}$$

where $X(k, j) \in \Gamma T M$ and $\theta(k, j) \in C^\infty(M, \mathbb{R}^n)$ both are uniquely determined by k, then we observe that $X^0(k, j)$, the divergence-free part of $X(k, j)$, is uniquely determined by $\mathrm{d}k$. This motivates us to introduce a structural viscosity $\nu(\mathrm{d}j, \mathrm{d}k) \in C^\infty(M, \mathbb{R})$ via

$$X(\mathcal{H}(\mathrm{d}j, \mathrm{d}k), j) = \nu(\mathrm{d}j, \mathrm{d}k)X^0(k, j) + \hat{X}(\mathcal{H}(\mathrm{d}j, \mathrm{d}k), j). \tag{26}$$

The second term in (26) is such that this equation holds. The following theorem describes a generalization of the *Navier-Stokes equation* (cf. [7]):

Theorem 8 *In absence of external force densities, the general equations of motion on $E(M, \mathbb{R}^n)$ of a deformable medium subjected to a constitutive law F are given by*

$$\nabla(\sigma(t))_{Z(t)} Z(t) + \dot{Z}(t) + 2 \cdot \epsilon(\sigma(t), \dot{\sigma}(t))W(\sigma(t))Z(t)$$
$$- \epsilon(\sigma(T), \dot{\sigma}(T))\mathrm{grad}_{\sigma(t)}\epsilon(\sigma(t), \dot{\sigma}(t))$$
$$= \rho^{-1}(\sigma(t))\big[\mathrm{grad}_{\sigma(t)}\tau_{int}(\sigma(t), \dot{\sigma}(t))$$
$$+ \Delta(\sigma(t))\big(\nu(\mathrm{d}\sigma(t), \mathrm{d}\dot{\sigma}(t))Z^0(t) + \hat{X}(\mathcal{H}(\mathrm{d}\sigma(t), \mathrm{d}\dot{\sigma}(t)), \sigma(t))\big)$$
$$- 2 \cdot W(\sigma(t))\mathrm{grad}_{\sigma(t)}\theta(\mathcal{H}(\mathrm{d}\sigma(t), \mathrm{d}\dot{\sigma}(t)), \sigma(t))$$
$$+ (W(\sigma(t)))^2 \big(\nu(\mathrm{d}\sigma(t), \mathrm{d}\dot{\sigma}(t))Z^0(t) + \hat{X}(\mathcal{H}(\sigma(t), \dot{\sigma}(t)), \sigma(t))\big)$$
$$- \theta(\mathcal{H}(\mathrm{d}\sigma(t), \mathrm{d}\dot{\sigma}(t)), \sigma(t)) \cdot \mathrm{grad}_{\sigma(t)}\mathcal{H}(\sigma(t))\big]$$

$$\dot{\epsilon}(\sigma(t),\dot{\sigma}(t)) = \rho^{-1}(\sigma(t))\big[-\tau_{int}(\mathrm{d}\sigma(t),\mathrm{d}\dot{\sigma}(t))H(\sigma(t))$$
$$- \mathrm{d}H(\sigma(t))(\nu(\mathrm{d}\sigma(t),\mathrm{d}\dot{\sigma}(t))Z^0(t) + \hat{X}(\mathcal{H}(\mathrm{d}\sigma(t),\mathrm{d}\dot{\sigma}(t)),\sigma(t)))$$
$$+ 2 \cdot \mathrm{div}_{\sigma(t)}\nu(\mathrm{d}\sigma(t),\mathrm{d}\dot{\sigma}(t))W(\sigma(t))Z^0(t)$$
$$+ 2 \cdot \mathrm{div}_{\sigma(t)}W(\sigma(t))\hat{X}(\mathcal{H}(\mathrm{d}\sigma(t),\mathrm{d}\dot{\sigma}(t)),\sigma(t))$$
$$+ \theta(\mathcal{H}(\mathrm{d}\sigma(t),\mathrm{d}\dot{\sigma}(t)),\sigma(t)) \cdot \mathrm{Tr}\, W(\sigma(t))^2$$
$$+ \Delta(\sigma(t))\theta(\mathcal{H}(\mathrm{d}\sigma(t),\mathrm{d}\dot{\sigma}(t)),\sigma(t))\big]$$
$$+ \mathbf{f}(\sigma(t))(Z(t),Z(t)) - \mathrm{d}\epsilon(\sigma(t),\dot{\sigma}(t))Z(t),$$

where $Z^0(t) := X^0(\dot{\sigma}(t),\sigma(t))$. *The motion of a deformable medium along a fixed surface* $i(M) \subset \mathbb{R}^n$ *is given by*

$$\nabla(i)_{V(t)}V(t) + \dot{V}(t) = \rho^{-1}(\sigma(t) \circ g(t)^{-1})\big[\mathrm{grad}_i \tau_{int}(\sigma(t),V(t))$$
$$+ \Delta(i)\left(\nu(\sigma(t),V(t))V^0(t) + \hat{V}(\mathcal{H}(\sigma(t),V(t)),V(t))\right)$$
$$- 2 \cdot W(i)\mathrm{grad}_i\theta(\mathcal{H}(\sigma(t),V(t)),V(t))$$
$$+ (W(i))^2\left(\nu(\sigma(t),V(t))V^0(t) + \hat{V}(\mathcal{H}(\sigma(t),V(t)),V(t))\right)$$
$$- \theta(\mathcal{H}(\sigma(t),V(t)),V(t)) \cdot \mathrm{grad}_i H(i)\big]$$

$$0 = \rho^{-1}(\sigma(t) \circ g(t)^{-1})\big[-\tau_{int}(\sigma(t),V(t)) \cdot H(i)$$
$$- \mathrm{d}H(i)\left(\nu(\sigma(t),V(t))V^0(t) + \hat{V}(\mathcal{H}(\sigma(t),V(t)),V(t))\right)$$
$$+ 2 \cdot (\mathrm{div}_i\nu(\sigma(t),V(t))W(i)V^0(t)$$
$$+ \mathrm{div}_iW(i)\hat{V}(\mathcal{H}(\sigma(t),V(t)),V(t)))$$
$$+ \theta(\mathcal{H}(\sigma(t),V(t)),V(t)) \cdot \mathrm{Tr}\, W(i)^2$$
$$+ \Delta(\sigma(t))\theta(\mathcal{H}(\sigma(t),V(t)),V(t))\big]$$
$$+ \mathbf{f}(i)(V(t),V(t)),$$

where $\sigma(t) = i \circ g(t)$ *with* $g(t) \in \mathrm{Diff}\, M$. *The vector fields* $V(t), V^0(t)$ *and* $\hat{V}(\mathcal{H}(\sigma(t),V(t)),V(t))$ *are respectively the push-forwards of* $Z(t), Z^0(t)$ *and* $\hat{X}(\mathcal{H}(\mathrm{d}\sigma(t),\mathrm{d}\dot{\sigma}(t)),\sigma(t))$ *by* $g(t)$. *The map* $\tau_{int} \in C^\infty(M,\mathbb{R}^n)$ *is given by decomposing the tangential component of the force density via Hodge's method into a gradient (of* τ_{int}*) and a divergence-free part.*

As to whether this Navier-Stokes type of equation is invariant under the substitution $t \mapsto -t$ depends on the map ν.

As an application of the above approach and of the remark to Theorem 4 let us consider the following special case, determined by the form

$$\alpha(\mathrm{d}j,\mathrm{d}k) = -\tau_{int}(\mathrm{d}j,\mathrm{d}k) \cdot \mathrm{d}j + \nu \cdot \mathrm{d}j \cdot L_{V^0(k,j)},$$

with a constant $\nu \in \mathbb{R}$ and where $L_{V^0(k,j)}$ denotes the Lie derivative on ΓTM in the direction $V^0(k,j)$. In the incompressible case, i.e. for $\mathrm{div}_iV(t) = 0$, the motion along the fixed manifold $i(M) \subset \mathbb{R}^n$ is governed by

$$\rho^{-1}(i) \cdot \mathrm{grad}_i \tau_{int}(\sigma(t), V(t)) + \nu \cdot \Delta(i)V(t) - \nu \cdot \mathrm{Ric}(i)V(t)$$
$$= \nabla(i)_{V(t)}V(t) + \dot{V}(t)$$

and

$$\rho^{-1}(i) \cdot (\tau_{int}(\sigma(t), V(t)) \cdot H(i)$$
$$- 2 \cdot \nu \cdot \mathrm{d}H(i)V(t) + 2 \cdot \nu \cdot \mathrm{div}_i W(i)V(t))$$
$$= \mathbf{f}(i)(V(t), V(t)).$$

In case of $i(M)$ being a large two-sphere in $\mathbb{R}^3$ and ν being small $\nu \cdot \mathrm{Ric}(i)$ may be neglected in this equation. If moreover $\nu = 0$ the equation turns over in Euler's equation.

Appendix 1 Geometric Preliminaries

Let M be as in Sect. 1 and N be a connected, smooth and oriented manifold with a Riemannian metric $(\ ,\ )$. In case $N = \mathbb{R}^n$ we write $<\ ,\ >$ instead of $(\ ,\ )$ and mean a *fixed scalar product*. The Levi-Civita connections of $(\ ,\ )$ and $<\ ,\ >$ are denoted by ∇ and d respectively. For $J \in E(M, N)$ and $j \in E(\partial M, N)$ we define a *Riemannian metric* on M and ∂M, respectively, by

$$m(J)(X, Y) := (TJX, TJY), \qquad \forall X, Y \in \Gamma TM$$

and

$$m(j)(X, Y) := (TjX, TjY), \qquad \forall X, Y \in \Gamma T\partial M.$$

One also uses the notations $m(J) = J^*(\ ,\ )$ and $m(j) = j^*(\ ,\ )$. Here ΓTQ denotes the collection of all smooth vector fields of any smooth manifold Q (with or without boundary). Both $m(J)$ and $m(j)$ depend smoothly on J and j.

Associated with the metrics $m(J)$ and $m(j)$, we have the respective *Levi-Civita connections* $\nabla(J)$ on M and $\nabla(j)$ on ∂M. They are determined by

$$TJ\nabla(J)_X Y = \nabla_X(TJY) - (\nabla_X(TJX))^\perp, \qquad \forall X, Y \in \Gamma TM$$

and

$$Tj\nabla(j)_X Y = TJ\nabla(J)_X Y - m(j)(W(j)X, Y) \cdot N(j), \forall X, Y \in \Gamma T\partial M.$$

By $W(j)$ we mean the Weingarten map given as follows: Let $N(j) = TJ \cdot \mathbf{n}$, where $\mathbf{n}$ is the positively oriented unit normal vector field (depending on $m(J)$), then

$$(\nabla_Z N(j))^\top = TJ \cdot W(j)Z, \qquad \forall Z \in \Gamma TM,$$

where "$^\top$" denotes the component tangential to $TJTM$. In case $N = \mathbb{R}^n$, we may replace TJ and Tj by $\mathrm{d}J$ and $\mathrm{d}j$ respectively. For any $J \in E(M, \mathbb{R}^n)$ and any $j \in E(\partial M, \mathbb{R}^n)$ let us denote by $\mu(J)$ and $\mu(j)$ the *Riemannian volume* form determined by $m(J)$ and the orientation of M respectively by $m(j)$ and the orientation of ∂M. We define in case of $N = \mathbb{R}^n$ and $\dim \partial M = n - 1$ the *second fundamental* form $\mathbf{f}(j)$ by

$$\mathbf{f}(j)(X,Y) = m(j)(W(j)X,Y)$$

for all $j \in E(\partial M, \mathbb{R}^n)$ and for all $X, Y \in \Gamma TM$. Moreover $H(j) := \mathrm{Tr}W(j)$ is called here the *mean curvature* and $\kappa(j) := \det W(j)$ the *Gaussian curvature* of $j(\partial M) \subset \mathbb{R}^n$ respectively. Reference for this section are [1],[2] and [10].

Appendix 2 Computing Force Densities

M and N are as in Append. 1. Computation of force densities is based on the following proposition:

Proposition A1 *Let* $\dim M = m \le \dim N$ *and let* $K \in C^\infty(M, TN)$, *then for each* $J \in E(M, N)$ *and each* $L \in C^\infty(M, TN)$ *the following equation holds :*

$$\int_M \nabla K \cdot \nabla L \mu(J) = \int_M (\Delta(J)K, L)\mu(J) + \int_{\partial M} (\nabla_\mathbf{n} K, L)i_\mathbf{n}\mu(J), \qquad (A1)$$

where

$$\nabla K \cdot \nabla L = -\frac{1}{2}\mathrm{Tr}\ c(\nabla K(J), TJ) \cdot c(\nabla(J)L, TJ) + \mathrm{Tr}\ A(\nabla K(J), TJ) \cdot \tilde{A}(\nabla L, TJ)$$

(with ∇ the Levi-Civita connection of the metric (,) on N) and $\Delta(J)$ denotes the Laplacian of $m(J)$.. This implies in turn a Green's equation

$$\int_M (\Delta(J)K, L)\mu(J) - \int_M (K, \Delta(J)L)\mu(J)$$

$$= \int_{\partial M} (\nabla_\mathbf{n} L, K)i_\mathbf{n}\mu(J) - \int_{\partial M} (\nabla_\mathbf{n} K, L)i_\mathbf{n}\mu(J).$$

Here $i_\mathbf{n}\mu(J)$ is the volume element on ∂M defined by $\mu(J)$.

Proof.   Writing any $L \in C^\infty(M, TN)$ relative to a given $J \in E(M, N)$ in the form

$$L = TJX(L, J) + L^\perp,$$

with a unique $X(L, J) \in \Gamma TM$ (and $L^\perp$ being such that $L^\perp(p)$ is the component normal to TJT_pM for all $p \in M$), we have the following formula at hand :

$$\nabla_X L = TJ\nabla(J)_X X(L, J) + (\nabla_X L)^\perp, \qquad \forall X \in \Gamma TM. \qquad (A2)$$

and write in turn

$$T_X L = c(\nabla L, TJ) \cdot TJ + TJA(\nabla L, TJ). \qquad (A3)$$

From (A2) and (A3) we read off :

$$c(\nabla L, T)TJ = (\nabla L)^\perp,$$

as well as

$$A(\nabla L, TJ) = \nabla X(L, J) + W(J, L) \quad \forall L \in C^{\infty}(M, TN) \quad \forall J \in E(M, N).$$

Here $W(J, L)$ is given by $TJW(J, L)X = (\nabla L^{\perp}X)^{\top}$, with " $^{\top}$" denoting the pointwise formed component tangential to $TJTM$. Following [18] we introduce the notions of $\Delta(J)K$ and $\Delta(J)\gamma$, the Laplacian, for any $K \in C_J^{\infty}(M, TN)$ and any $\gamma \in A_J^1(M, TN)$ (where the index J means that K and γ respectively cover J). Introducing $\nabla^{\star}$ we set for any $K \in C_J^{\infty}(M, TN)$, any $\gamma \in A_J^1(M, TN)$ and any $J \in E(M, N)$

$$\nabla^{\star}K := 0$$

and

$$\nabla^{\star}\gamma := -\sum_{r=1}^{m} \nabla(J)_{E_r}(\gamma)(E_r),$$

with $\nabla(J)_X(\gamma)(Y) = \nabla_X(\gamma Y) - \gamma(\nabla(J)_X Y)$ for any pair $X, Y \in \Gamma TM$. Clearly if $\gamma \in A^1(M, \mathbb{R})$ and $\nabla = d$ then

$$d^{\star}\gamma = -\text{div}_J Y,$$

provided that $\gamma(X) = m(J)(Y, X)$, for all $X \in \Gamma TM$. The Laplacian $\Delta(J)$ is then defined by

$$\Delta(J) := \nabla\nabla^{\star} + \nabla^{\star}\nabla,$$

Consequently we have

$$\Delta(J)K = \nabla^{\star}\nabla K = -\sum_{r=1}^{m} \nabla_{E_r}(\nabla K)(E_r). \tag{A4}$$

For each $L \in T_J E(M, N)$ and for each $J \in E(M, N)$ we write

$$\nabla L = \bar{A}(\nabla L, TJ)TJ,$$

with $\bar{A}(\nabla L, TJ) : TN|J(M) \longrightarrow TN|J(M)$ a smooth bundle endomorphism. Then for any moving frame $E_1, ..., E_m$ on M orthonormal with respect to $m(J)$ we deduce

$$\nabla K \cdot \nabla L = \sum_{r=1}^{m} (\bar{A}^{\star}(\nabla K, TJ) \cdot \bar{A}(\nabla L, TJ)(TJE_r), TJE_r)$$

$$= \sum_{r=1}^{m} (\bar{A}^{\star}(K, TJ) \cdot \nabla_{E_r} L, TJE_r),$$

with $\bar{A}^{\star}(\nabla K, TJ)$ being the adjoint of $\bar{A}(\nabla K, TJ)$ formed with respect to $(\ ,\ )$. Hence

$$\nabla K \cdot \nabla L = \sum_{r=1}^{m} (\nabla_{E_r}(\bar{A}^{\star}(\nabla K, TJ)L), TJE_r)$$

$$- \sum_{r=1}^{m} (L, \nabla_{E_r}(\bar{A}(\nabla K, TJ))TJE_r)$$

holds true. We derive from (A4)

$$\nabla K \cdot \nabla L = \sum_{i=1}^{m} (\nabla_{E_r}(\bar{A}^*(\nabla K, TJ)L), TJE_r)$$

$$+ (\Delta(J)K, L) + \sum_{i=1}^{m} (\bar{A}(\nabla K, TJ)\nabla_{E_r}(TJ)E_r, L).$$

Since $(\bar{A}^*(\nabla(K, TJ)L)^{\top} = TJZ(K, L, J)$ for some well-defined $Z(K, L, J)$ and $\nabla_{E_r}(TJ)E_r$ is pointwise normal to $TJTM$ the following series of equations hold true :

$$\nabla K \cdot \nabla L = -\sum_{i=1}^{m} (\nabla_{E_r}(c(\nabla K, TJ)L), TJE_r) + \mathrm{div}_J Z(K, L, J)$$

$$+ (\Delta(J)K, L) + \sum_{i=1}^{m} (c(\nabla K, TJ)\nabla_{E_r}(TJ)E_r, L^{\top})$$

$$= -\sum_{i=1}^{m} (\nabla_{E_r}(c(\nabla K, TJ)L^{\perp}), TJE_r)$$

$$- \sum_{i=1}^{m} (\nabla_{E_r}(c(\nabla K, TJ)L^{\top}), TJE_r)$$

$$+ \mathrm{div}_J Z(K, L, J) + (\Delta(J)K, L)$$

$$+ \sum_{i=1}^{m} (c(\nabla K, TJ)\nabla_{E_r}(TJ)E_r, L^{\top}),$$

with div_J the divergence operator associated with $m(J)$.

Writing $c(\nabla K, TJ)L^{\perp} = TJU(K, L, J)$, for some well defined $U(K, L, J) \in \Gamma TM$, we obtain

$$\nabla K \cdot \nabla L = -\mathrm{div}_J U(K, L, J) + \mathrm{div}_J Z(K, L, J) + (\Delta(J)K, L).$$

Here $U(K, L, J)$ is given by $TJU(K, L, J) = c(\nabla K, TJ)L^{\perp}$. The theorem of Gauss yields the desired equation (A1). □

Appendix 3 Simple Examples

Here we will study well known one-forms on $E(M, \mathbb{R}^n)$ in the light of our formalism developed above. For simplicity let dim $M = n$. However, (A5) and (A6) below hold also for dim $M \leq n$. Again $j := J|\partial M$ and $l := L|\partial M$. The index "0" means here the component in $E_0(M, \mathbb{R}^n)$.

(i.) In our first example we specify a constitutive map $\mathcal{H}$ by $\mathcal{H}(dJ) = J_0$ for all $J \in E(M, \mathbb{R}^n)$, with J_0 the component of J in $E_0(M, \mathbb{R}^n)$. Then if $X(L, J)$ is as in (A2)

$$\int_M dJ \cdot dL\mu(J) = \int_M \text{Tr}\, \bar{A}(dL, dJ)\mu(J) = \int_M \text{Tr}\, \nabla(J)X(L,J)\mu(J)$$

$$= \int_M \text{div}_J X(L,J)\mu(J) = \int_{\partial M} < N(j), L >_{i_\mathbf{n}}\mu(j)$$

$$= \text{D}(\int_M \mu(J))(L),$$

holds for all $J \in E(M, \mathbb{R}^n)$ and for all $L \in C^\infty(M, \mathbb{R}^n)$. Moreover $N(j) = TJ\mathbf{n}$. Introducing the *volume function* $V : E(M, \mathbb{R}^n) \longrightarrow \mathbb{R}$, which assigns to any $J \in E(M, \mathbb{R}^n)$ the *volume*

$$V(J) = \int_M \mu(J),$$

yields

$$DV(J)(L) = \int_M dJ \cdot dL\mu(J). \tag{A5}$$

Let $\hbar_\mathbf{n}$ be given by $\Delta(j)\hbar_\mathbf{n}(dj) = N(j)$ for all $j \in E(\partial M, \mathbb{R}^n)$. Hence

$$DV(J)(L) = \int_{\partial M} \hbar_\mathbf{n}(dj) \cdot dl\mu(j)$$

$\forall J \in E(M, \mathbb{R}^n)$ and $\forall L \in C^\infty(M, \mathbb{R}^n)$.

(ii.) Next let us turn our attention to $\hbar_\partial$ defined by the formula $\hbar_\partial(dj) = j_0$ for all $j \in E(\partial M, \mathbb{R}^n)$ with $j_0 \in E_0(\partial M, \mathbb{R}^n)$. For any $J \in E(\partial M, \mathbb{R}^n)$ and any $l \in C^\infty(\partial M, \mathbb{R}^n)$ one easily verifies the following (cf. (25)) :

$$\int_{\partial M} < \Delta(j)j, l > \mu(j) = \int_{\partial M} dj \cdot dl\mu(j)$$

$$= \int_{\partial M} (\text{div}_j X(l,j) + \theta(l,j) \cdot H(j))\mu(j)$$

$$= \int_{\partial M} < H(j) \cdot N(j), l > \mu(j)$$

$$= \text{D}(\int_{\partial M} \mu(j))(l).$$

(Thus $\Delta(j)j = H(j) \cdot N(j)$ for each $j \in E(\partial M, \mathbb{R}^n)$.) If we define the *area function* $\mathcal{A} : E(\partial M, \mathbb{R}^n) \longrightarrow \mathbb{R}$, which sends any $j \in E(\partial M, \mathbb{R}^n)$ into $\mathcal{A}(j) = \int_{\partial M} \mu(j)$ we get

$$D\mathcal{A}(j)(l) = \int_{\partial M} dj \cdot dl\mu(j) \tag{A6}$$

for all variables of $D\mathcal{A}$.

(iii.) Next let us consider the map $\hbar'_\partial : E(\partial M, \mathbb{R}^n) \longrightarrow C^\infty(\partial M, \mathbb{R}^n)$ given by $\hbar'_\partial(dj) = N(j)$ for all $j \in E(\partial M, \mathbb{R}^n)$. Then the formula

$$\Delta(j)N(j) = -dj \; \mathrm{grad}_j H(j) + (\mathrm{Tr}W(j)^2) \cdot N(j)$$

holds for any $j \in E(\partial M, \mathbb{R}^n)$. In the special case of dim $\partial M = 2$ a topological constant, the *Euler characteristic* $\chi(\partial M)$, enters the constitutive law F. It is hidden in the virtual work obtained by destorting ∂M in the unit normal direction:

$$F(j)(N(j)) = \int_{\partial M} < \Delta(j)N(j), N(j) > \mu(j) = \int_{\partial M} \mathrm{Tr}W(j)^2 \mu(j).$$

Applying the theorems of Cayley-Hamilton (cf. [11]) and of Gauss-Bonnet (cf. [10]) to the very right hand side of equation (A6) yields :

$$F(j)(N(j)) = -4\pi\chi(\partial M) + \int_{\partial M} H(j)^2 \mu(j).$$

Appendix 4 A Metric on $E(M, N)$ Defined via Mass Density and Its Geodesics

Here N is as in Append. 1. We first introduce a metric $\mathcal{B}$ on $E(M, N)$, of which the geodesics represent free motions. This metric is based on a density map ρ of which we may think as a mass density. A map $\rho \in C^\infty(E(M, N), C^\infty(M, \mathbb{R}))$ is called a *density map* if $\rho(J)(p) > 0$ for all $p \in M$ and for $J \in E(M, N)$ and in addition satisfies the *continuity equation*

$$\mathbf{d}\rho(J)(L) = -\frac{\rho(J)}{2}\mathrm{Tr}_{m(J)}\mathbf{d}m(J)(L) \quad \forall J \in E(M, N) \text{ and } \forall L \in C^\infty(M, TN).$$
$$(A7)$$

For some simple proof of the existence of density maps we refer to [7]. Here $\mathbf{d}$ denotes the differential of maps from Fréchet manifolds into Fréchet spaces and $\mathbf{d}m(J)$ is given for all $J \in E(M, N)$ by

$$\mathbf{d}m(J)(L)(X, Y) = (\nabla_X L, TJY) + (TJX, \nabla_Y L),$$

for all $L \in C^\infty(M, TN)$ and all $X, Y \in \Gamma TM$, cf. [8]. We now define the metric $\mathcal{B}$ on $E(M, N)$ by

$$\mathcal{B}(J)(L_1, L_2) := \int_M \rho(J)(L_1, L_2)\mu(J),$$

for any $J \in E(M, N)$ and any pair $L_1, L_2 \in C^\infty_J(M, TN) = T_J E(M, N)$. Let us next determine the *geodesic spray* of $\mathcal{B}$. The cotangent bundle of $E(M, N)$ is denoted by $T^*E(M, N)$. For any $J \in E(M, N)$ its fiber $T^*_J E(M, N)$ consists of the smooth linear maps from $T_J E(M, N)$ into $\mathbb{R}$. The metric $\mathcal{B}$ yields the injective map $\mathcal{B}^\sharp$ defined by

$$\mathcal{B}^\sharp(J)(L) := \mathcal{B}(J)(L, ..) = \int_M \rho(J)(L, ..)\mu(J) \qquad (A8)$$

$\forall L \in C^\infty_{\mathcal{F}}(M,N)$ and $\forall J \in E(M,N)$. We will work on Im $\mathcal{B}^\sharp$ only (the topology is such that $\mathcal{B}^\sharp$ is a diffeomorphism). Let $\Pi^* : \text{Im } \mathcal{B}^\sharp \longrightarrow E(M,N)$ be the projection. Then the canonical one-form on Im $\mathcal{B}^\sharp$ is given by

$$\theta_E(\mathcal{W}_{\mathcal{B}^\sharp(L)}) = -\mathcal{B}^\sharp(L)(T\Pi^*\mathcal{W}_{\mathcal{B}^\sharp(L)}),$$

where $\mathcal{W}_{\mathcal{B}^\sharp(L)} \in T_{\mathcal{B}^\sharp(L)}\text{Im}\mathcal{B}^\sharp$ for all $L \in TE(M,N)$ and all $K \in T_L TC^\infty(M,N)$. Let $C^\infty_L(M,T^2N) := \{K \in C^\infty(M,TN)|\pi_{TN} \circ K = L\}$, for a given map $L \in C^\infty(M,TN))$. Thus for all $K \in C^\infty_L(M,T^2N)$ for any $L \in C^\infty_{\mathcal{F}}(M,TN)$ equation (A8) turns into

$$(\mathcal{B}^\sharp\theta_E)(L)(K) = -\int_M \rho(J)(T\pi_N \circ K, L)\mu(j).$$

Due to the continuity equation (A7) the symplectic form ω_E is given by

$$\omega_E(L)(K_1, K_2) = \int_M \rho(J)\omega(L)(K_1, K_2)\mu(J) \qquad \text{(A9)}$$

where $\omega := d\theta_N$ and where θ_N is the pull-back (by the metric on N) of the canonical one-form on T^*N. Thus the *geodesic spray*

$$s : TE(M,N) \longrightarrow T^2E(M,N)$$

is determined by

$$\omega_E(L)(s(L), K) = d\mathcal{E}(L)(K),$$

where

$$\mathcal{E}(L) := \frac{1}{2}\mathcal{B}(L,L), \qquad \forall L \in C^\infty(M,TN) \text{ and } \forall J \in E(M,N).$$

The *energy* $e : TN \longrightarrow \mathbb{R}$ is defined by

$$e(v_q) := \frac{1}{2}(v_q, v_q), \qquad \forall v_q \in T_q N \text{ and } \forall q \in N.$$

Let $e_\star$ be given by

$$(e_\star)(J)(L)(p) := e(L(p))$$

for all $L \in C^\infty_{\mathcal{F}}(M,TN)$, all $J \in E(M,N)$ and all $p \in M$. Since

$$\mathcal{E}(L) = \int_M \rho(J)e_\star(L)\mu(J), \qquad \forall L \in C^\infty_{\mathcal{F}}(M,TN) \text{ and } \forall J \in E(M,N), \qquad \text{(A10)}$$

we immediately deduce from (A9) and (A10) the following:

Lemma A2 *The geodesic spray s of $\mathcal{B}$ is $s_\star : C^\infty(M,TN) \longrightarrow C^\infty(M,T^2N)$, given by*

$$s_\star(L) = s \circ L, \qquad \forall L \in C^\infty(M,N), \text{ and } \forall J \in E(M,N)$$

where $s : TN \longrightarrow T^2N$ is the geodesic spray of the metric $(\ ,\ )$ on N. Hence a geodesic curve $\sigma : (-\lambda, \lambda) \longrightarrow E(M, N)$ with $\lambda \in \mathbb{R}^+$ and the initial conditions $\sigma(0) = J$ and $\dot{\sigma}(0) = L$ on $E(M, N)$ is uniquely determined by the requirement that $\sigma_p : (-\lambda, \lambda) \longrightarrow N$ given by

$$\sigma_p(t) := \sigma(t)(p), \qquad \forall t \in (-\lambda, \lambda)$$

is a geodesic curve in N with the initial data $\sigma_p(0) = p$ and $\dot{\sigma}(0) = L(p)$ for any fixed $p \in M$. Clearly if $N = \mathbb{R}^n$ and $(\ ,\ ) = <\ ,\ >$, then σ is a straight line segment.

One now introduces the covariant derivative of $\mathcal{B}$ in the obvious fashion.

Acknowledgment

I am indebted to T. Ackermann and G. Schwarz for many helpful discussions.

References

1.   R. Abraham, J.E. Marsden, T. Ratiu: *Manifolds, Tensor Analysis, and Applications* (Addison Wesley, Global Analysis, 1983)
2.   M. Berger, B. Gostiaux: *Differential Geometry: Manifolds Curves and Surfaces* (Graduate Texts in Mathematics **115**, Springer-Verlag, New York, Berlin, Heidelberg, 1987)
3.   E. Binz: "On the Notion of the Stress Tensor Associated with $\mathbb{R}^n$-invariant Constitutive Laws Admitting Integral Representations", Rep. Math. Phys. **87** 49-57 (1989)
4.   E. Binz: "Constitutive Laws of Bounded Smoothly Deformable Media", to appear in Proceedings of the Winter School of the Deutsche Physikalische Gesellschaft, ed. by A.Hirshfeld and J.Debrus (Lecture Notes in Physics, Springer-Verlag, Berlin Heidelberg New York, 1991)
5.   E. Binz: "Symmetry, Constitutive Laws of Bounded Smoothly Deformable Media, and Neumann Problems", to appear in *Symmetries in Science V*, ed. by B.Gruber (Plenum Press)
6.   E. Binz, H. Fischer: "On the Manifold of Embeddings of a Closed Manifold", in Proceedings of the Conference on Differential Geometric Methods in Mathematical Physics, Techn. Universität Clausthal, (1978) (Lecture Notes in Physics **134** 310-324 (1981))
7.   E. Binz, G. Schwarz, D. Socolescu: "On a Global Differential Geometric Description of the Motions of Deformable Media", in *Infinite Dimensional Manifolds, Groups, and Algebras*, Vol. II, ed. by H.D.Doebner, J.Hennig (1990)
8.   E. Binz, J. Śniatycki, H.Fischer: *Geometry of Classical Fields, Mathematics Studies* **154** (North-Holland Verlag, Amsterdam, 1988)
9.   A. Frölicher. A. Kriegl: *Linear Spaces and Differentiation Theory* (John Wiley, Chichester, England, 1988)

10. W. Greub, S. Halperin, J. Vanstone: *Connections, Curvature and Cohomology*, Vols. I and II (Acad. Press, New York, 1972-73)

11. W. Greub: *Lineare Algebra I*, Graduate Texts in Mathematics, Vol. 23 (Springer-Verlag, Berlin, Heidelberg, New York, 1981)

12. M. Epstein, R. Segev: "Differentiable Manifolds and the Principle of Virtual Work in Continuum Mechanics", J. Math. Phys. **21** 1243-1245 (1980)

13. E. Hellinger: "Die allgemeinen Ansätze der Mechanik der Kontinua", Enzykl. Math. Wiss. **4/4** (1914)

14. M.W. Hirsch: *Differential Topology* (Springer GTM, Berlin, 1976)

15. L. Hörmander: *Linear Partial Differential Operations*, Grundlehren der mathematischen Wissenschaften, Vol.116 (Springer Verlag, Berlin, Heidelberg, New York, 1976)

16. F. John: *Partial Differential Equations*, (Applied Mathematical Science, Vol.1, 1978)

17. L.D. Landau, E.M. Lifschitz: *Elastizitätstheorie*, Lehrbuch der theoretischen Physik, Vol. VII (4. Auflage, Akademie Verlag, Berlin, 1975)

18. Y. Matsushima: "Vector Bundle Valued Canonic Forms", Osaka J. Math. **8** 309-328 (1971)

19. R. Montgomery: "Isoholonomic Problems and Some Applications", Commun. Math. Phys. **128** 562-592 (1990)

20. J. Serrin: *Mathematical Principles of Classical Fluid Mechanics*, Encyclopedia of Physics, Vol. VIII/1, ed. by C. Truesdell and S.Flügge (Springer-Verlag, Berlin, 1959)

21. M. Spivak: *A Comprehensive Introduction to Differential Geometry*, Vol. V (Publish and Perish Inc., Boston Ma.)

22. C. Truesdell, W. Noll: *The Non-Linear Field Theories of Mechanics*, Encyclopedia of Physics, Vol. III/3, ed. by C.Truesdell and S.Flügge (Springer-Verlag, Berlin, 1973)

GL(n, ℝ), Tetrads and Generalized Space-Time Dynamics

Jan J. Sławianowski

Institute of Fundamental Technological Research, Polish Academy of
Sciences, Świętokrzyska 21, 00-049 Warsaw, Poland

Abstract: We discuss the possibility of eliminating the concepts of internal scale and internal geometry from generally-covariant field theories. By internal scale and internal geometry we mean objects like scalar products, norms, etc, introduced by hand, as primary entities, to internal spaces of field multiplets. Typical examples are: Minkowskian metric in the internal space of gravitational co-tetrad, two-dimensional symplectic form used in spinor theory, and Hermitian or Euclidean scalar products in internal spaces of gauge models. We suggest dynamical models free of such absolute objects, thereby, invariant under full linear groups, not only under "rigid" subgroups preserving bilinear forms. In particular, constructed are generally-covariant Lagrangians for tetrads invariant under GL(4, ℝ), and Lagrangians for the Weyl spinor - tetrad, invariant under the homomorphically correlated actions of GL(2, ℂ) and the linear Lorentz-conformal group CO(1, 3) = ℝ⁺ SO(1, 3); the tetrad part is then separately invariant under GL(4, ℝ). In usual models, gravitational Lagrangians are built in a SO(1, 3)-invariant way, and Lagrangians for spinor-tetrad systems are invariant under the homomorphically correlated action of SL(2, ℂ) and SO(1, 3); dilatations are not symmetries. Discussed are perspectives of physical applications of our models. We also suggest some mechanism of deriving absolute objects like structure constants of gauge groups and spatio-temporal fibration of Kaluza worlds from dynamical principles, e.g. from GL(n, ℝ)-invariant and generally-covariant Lagrangians for the field of linear frames.

1 Introduction

Tetrads have been used for a long time in general relativity as gravitational potentials. The tetrad-based formulation has an advantage of using suggestive pseudo-Euclidean frames in all tangent spaces; besides, it is unavoidable when one deals with spinor fields. More recently, it turned out that the use of tetrads as potentials, or as reference frames, enables one to consider alternative theories of gravitation, e.g. metric-teleparallel models and gauge-like treatments.

In typical theories Lagrangians are invariant under the matrix Lorentz group O(1, 3) acting on the tetrad legs. In this paper I am trying to present arguments for conformal and $GL(4, \mathbb{R})$-invariant models for tetrads, and more generally, for $GL(n, \mathbb{R})$-invariant models to be applied in multidimensional Kaluza-like approaches. Such models have certain promising features, e.g. for $n = 4$ the normal-hyperbolic signature may be, in a sense, deduced from differential equations; the velocity of light is then an integration constant. There are no internal metrics and internal scale standards introduced by hand. Concepts like internal symmetry groups, their structure constants, internal metrics, and perhaps even the very spatio-temporal fibration of Kaluza-like worlds, may appear in a dynamical way as parameters characterizing certain special vacuum-like solutions, without being a priori introduced by fundamental Lagrangians.

The models I suggest are different from those investigated by Ne'eman, Šijacki [6] and others, nevertheless, they are motivated by the same belief in $GL(n, \mathbb{R})$.

Let M be a manifold and W a real vector space; $\dim M = \dim W = n$. Our dynamical variables will be W-valued differential one-forms e on M. If e is a co-frame, i.e. $e_x : T_x M \to W$ is an isomorphism for any x, we define the dual contravariant frame $\check{e}$ by $\check{e}_x := e_x^{-1} : W \to T_x M$. Let $p \in W^*$, $u \in W$; the symbols $e[p]$, $\check{e}[u]$ denote, respectively, the differential one-form and vector field defined by $e[p]_x := e_x^* \cdot p = p \circ e_x$, $\check{e}[u]_x := \check{e}_x \cdot u$.

If $W = \mathbb{R}^n$, $\check{e}$ and e become n-tuples of vectors e_K and covectors e^K, $K = 1, \ldots, n$. Using electromagnetic and gauge language: e is a multiplet of potentials e^K, W is its internal space, and the non-holonomic label K is an internal index. The action of the internal group $GL(W)$ is defined by

$$(Ue)_x = U \circ e_x, \quad \text{i.e.} \quad (Ue)^K = U^K{}_M e^M. \tag{1}$$

Lagrangians are Weyl densities of weight one built algebraically of e and de, $L[e] = L(e, de)$. L is to be generally-covariant, i.e. for any $f \in \text{Diff}\, M$, $L[f^*e] = f^*L[e]$ modulo divergence. If $G \subset GL(W)$ is the maximal subgroup preserving L, we say that L is G-suited. Let us review the general structure of G-suited models for natural groups G.

2 The Group Hierarchy of Models

1. $GL(W)$-suited Lagrangians. There is no additional geometry in W. Lagrangians depend on e through the torsion tensor S of the e-teleparallelism connection, $L[e] = L(S)$, where $S^a{}_{bc} = \Gamma^a{}_{[bc]}$, $\Gamma^a{}_{bc} = e^a{}_A e^A{}_{b,c}$. The defining property of Γ is that $\nabla e^A = 0$, $A = 1, \ldots, n$. S may be represented as

$$S = -e_A \otimes de^A = \frac{1}{2} F^A{}_{BC} e_A \otimes e^B \otimes e^C, \tag{2}$$

where F denotes the non-holonomy object of e,

$$[e_A, e_B] = F^C{}_{AB} e_C, \qquad de^A = \frac{1}{2} F^A{}_{BC} e^C \wedge e^B.$$

The field e is holonomic iff $S = 0$, i.e. $F = 0$. All GL(W)-Lagrangians are homogeneous of degree n in S, thus non-quadratic in derivatives. They result in essentially nonlinear equations. The simplest models are obtained by square-rooting second-order tensors quadratic in derivatives

$$L = \sqrt{|T|} = \sqrt{|\det[T_{ab}]|}, \tag{3}$$

$$T_{ab} = 4A\, S^c{}_{ad}\, S^d{}_{bc} + 4B\, S^c{}_{ac}\, S^d{}_{bd} + 4C\, S^c{}_{dc}\, S^d{}_{ab},$$

A, B, C being real constants. $T_{(ab)}$ might be used as a metric tensor of M. The most natural choice corresponds to the first term, i.e. to the Killing-Cartan tensor

$$g_{ab} = 4S^c{}_{ad}\, S^d{}_{bc}. \tag{4}$$

General Diff$M$ × GL(W)-invariant Lagrangians may be written as

$$L(S) = a(S)\sqrt{|g|} = b(S)\sqrt{|T|}, \tag{5}$$

a, b being scalars built algebraically of S alone. They are homogeneous of degree zero in S, e.g. $g_{ai}\, g^{bj}\, g^{ck}\, S^a{}_{bc}\, S^i{}_{jk}$. The simplest nontrivial model is $L = \sqrt{|g|}$, action equal to the g-volume. The square-root structure of all GL(W)-Lagrangians has an analog in Born-Infeld electrodynamics.

2. SL(W)-suited models. The symmetry group is reduced, but W may be still structureless, because SL(W) preserves all standards of volume. Lagrangians have the form $L[e] = L(S, |e|)$, where $|e| := \det[e^A{}_a]$. One can represent them as $L = a(S, |e|)\sqrt{|g|} = b(S, |e|)\sqrt{|T|} = c(S, |e|)|e|$; a, b, c being scalar functions.

3. $\mathbb{R}_+$O(W, n) = CO(W, n)-suited Lagrangians. W is now endowed with conformal geometry, i.e. a one-dimensional subspace $\mathbb{R}n$ is fixed in $W^* \otimes W^*$, n being a metric of non-specified signature. For any $n \in W^* \otimes W^*$ the Dirac-Einstein metric tensor $h[e, n]$ on M is defined as

$$h[e, n]_x = e_x^* \cdot n, \quad \text{i.e.} \quad h[e, n]_{ab} = n_{AB}\, e^A{}_a\, e^B{}_b. \tag{6}$$

Unlike $g[e]$, $h[e, n]$ is e-parallel, $\nabla h = 0$. Its signature is introduced by hand through n. h depends on e in a Diff$M$-covariant and O(W, n)-invariant way; O(W, n) denotes the group of linear isometries of n. Conformal Lagrangians have the form $L[e] = L(S, h)$, the dependence on h being homogeneous of degree zero, $L(S, h) = L(S, lh)$ for any $l > 0$. Having fixed some $n \in W^* \otimes W^*$ we can define three Weitzenböck invariants, quadratic in S,

$$
\begin{aligned}
J_1[e, n] &= h_{ai}\, h^{bj}\, h^{ck}\, S^a{}_{bc}\, S^i{}_{jk}, \\
J_2[e, n] &= h^{ab}\, S^m{}_{an}\, S^n{}_{bm} = \frac{1}{4} h^{ab}\, g_{ab}, \\
J_3[e, n] &= h^{ab}\, S^k{}_{ak}\, S^m{}_{bm}.
\end{aligned}
\tag{7}
$$

They are homogeneous of degree -1 in n, i.e. in h, thus when inserted into Lagrangians, they must be combined multiplicatively with other h-dependent quantities. The simplest models are obtained by square-rooting second order tensors quadratic in S,

$$L = \sqrt{|Z|} = \sqrt{|\det[Z_{ab}]|}, \tag{8}$$

$$Z_{ab} = T_{ab} + 4D\, S^i_{\ ak}\, S^j_{\ bl}\, h_{ij}\, h^{kl} + (EJ_1 + FJ_2 + GJ_3)h_{ab}, \tag{9}$$

D, E, F, G being constant and T given by (3). This is a general twice covariant tensor quadratic in S and conformally-invariant. More general conformal models may be represented as

$$L = a(S, h)\sqrt{|g|} = b(S, h)\sqrt{|Z|},$$

a, b being scalars homogeneous of degree zero in h.

4. $O(W, n)$-suited Lagrangians. W is now endowed with a fixed $n \in W^* \otimes W^*$. The most natural candidate for the metric of M is now $h[e, n]$. It is flat iff e is holonomic, $S = 0$. Lagrangians have the form $L[e] = L(S, h)$, now without any restrictions on the dependence of L on h. There are various scalar-density factorizations,

$$L(S, h) = a(S, h)\sqrt{|g|} = b(S, h)\sqrt{|T|} = c(S, h)\sqrt{|h|} = c(S, h)|e|. \tag{10}$$

The most convenient form is $c\sqrt{|h|}$. Usually a, b, c are simple functions of basic scalars, e.g. of (7). Two kinds of models are distinguished by the use of expressions quadratic in S, namely, the quadratic forms, [3], [5],

$$L = c_1 L_1 + c_2 L_2 + c_3 L_3 = (c_1 J_1 + c_2 J_2 + c_3 J_3)\sqrt{|h|}, \tag{11}$$

and the square-root model

$$L = \sqrt{|Y|} = \sqrt{|\det[Y_{ab}]|}, \quad Y_{ab} = Z_{ab} + N\, h_{ab}, \quad N = \text{const.}, \tag{12}$$

smoothly corresponding with pseudo-Euclidean, conformal, and $GL(W)$-Lagrangians. Models (9), (12) differ by the "cosmological" term Nh_{ab}.

3 Homogeneous Solutions

We are interested in $O(W, n)$-, $CO(W, n)$- and $GL(W)$-models. Their Lagrangians may be generally written as $L[e] = L(S, h)$. It is convenient to use tensor densities $H := \frac{\partial L}{\partial S}$, $Q = \frac{\partial L}{\partial h}$ – field momentum and Dirac-Einstein stress. Models $CO(W, n)$ and $GL(W)$ correspond, respectively, to conditions $h_{ij}Q^{ij} = 0$, $Q^{ij} = 0$. Field equations have the form

$$K_i^{\ j} := \nabla_k H_i^{\ jk} + 2S^m_{\ mk} H_i^{\ jk} - 2h_{ik} Q^{kj} = 0. \tag{13}$$

There is a family of geometrically distinguished solutions of (13). To present them we must introduce a few additional concepts. Let $LA[e]$ denote the Lie algebra of vector fields generated by the space $\check{e}[W] := \{\check{e}[u] : u \in W\}$. If $LA[e] = \check{e}[W]$, we say that e is a Lie form. This occurs iff $F^A_{\ BC} = \text{const.}$, i.e. $\nabla S = 0$. We say that a Lie form e is abelian, semisimple, etc., if the algebra $LA[e] = \check{e}[W]$ is abelian, semisimple, etc. A Lie form e is said to be splitting if there is such a pair of complementary subspaces T, S of W such that the linear spaces of vector fields $\check{e}[T] := \{\check{e}[t]: t \in T\}$, $\check{e}[S] := \{\check{e}[s]: s \in S\}$ are mutually commuting Lie subalgebras of $\check{e}[W]$, thus $\check{e}[W] = \check{e}[T] \oplus \check{e}[S]$. The vector fields $\check{e}[s]$ span an integrable distribution annihilated by the Pfaff forms $e[f]$, $f \in T^*$, and vice versa. We say that e is hypersplitting if dim $T = 1$. Obviously, hypersplitting Lie forms are never semisimple. Now, let E be a hypersplitting Lie form and l a function on M constant on integral surfaces of $E[T^*]$, i.e., on leaves of the $\check{E}[S]$-distribution; $\check{E}[s]l = \, <dl, \check{E}[s]> \, = 0$ for any $s \in S$. We define a new co-frame $e = {}_lE$ such that $\check{e}[t] = \check{E}[t]$, $\check{e}[s] = l\check{E}[s]$ for any $t \in T$, $s \in S$. If $l \neq$ const., e is not a Lie form. Co-frames of this type will be called Galilean; T refers to "time" and S to "space". We say that a Galilean form is regular if the corresponding Lie algebra $\check{E}[S] = \{\check{E}[s]: s \in S\}$ is semisimple, and that it is normal if this semisimple Lie algebra is compact.

Theorem *Let e be a semisimple Lie form or a regular Galilean form with an arbitrary dilatational factor l without critical points. It solves* GL(W)*-invariant equations (13) derived from any Lagrangian $L(S) = a(S)\sqrt{|g|}$, provided a is smooth at $S[e]$. It is also a solution for any Lagrangian $L(S) = b(S)\sqrt{|T|}$ if b is smooth at $S[e]$ and B, C are sufficiently small. In particular, e solves equations derived from $L = \sqrt{|g|}$ and $L = \sqrt{|T|}$, at least for sufficiently small B, C, cf. [7].*

There is no place for the proof of this statement. Let me only quote the arguments showing that semisimple Lie forms are solutions. Namely, equations (13) may be rewritten as

$$G_i{}^{jk}{}_a{}^{bc}\nabla_k S^a{}_{bc} + 2S^m{}_{mk}H_i{}^{jk} - 2h_{ik}Q^{kj} = 0, \qquad (14)$$

where $G := \frac{\partial H}{\partial S} = \frac{\partial^2 L}{\partial S \partial S}$. For GL($W$)-models $Q = 0$. Besides, we have $\nabla S = 0$, because e is a Lie form, and $S^m{}_{mk} = 0$, because the structure constants of semisimple Lie algebras are traceless. The objects G, H are finite under substituting $S[e]$ for S, because the Killing metric g is non-degenerate for semisimple Lie algebras, and, therefore, L is differentiable at $S[e]$, without differential $\sqrt{0}$-singularities. Thus, all terms in (14) vanish, being products of vanishing and finite expressions.

The theorem does not work for non-semisimple Lie forms, in particular for abelian ones. We are dealing then with $\sqrt{0}$-differential singularities, G and H are infinite and the left-hand side of (14) is non-determined, although, by abuse of language, one could claim it vanishes for $n > 2$, because it consists of two terms homogeneous of degree $n-2$ and $n-1$ in S. However, this would be mathematically meaningless, and besides, there would be no GL(W)-invariant metrical interpretation of the formalism, because $g = 0$ for holonomic frames. On the other hand, abelian Lie forms provide geometric solutions for a wide class of O(W, n)-suited models. Namely, they satisfy equations (14) if $Q(0, h) = 0$ and $H(0, h), G(0, h)$ are finite and well defined. This occurs for physically useful polynomial models free of the "cosmological" term const.$\sqrt{|h|}$. Semisimple Lie forms are universal solutions for all GL(W)-models. They are also possible, but no longer universal, in O(W, n)-

suited models; for a given Lagrangian the arbitrariness of structure constants is then restricted by field equations.

The theorem above provides a constructive proof of the formal consistency of field equations. The problem is non-academic, because equations (13) are generally-covariant, thus, overdetermined. Among the n^2 field variables $e^A{}_i$ there are n non-dynamical quantities depending on the choice of coordinates. The strong non-linearity prevented me from performing the Dirac reduction procedure. Everything that may be said a priori is that the n-tuple of equations $K_a{}^0 = 0$, $a = 0, \ldots, n-1$ is free of the second derivatives, thus, it represents a subsystem of Dirac's secondary constraints; the label a runs here from 0 to $n-1$, and the coordinate x^0 is interpreted as the time variable. There is an obvious analogy to equations $R_a{}^0 = 0$ in Einstein's theory of gravitation. However, now it is impossible to decide a priori whether there are no additional secondary constraints. Thus, the mathematical significance of the above quoted solutions consists in providing the constructive proof that secondary constraints are not so strong to exclude any solutions at all.

The above discussion was purely formal and yet far from physics. Let us now try to explain its physical motivation and the physical message of the presented results. I expect that really new ideas in microscopic theory of space-time and perhaps in unifying multidimensional theories should arise from $GL(W)$-invariant models. Before presenting arguments for this view, it is convenient to start with some comments concerning $O(W, n)$-invariant theories. The reason is that this class comprises well established models, including Einstein's theory. Quadratic Lagrangians (11) describe so-called metric-teleparallel theories of gravitation. The tensor $h[e, n]$ is interpreted as a macroscopic metric, $n = \mathrm{diag}(1, -1, -1, -1)$, $\dim M = 4$. Einstein's theory corresponds to $c_1 : c_2 : c_3 = 1 : 2 : -4$; Lagrangian is then invariant modulo divergence under local Lorentz rotations $e \mapsto Ue$, where U depends on $x \in M$, $(Ue)_x = U(x)e_x$, and $U(x) \in O(1, 3)$ for any x. This local invariance means that it is $h[e, n]$ that is a true dynamical variable, whereas the other degrees of freedom contained in e are redundant and describe only local reference frames, h-orthonormal and necessary for introducing spinors. For other ratios $c_1 : c_2 : c_3$ the local Lorentz symmetry is reduced to the global one; e acquires an autonomous physical meaning. Abelian Lie forms are self-evident vacuum solutions of the resulting equations, independently of the choice of c's; the space-time $(M, h[e, n])$ is then flat. Möller [5] hoped to avoid singularities appearing in gravitation theory, by the use of non-quadratic models (10). In all such treatments one assumes that L is $O(1, 3)$-suited and takes as a dogma that the internal tetrad symmetry cannot be extended to a larger subgroup of $GL(4, \mathbb{R})$. This is motivated by the crucial role of the spatio-temporal Lorentz group in special relativity. However, this analogy is wrong. $O(W, n)$ acting as in (1) is an internal symmetry, whereas the Lorentz group is in generally-covariant theories replaced by $\mathrm{Diff}\, M$. If we once suspend the local Lorentz invariance and admit models (10) more general than Einsteinian, there are no physical reasons for the demand of global $O(W, n)$ as a maximal subgroup of $GL(W)$ preserving L. Instead one should consider seriously all dynamical models for tetrads, invariant under the natural subgroups of $GL(W)$ as quoted above. The full $GL(W)$ is the most important candidate, suggested by geometry of degrees of freedom. There are also physical indications of the lack of scale on fundamental level; let us mention, e.g., the Bjorken scaling. Together with certain philosophical arguments this suggests that quantities like internal metrics in isotropic spaces, e.g., n, should be eliminated or derived from something more elementary, e.g., as integration constants characterizing particular vacuum solutions. Their status as primary parameters of fundamental Lagrangians is rather doubtful. This philosophy, evidently related to the spontaneous symmetry breaking and mass generation mechanism, motivated my interest in conformal and GL(W)-invariant Lagrangians. The program of eliminating the internal scale is convergent with the idea of nonlinear regularization and non-quadratic models due to Möller, because conformal Lagrangians (8) resemble Born-Infeld electrodynamics and are essentially non-quadratic in derivatives [5], [7].

Lie-algebraic solutions of (13) are interesting as "vacua" of our models. Ground states, i.e., vacuums in classical field theory are solutions minimizing the energy functional. In generally-covariant theories this definition is inconvenient because of problems with the energy concept. The formal Hamiltonian resulting from a $1 + (n-1)$-splitting of M vanishes on all solutions. The effective ADM-like Hamiltonian following from the reduction procedure is nontrivial, however, in our strongly nonlinear models there is no chance for the explicit construction. One can use, however, a qualitative concept of vacuum, based on some rough, but physically intuitive ideas. Namely, vacua are expected to be as non-excited, thus, homogeneous, constant, as possible. Usually they are self-evident without any calculations. Canonical examples are: flat metric for the Einsteinian gravitational field and $\Psi = 0$ for the Klein-Gordon field.

Lie-algebraic co-frames are most natural candidates for vacuum-like W-valued forms. Indeed, we cannot define the constancy of e by $\nabla e = 0$, because all co-frames fulfill it. As we have seen, the condition $de = 0$, satisfied by holonomic frames, is too restrictive and useless in GL(W)-models. Thus, the only natural "constancy" of e is $\nabla de = 0$, i.e., $\nabla S = 0$, $F^A{}_{BC} = $ const. As S is linear in derivatives of e, $\nabla S = 0$ is analogous to the definition of symmetric spaces, $DR = 0$, D being a covariant differentiation, and R the corresponding curvature tensor.

If e is a Lie form, (M, e) becomes a local group space, i.e., homogeneous space, with trivial isotropy groups, of a Lie group G corresponding to the structure constants $F^A{}_{BC}$. Vector fields e_K are generators of G. Fixing an origin $o \in M$ we identify M with the group G, o becomes its identity, W a Lie algebra, and e the canonical right-invariant form. Vector fields e_K are right-invariant and generate left regular translations; their duals e^K are right-invariant differential forms. Any choice of $o \in M$ gives rise to the adjoint left-invariant W-valued canonical one form ${}^o e$. The corresponding vector fields ${}^o e_K$ are left-invariant and generate right regular translations. Commutation relations have the form

$$[e_K, e_L] = F^M{}_{KL}\, e_M, \quad [e_K, {}^o e_L] = 0, \quad [\,{}^o e_K, {}^o e_L] = -F^M{}_{KL}\, e_L. \tag{15}$$

The field $g[e]$ becomes the Killing-Cartan metric of G, $g[e] = h[e, n]$, where $n_{AB} = F^C{}_{AD} F^D{}_{BC}$. If e is semisimple, $g[e]$ is non-degenerate and $(M, g[e], \Gamma[e])$ is a pseudo-Riemann-Cartan space. There is a $2n$-dimensional isometry group generated by Killing vectors e_K, $\,^o e_K$, $K = 1, \ldots n$, thus, the state described by e really is highly homogeneous.

The idea of group spaces as vacua was independently suggested by many authors [1], [2], [8]. A drawback of many models I saw was that structure constants F were introduced by hand to the Lagrangians. In my $GL(W)$-models semisimple Lie forms are universal and canonical solutions; structure constants are something like integration constants. There is a difficulty, however, because certain naturals are impossible as dimensions of semisimple Lie algebras, and, unfortunately, this is just the case with $n = 4$. In view of desirable properties of group-space vacua this seems to be an argument against $GL(W)$-models. However, one can save the idea with the help of multidimensional Kaluza approaches. Besides, from the theorem above it follows that dimensions "semisimple + one", e.g. $n = 4$, are also not so bad. Moreover, they furnish interesting arguments supporting the idea of $GL(W)$-invariance.

If e is a regular Galilean form, and the extension factor l has no critical points, then $g[e]$ is non-singular and the contribution of $\check{e}[T]$ to signature is positive. The pseudo-Riemannian structure $(M, g[e])$ has a $(2n-1)$- dimensional isometry group; the Killing vectors have the form $\check{e}[t] = \check{E}[t]$, $\check{E}[s] = \frac{1}{l}\check{e}[s]$, $\,^o\check{E}[s]$, where $t \in T$, $s \in S$ and $o \in M$ is a fixed origin. Let us fix a basis $\{t \in T, s_R \in S : R = 1, \ldots n-1\}$ in W and denote $E_O = \check{E}[t]$, $E_R = \check{E}[s_R]$. Commutation rules for Killing vectors read

$$[E_R, E_S] = C^Z{}_{RS} E_Z, \quad [E_R, \,^o E_S] = 0, \quad [\,^o E_R, \,^o E_S] = -C^Z{}_{RS} \,^o E_Z,$$

$$[E_O, E_R] = [E_O, \,^o E_S] = 0, \qquad R, S, Z = 1, \ldots n-1, \tag{16}$$

$C^Z{}_{RS}$ being constant. For comparison let us also quote the rules for e,

$$[e_O, e_R] = e_O \log l e_R, \qquad [e_R, e_S] = l C^Z{}_{RS} e_Z, \tag{17}$$

where $e_O = E_O$, $e_R = l E_R$. It is seen that e becomes a Lie form iff $l = $ const. If e is normal, i.e., $\check{E}[S]$ is compact, then $(M, g[e])$ is a normal- hyperbolic space-time, $\check{e}[t] = e_O$ is time-like, $\check{e}[s]$, $s \in S$ are space-like and orthogonal to $\check{e}[t]$, and the metric $g[e]$ is stationary and static in the above sense of the term "time". Let $k \in S^* \otimes S^*$ denote the Killing metric of the Lie algebra $\check{E}[S]$, thus, $k_{RS} = C^U{}_{RZ} C^Z{}_{SU}$. Then we have

$$g[e] = (n-1) \mathrm{d} \log l \otimes \mathrm{d} \log l + k_{RS} E^R \otimes E^S. \tag{18}$$

It is seen that $g[e]$ is stationary and static and $\log l$ is proportional to the time variable, $l = \mathrm{d} \exp(at)$, d, a being real constants,

$$g[e] = (n-1) a^2 \mathrm{d}t \otimes \mathrm{d}t + k_{RS} E^R \otimes E^S. \tag{19}$$

The pseudo-Riemannian structure $(M, g[e])$ is locally isomorphic with the group space of $G = \mathbb{R} \times H$, H being a compact Lie group corresponding to the structure constants $C^T{}_{RZ}$. The group G is endowed with the natural invariant metric reducing on $\mathbb{R}$- and H-fibers respectively to the natural metric of real axis and the Killing metric k of H; IR- and H-fibers assumed orthogonal. The isometry between $(M, g[e])$ and canonical pseudo-Riemannian structure of IR $\times H$ is unique up to the choice of the velocity of light, $c = \sqrt{n-1}\, a$. The group parameter of the vector field $e_O = E_O$ coincides with the time variable iff $E_O \log l = c/\sqrt{n-1}$, i.e., iff l is exponential. Analytically, the exponential shape of l appears in solutions of GL(W)-invariant equations when we demand the separation of variables, i.e., factorization into time-dependent and space-dependent terms. Nevertheless, it must be stressed that Galilean forms are universal solutions of GL(W)-models for any l without critical points.

As mentioned, $g[e]$ is stationary and static in spite of the very strong time-dependence of e itself. More generally, all quantities built out of a normal Galilean form e in a GL(W)-invariant way do not depend on "time". On the contrary, prescriptions breaking affine symmetry introduce the explicit time dependence. As an instructive example let us consider the Dirac-Einstein metric $h[e, n]$ built of the above Galilean form, $n \in W^* \otimes W^*$ being the direct sum of a one dimensional positive metric b^2 on IR and the Killing metric k of $\breve{E}[S]$, thus $n_{OO} = b^2$, $n_{OR} = 0$, $n_{RS} = k_{RS}$,

$$h[e, n] = b^2 dt \otimes dt + d^{-2} \exp(-2at) k_{RS} E^R \otimes E^S. \tag{20}$$

This metric is not stationary, namely, its spatial part undergoes a uniform expansion or contraction with the time dependent factor $\exp(-2at)$. Thus, $(M, h[e, n])$ is an empty de Sitter world. The exponential multiplier of expansion implies that in the Galilean vacua there is a kind of arrow of time hidden in internal variables and in GL(W) non-invariant quantities. This arrow may be detected by phenomena ruled by Lagrangians breaking the GL(W)-symmetry. For example, let us assume that we have at disposal a test matter interacting with the gravitational tetrad e through $h[e, n]$, and inject this matter into the space-time (M, e), where e is a Galilean form and n is given as above. Although $(M, g[e])$ is stationary, it is the nonstationary tensor $h[e, n]$ that will be experienced by such matter as a spatio-temporal metric, thus, our experiments and observations will prove something like cosmological expansion, or contraction, depending on the sign of a.

4 Spinors and Conformal Models

A non-academic example is provided by spinor fields. Namely, in standard theories, Lagrangians describing the interacting gravitational and Weyl or Dirac fields in four-dimensional space-time are not invariant under the action of GL(4, IR) on the gravitational tetrad. Instead, they are invariant under the co-tetrad action of the Lorentz group $SO(1, 3) \subset$ GL(4, IR) homomorphically correlated with the simultaneous action of SL(2, $\mathbb{C}$) $\subset$ GL(2, $\mathbb{C}$) on the Weyl spinors. It is possible to modify the theory in such a way that, retaining some correspondence with the standard models, it becomes invariant under the correlated action of the Lorentz conformal group CO(1, 3) on e and the complex linear group GL(2, $\mathbb{C}$) on the Weyl field f. Moreover, one can do it in such a way that for any solution e of (13) the pair $(e, f) = (e, 0)$ is a solution for the coupled system co-tetrad + spinor. In particular, there are models with GL(4, ℝ)- invariant dynamics of the matter-free background. A simple example is

$$L[e, f] = \sqrt{|\det[g[e]_{ij} + Gm[e, f]_{ij}]|}, \qquad (21)$$

f denoting the Weyl spinor, G – the coupling constant, and m the material tensor given by

$$m_{kl} = \frac{1}{2i}(e^A{}_k \, \sigma_A{}^{\dot{a}b} \, \bar{f}_{\dot{a}} f_{b,l} - e^A{}_l \, \sigma_A{}^{\dot{a}b} \, \bar{f}_{\dot{a},k} f_b). \qquad (22)$$

σ's in (22) denote the relativistic quadruplet of Pauli matrices. It is possible to modify (21) by substituting T instead of g. Replacing g in (21) by Z we restrict the symmetry of the matter-free background to CO(1, 3). It is known that without far-reaching and rather drastic [4] changes like infinite-dimensional spinors or nonlinear representations, it is impossible to extend the symmetry group of (21) to the total GL(4, ℝ) correlated with an appropriate spinorial group. Thus, even in models with conformally-invariant or GL(4, ℝ)-invariant matter-free dynamics, the spinorial matter will detect the expansion/contraction of (20), although, seemingly, $h[e, n]$ is absent in (21). This is interesting from the point of view of the above mentioned cosmological speculations like escaping of galaxies, because the genuine heavy matter is of fermionic origin.

5 Space-time Dimension, Signature, and "Elastic" Kaluza Worlds

In formal schemes of typical theories the space-time dimension and the signature of macroscopic metric are primary elements, taken from experiment. Nevertheless, philosophically oriented physicists always aimed at deriving them formally from something more elementary. In some sense, our GL(W)-invariant dynamics for the field of frames provides a small step in this direction. Dimensions $n = 1, 2$ are in such models excluded, because for $n = 1$ we have $S = 0$, and for $n = 2$ $\det[g_{ij}] = \det[T_{ij}] = 0$. If $n = 3$, there are no Galilean solutions, because in two dimensions there are no semisimple Lie algebras. There are Lie-form solutions corresponding to the Lie-algebras su(2) = so(3, ℝ), sl(2, ℝ) = so(1, 2). For su(2) the Killing metric is elliptic, whereas for sl(2, ℝ) it is normal-hyperbolic $(- + +)$, the minus sign corresponding to the compact dimension. Thus, in the group manifold of SL(2, ℝ) there are closed time-like curves. However, the manifold of the universal covering group $\overline{\text{SL}(2, ℝ)}$ is free of this causal defect; I am grateful to Professor D. Finkelstein for calling my attention to that during the Gauge Conference in Jabłonna, September 1989. In four dimensions there are no semisimple Lie algebras, thus, there are no Lie solutions to GL(4, ℝ)-invariant equations for tetrads. However, there are two Galilean solutions corresponding to three-dimensional Lie algebras su(2), sl(2, ℝ). For the su(2)-based solution the Killing tensor has the signature $(+ - - -)$ and locally $M = ℝ \times S^3 = ℝ \times SU(2)$; ℝ is the time axis and S^3 the compact three-dimensional space. This Galilean solution is normal and the velocity of light is an integration constant, cf. (19). The sl(2)-based solution has signature $(+ + +-)$ and is locally identical with $\mathbb{R} \times \overline{\mathrm{SL}(2, \mathbb{R})}$. Although the resulting pseudo- Riemannian structure is normal-hyperbolic, the corresponding Galilean solution is not normal, because the $\mathbb{R}$-axis is space-like; the time-like directions correspond to compact subgroups of $\mathrm{SL}(2, \mathbb{R})$. Thus the GL($W$)-invariant formalism is nontrivial for dimensions $n \geq 3$. If we assume the normal-hyperbolic signature as a dogma, then $n = 3$ is admissible, however, its drawback is that the $(n - 1) = 2$-dimensional space is not homogeneous. Thus, $n = 4$ is the lowest dimension admissible in GL(W)-models and fully compatible with our intuitions concerning space-time. On the other hand, if we assume $n = 4$, then the normal-hyperbolic signature is in some sense implied by differential equations, because it is characteristic for the self-evident vacuum solutions. This is a promising structure of my GL(W)-invariant formalism as possible candidate for the correct microscopic theory of space-time. Another argument is based on the existence of interesting group-space solutions. Namely, there is a hope that GL(W)-models may eliminate scalar products and the concept of scale from internal spaces of field multiplets, e.g., in gauge theories. In all theories of this kind there exist internal metrics, usually hermitian; these metrics together with structure constants of their symmetry groups are primitive elements of the theory, introduced by hand and verified a posteriori by experiments. Thus, just as the above n_{AB} or symplectic metric in spinorial space, they evoke philosophical objections. It would be much better if they followed from something more fundamental, e.g., from differential equations as integration constants of special solutions. I suppose that my GL(W)-models applied to multidimensional Kaluza-like universes may provide a proper way, because, as we have seen, things like Lie groups, structure constants and internal metrics appear in such models as properties of vacuum solutions. Moreover, such an approach could help us to overcome an unpleasant feature of traditional Kaluza theories, namely, the lack of the full n-dimensional covariance; $n = 4 + m$ being the dimension of Kaluza universe. Those theories are generally-covariant in four-dimensional sense, however the total manifold is endowed with absolute objects, namely, the fibration over the four-dimensional space-time and the particular structure group. The idea of eliminating them is to assume that it is the co-frame field that is a proper unifying field, and to search for special solutions of the form

$$e = (\dots, a^A, \dots; \dots, e^Z, \dots), \qquad A = 0, 1, 2, 3, \quad Z = 4, \dots, n - 1,$$

where the Pfaff problem induced by e^A's is involutive. The maximal $(n - 4)$-dimensional integral surfaces of the system e^A, $A = 0, 1, 2, 3$ would generate the four-dimensional space-time as the quotient manifold of the corresponding foliation. The remaining forms e^Z, $Z = 4, \dots n-1$ are expected to describe connections on the resulting bundle structure, and the gauge group would be generated by the contravariant vectors e_Z, $Z = 4, \dots n - 1$ from the dual frame

$$\check{e} = (\dots, e_A, \dots; \dots, e_Z, \dots).$$

It is seen that the field of frames has exactly as many degrees of freedom as needed to provide the kinematical construction of fibrations, connections and structural groups. But, at this stage, it is yet an open question whether this program is compatible with the dynamics of GL(W)-models.

My motivation was rather microscopic and concerned the fundamental level. Nevertheless, it is reasonable to ask whether GL(4, $\mathbb{R}$)-invariant formalism may provide an alternative description of macroscopic gravitation. A reasonable way would be to investigate stationary spherically symmetric solutions. I have done some steps in this direction, however, the very strong nonlinearity prevented me as yet from obtaining a convincing answer.

Acknowledgments

The gratitude and esteem to Professor H.D. Doebner is obvious from the very contributing to this volume. Nevertheless, I would like also to express explicitly how indebted I am to him for his support and hospitality in Clausthal, which gave me so much opportunity for scientific discussions and exchange of views. The ideas and results reported here have also their origin in my visits to Clausthal. I am also indebted to Alexander von Humboldt Foundation for supporting this collaboration during many years.

Thanks are also due to colleagues from the Arnold Sommerfeld Institute of Mathematical Physics in Clausthal for their effort in edition of this volume and technical help with preparation of my manuscript.

References

1. A. D'Adda, J.E. Nelson, T. Regge: Ann. Phys. (N.Y.) **165** 384 (1985)
2. L. Halpern: In: *Quantum Theory of Gravity*, ed. by S. M. Christensen (Adam Hilger Ltd., Bristol, 1984)
3. W. Kopczyński: J. Phys. A. Math. Gen. **15**, 493, (1982)
4. F. Miglietta: Nuovo Cimento A**52**, 2, 151, (1979)
5. C. K. Möller: Danske Vidensk. Selsk., Mat-Fys Medd. **39** 13 (1978)
6. Y. Ne'eman, D. Šijacki: Phys. Rev. D**37** 11, 3267 (1988)
7. J. J. Sławianowski: Rep. Math. Phys. **23** 177 (1986)
8. M. Toller: Nuovo Cimento B**58** 2, 181 (1980)

On Boundary Conditions for
Yang-Mills Fields
in Spatially Bounded Domains

Jędrzej Śniatycki

Department of Mathematics and Statistics
University of Calgary
Calgary, Alberta, Canada

Abstract: The role of boundary conditions for Yang-Mills fields in spatially bounded domains is examined. It is shown that the conservation laws and the structure of the constraints depend on the choice of boundary conditions. The difficulties due to lack of any existence and uniqueness theorems for mixed problems in Yang-Mills theory are discussed.

1 Motivation

The macroscopic phenomena in physics are fairly well described by classical theories, while the microscopic phenomena are in the domain of quantum mechanics. Moreover, according to the Copenhagen interpretation quantum mechanical measurements are made in terms of classical apparata used to prepare and detect quantum states [1]. An exception to this rule is provided by quantum cosmology, however, it suffers some interpretational difficulties [2].

According to the present state of physical theories, Yang-Mills fields are carriers of the strong and the weak interactions. These interactions are quantum in nature and they have short range. Most controlled laboratory experiments in high energy physics are performed in bounded regions of space-time. The data of these experiments could be roughly decomposed into the initial and final Cauchy data and the boundary data.

Therefore, it is desirable to study Yang-Mills fields, as well as other physical fields, in bounded regions of space-time. Ideally, one would like to have a theory with a quantum description of the behaviour of the system inside a bounded domain of space-time, and a classical description of the measuring apparata. The interaction between the classical and the quantum components would be described by the boundary conditions. Such a theory would give a realistic description of existing experiments. Moreover, it would provide a natural length scale given by the size of the space-time region under consideration.

The program of analysing classical fields in terms of their boundary data on compact regions in space-time is due to Tulczyjew [3,4]; it was continued in [5,6,7]. The aim of this paper is to review the existing results and to discuss unsolved problems.

2 Field Equations

Yang-Mills fields are connections in a principal G bundle P over the base manifold X. Here G is a connected Lie group describing the internal degrees of freedom. The base manifold X represents a spatially bounded region in space-time. It is an oriented Lorentzian manifold with a time-like boundary ∂X. In terms of a local trivialization of P we can represent a Yang-Mills field as a 1-form A on X with values in the Lie algebra g of G. The curvature form of A, given by

$$F = \mathrm{d}A + [A, A] , \tag{1}$$

where $[\,,\,]$ denotes the Lie bracket in g, is called the field strength. The Yang-Mills equations in absence of sources are

$$D_A * F = 0 , \tag{2}$$

where D_A is the covariant differential defined in terms of the connection A in P, and $*$ is the Hodge operator defined in terms of the orientation of X and the Lorentzian metric g_X in X. In terms of local coordinates (x^μ) in X equations (1) and (2) take on the form

$$F_{\mu\nu} = \partial_\mu A_\nu - \partial_\nu A_\mu + [A_\mu, A_\nu] ,$$

and

$$\partial^\mu F_{\mu\nu} + [A^\mu, F_{\mu\nu}] = 0 .$$

The field equations are invariant under gauge transformations, that is automorphisms of the principal bundle P covering the identity transformation in X. The group of gauge transformation is isomorphic to the group of sections of the bundle $P[G]$ associated to P with typical fibre G. Its Lie algebra is the algebra of sections of the adjoint bundle $P[g]$. The action on the field variables of a section ξ of $P[g]$ is given by $A \to A + \delta_\xi A$, $F \to F + \delta_\xi F$, where

$$\delta_\xi A = D_A \xi \quad \text{and} \quad \delta_\xi F = [F, \xi] .$$

For a given $3 + 1$ splitting of X as the product $M \times \mathbb{R}$, where M is a Cauchy surface, and $\mathbb{R}$ corresponds to the time axis, the Cauchy data for the field equations consist of triplets $(A_0, \boldsymbol{A}, \boldsymbol{E})$, where A_0 is the component of A normal to M, $\boldsymbol{A}$ is the pull-back of A to M, and $\boldsymbol{E}$ is the pull back to M of the left interior product of F with the unit normal vector η_M of M,

$$E = \eta_M \rfloor F \,.$$

The normal component A_0 can be eliminated by an apropriate choice of a gauge transformation. The condition

$$A_0 = 0$$

is called a temporal gauge condition. In a temporal gauge the field equations decompose into the evolution equations and the constraint equations for the Cauchy data A and E :

$$\dot{A} = E \,,$$

$$\dot{E} = D_A * F \,,$$

$$D_A * E = 0 \,,$$

where dot denotes the derivative with respect to time and F is the curvature of A,

$$F_{ij} = \partial_i A_j - \partial_j A_i + [A_i, A_j] \,.$$

3 Existence and Uniqueness of Solutions

Since we are dealing with a system of evolution equations in a spatially bounded domain we have to consider what boundary conditions we should impose to ensure the existence and the uniqueness (up to a gauge transformation) of solutions. A linearization of the Yang-Mills equations is a system of uncoupled Maxwell equations. In this case the usual classical boundary conditions consist of a subset of the Cauchy data on the boundary. Among the boundary conditions leading to a well posed problem are the Dirichlet conditions specifying the pull back to the boundary of the electromagnetic potential A, and the Neumann boundary conditions specifying the normal component of the electric field and the tangential component of the magnetic field.

In order to describe the corresponding boundary conditions for Yang-Mills theory we have to discuss the geometric structure given on the boundary. Let ∂P be the inverse image of ∂X under the principal fibre bundle projection $\pi_{XP} : P \to X$. It is a G principal fibre bundle over ∂X. A Yang-Mills field A defines by pull back a connection $A_{\partial X}$ in ∂P. The curvature form $F_{\partial X}$ of this connection coincides with the pull back of F to ∂P. Let $E_{\partial X}$ be the g-valued 1-form on ∂X obtained by the contraction of F with the inward pointing unit normal $\eta_{\partial X}$ of ∂X in X,

$$E_{\partial X} = \eta_{\partial X} \rfloor F \,.$$

The field strength F on ∂X is completely determined by $F_{\partial X}$ and $E_{\partial X}$. The Cauchy data on the boundary are given by $A_{\partial X}$, $E_{\partial X}$, and the normal component $\langle A, \eta_{\partial X} \rangle$ of A which can be eliminated, at least locally, by gauge transformations fixing all points of ∂P. They satisfy the equations

$$F_{\partial X} = \mathrm{d}A_{\partial X} + [A_{\partial X}, A_{\partial X}] \,,$$

$$D_{A_{\partial X}} * E_{\partial X} = 0 \,,$$

$$D_{A_{\partial X}} * F_{\partial X} = J_{\partial X} \,, \tag{3}$$

where $J_{\partial X}$ involves the normal derivative of $E_{\partial X}$.

A Dirichlet type boundary condition in Yang-Mills theory consist of specifying $A_{\partial X}$. Similarly, a Neumann type boundary condition consists of specifying $E_{\partial X}$. In particular, the homogeneous Neumann boundary condition

$$E_{\partial X} = 0$$

is known as the natural boundary condition; in the variational formalism this boundary condition yields the action integral invariant under the variations which need not vanish on the boundary, see Sect. 4.

The mixed problem in partial differential equations deals with determining the solutions of the evolution equations, which satisfy given boundary conditions an compatible initial conditions. At present we do not have existence theorems for mixed problems for Yang-Mills equations. The only existence results, due to Segal [8], Ginibre and Velo [9], and Eardley and Moncrief [10], ensure the existence of solutions of the Cauchy problem in Minkowski space for the Cauchy data in an appropriate Sobolev space. Their method consists of splitting the evolution equations into a linear part, giving a one parameter semi-group, and a non-linear correction term which satisfies assumptions of Segal's theory of non-linear evolution equations [11].

For Yang-Mills equations on bounded domains this method breaks down. The existence of the semi-group defined by the linear part of the equations requires that appropriate boundary conditions should be imposed. However, a wide range of choices of boundary conditions was found incompatible with the non-linear part of the evolution equations.

For example, consider the temporal gauge condition $A_0 = 0$ compatible with a homogeneous Dirichlet boundary condition $A_{\partial X} = 0$. The boundary condition implies that $F_{\partial X} = 0$, and (3) is satisfied only if $J_{\partial X} = 0$. This leads to an additional boundary condition since $J_{\partial X}$ involves the normal derivative of $E_{\partial X}$. A requirement that this new boundary condition should be preserved by the evolution leads to further boundary conditions involving higher normal derivatives. In this way we get a sequence of boundary conditions which terminates in a Sobolev space when we reach the derivatives of the order for which the trace on the boundary does not exist. Thus, if we deal with fields of a high order of differentiability, we end up with boundary conditions given by differential operators of higher order. This difficulty is avoided in Sobolev spaces with a sufficiently low order of differentiability, but in this case the non-linear part of the evolution equations is defined only in the sense of distributions and the Segal's theory does not apply.

One could approach the mixed problem for Yang-Mills equations through the perturbation expansion. For homogeneous Dirichlet data we have shown the existence of solutions in all orders and obtained estimates for their norms. However, the estimates indicated that power series representing the exact solution has zero radius of convergence [12].

Thus, some more work will be neeeded to solve the existence problem for mixed data for Yang-Mills equations in spatially bounded domain. The knowledge of the Cauchy data which admit evolution is essential for the study a canonical quantization of the theory.

4 Variational Formalism

The Yang-Mills equations are the Lagrange-Euler equations corresponding to the Lagrangian

$$L[A] = \tfrac{1}{4}\|F\|^2 \ ,$$

where $\|F\|$ is the norm of F defined by the Lorentzian metric g_X in X and a metric g_g in g invariant under the adjoint action of G. The action integral is

$$I[A] = \int_V \tfrac{1}{4}\|F\|^2 \sqrt{\det g}\, \mathrm{d}_4 x \tag{4}$$

where V is a relatively compact open domain in X, and $\sqrt{\det g}\,\mathrm{d}_4 x$ denotes the Riemannian volume form on X, that is $\det g$ is the determinant of the metric tensor $g_{\mu\nu}$ and $\mathrm{d}_4 x = \mathrm{d}x^1 \wedge \mathrm{d}x^2 \wedge \mathrm{d}x^3 \wedge \mathrm{d}x^4$. Since X is spatially bounded, one can take V to be bounded by $\bar{V} \cap \partial X$ and two Cauchy surfaces M_1 and M_2.

The Hamiltonian structure of the theory is determined by the variational principle

$$\delta I[A] = 0 \tag{5}$$

for all variations δA vanishing on ∂V. Using (4) one obtains the Hamiltonian formalism on the chosen Cauchy surfaces, in which E and A are canonically conjugate. In order to analyze the covariance of the resulting Hamiltonian theory under the group of symmetries which act nontrivially on Cauchy surfaces, I find it convenient to pass through the intermediate step, due to De Donder [13].

Let Y denote the bundle of connections of P and $Z = J^1 Y$ be the first jet extension of Y. Connections in P correspond to sections σ_{YX} of the projection map $\pi_{XY} : Y \to X$. Since $L[A]$ depends on first derivatives of A, it defines a function $L : Z \to \mathbb{R}$. If σ_{YX} is a section corresponding to A, then

$$L[A] = L \circ j^1 \sigma_{YX} \ .$$

The De Donder form Ω corresponding to the Lagrangian L can be defined intrinsically, either as a unique form satisfying certain conditions [14], or in terms of the canonical strcture of the dual bundle to the first jet bundle [15]. In terms of local coordinates $(x^\mu, y^a_\mu, z^a_{\mu\nu})$, adapted to the jet bundle structure it is given by

$$\Omega = \sqrt{\det g}\left(\frac{\partial L}{\partial z^a_{\mu\nu}} \mathrm{d}y^a_\mu \wedge \mathrm{d}_3 x_\mu - \left(\frac{\partial L}{\partial z^a_{\mu\nu}} z^a_{\mu\nu} - L \right) \mathrm{d}_4 x \right), \tag{6}$$

where $\mathrm{d}_3 x_\mu = \frac{\partial}{\partial x^\mu} \rfloor \mathrm{d}_4 x$. For every Yang-Mills field A,

$$I[A] = \int_V j^1 \sigma_{YX}^* \Omega, \tag{7}$$

where σ_{YX} is the section of the connection bundle corresponding to A. Moreover, using the expression (7) in the variational principle (5) we need not require that the variations should vanish on ∂V. Hence, we obtain the following result.

Theorem 1 *A Yang-Mills field A corresponding to a section σ_{YX} of the connection bundle satisfies the Yang-Mills equations if and only if σ_{YX} is the projection by the target map $\pi_{YZ} : Z \to Y$ of a section σ_{ZX} of the source map $\pi_{XZ} : Z \to X$, $\sigma_{YX} = \pi_{YZ} \circ \sigma_{ZX}$, such that*

$$\sigma_{ZX}^*(\xi_Z \rfloor \mathrm{d}\Omega) = 0$$

for every vector field ξ_Z on Z.

The proof of this theorem is given in [6].

5 Covariant Hamiltonian Formalism

The Hamiltonian formalism in field theory is concerned with the evolution of the Cauchy data for the field equations. Since we are dealing with spatially bounded domains, we have to consider what boundary conditions we want to impose on our solutions. Boundary conditions can be presented as a submanifold B of ∂Z projecting onto ∂X. The Hamiltonian formalism in the space of Cauchy data has been studied under the assumption that Ω pulls back to an exact form on B,

$$\iota_B^* \Omega = \mathrm{d}\Pi, \tag{8}$$

for some 3-form Π on B, where $\iota_B : B \to \partial Z$ is the inclusion map [5,6]. This condition is satisfied for all classical boundary conditions studied in literature. In particular, $\Pi = 0$ for Dirichlet boundary conditions, which specify $A_{\partial X}$, and also for the natural boundary conditions $E_{\partial X} = 0$. We shall see presently that Yang-Mills theory admits a Hamiltonian formalism also for boundary conditions which do not satisfy condition (8). For example, the choice $B = \partial Z$ corresponds to the study of all Cauchy data.

The presymplectic structure of the space of Cauchy data is obtained by integration of the exterior derivative of the De Donder form. More precisely, if M is a typical Cauchy surface, the space $\mathcal{C}$ of parametrized Cauchy surfaces consists of embeddings $\sigma : M \to X$, and the space $\mathcal{P}$ of parametrized Cauchy data consists of embeddings $\kappa : M \to Z$ such that $\sigma = \pi_{XZ} \circ \kappa \in \mathcal{C}$ and $\kappa(\partial M) \subset B$. Each κ in $\mathcal{P}$ can be presented as a triplet $(A_0, \boldsymbol{A}, \boldsymbol{E})$, where $\boldsymbol{A}$ is a connection in the pull back bundle $\sigma^* P$, while $\boldsymbol{E}$ is a $\mathfrak{g}$-valued 1-form and A_0 is a function from M to $\mathfrak{g}$. Vectors $\xi \in T_\kappa \mathcal{P}$ can be identified with maps $\xi : M \to TZ$ covering $\kappa : M \to Z$.

Let Θ be a 1-form on $\mathcal{P}$ defined by the integration of Ω,

$$\Theta(\xi) = \int_M \kappa^*(\xi \rfloor \Omega),$$

for every κ in $\mathcal{P}$ and every ξ in $T_\kappa \mathcal{P}$. By direct computation one obtains

$$d\Theta(\xi,\zeta) = \int_M \kappa^*(\xi \rfloor \zeta \rfloor d\Omega) + \int_{\partial M} \kappa^*(\xi \rfloor \zeta \rfloor \Omega). \qquad (9)$$

Since, for manifolds with boundary, vectors at the boundary points must be tangent to the boundary, it follows from (6) that the integral over ∂M vanishes. Hence,

$$d\Theta(\xi,\zeta) = \int_M \kappa^*(\xi \rfloor \zeta \rfloor d\Omega)$$

for every κ in $\mathcal{P}$ and every ξ, ζ in $T_\kappa \mathcal{P}$. For a fixed Cauchy surface the form $d\Theta$ defines the presymplectic structure in the space of Cauchy data. The symplectic reduction is obtained by ignoring the variable A_0. Thus, the symplectic space of Cauchy data consists of pairs (A, E).

It should be noted that the integral over ∂M in (9) vanishes for all theories for which

$$\xi \rfloor \zeta \rfloor \Omega = 0 \text{ whenever } \pi_{YZ*}\xi = \pi_{YZ*}\zeta = 0, \qquad (10)$$

where $\pi_{YZ} : Z \to Y$ is the target map. This condition characterizes the De Donder form in the family of modifications of the Lagrangian introduced by Lepage [16]. On the other hand, in general relativity condition (8) on the boundary data is essential for the Hamiltonian formulation of the theory because (10) is not satisfied.

6 Symmetries and Noether Theorems

There are various notions of symmetry, depending on which part of the theory is to be preserved by a symmetry transformation. Here we use the most restrictive notion, requiring that a symmetry should preserve all the structure of the theory under consideration, including the boundary conditions $B \subset \partial Z$. With this definition we get easily Noether Theorems relating infinitesimal symmetries, conservation laws and constraints.

Theorem 2 (First Noether Theorem) *Each infinitesimal symmetry ξ gives rise to a conserved momentum*

$$J_\xi = \Theta(\xi_\mathcal{P}),$$

where $\xi_\mathcal{P}$ is the vector field on $\mathcal{P}$ corresponding to ξ.

In Yang-Mills theory of special importance are infinitesimal gauge transformations ξ given by sections of the adjoint bundle $P[\mathfrak{g}]$. Whether an infinitesimal gauge transformation is an infinitesimal symmetry or not depends on the choice of boundary conditions. For Dirichlet boundary conditions, given by prescribing the boundary connection $A_{\partial X}$, an infinitesimal gauge transformation is an infinitesimal symmetry if it preserves the boundary connection. The natural boundary condition, given by $E_{\partial X} = 0$, is preserved by all gauge transformations. Similarly, if no boundary conditions are imposed, $B = \partial Z$, all gauge transformations are symmetries. The conserved momenta J_ξ corresponding to infinitesimal gauge symmetries can be interpreted as charges. For each infinitesimal gauge symmetry ξ, and $\kappa \equiv (A_0, \boldsymbol{A}, \boldsymbol{E}) \in \mathcal{P}$,

$$J_\xi(\kappa) = \int_{\partial M} (*\boldsymbol{E}|\xi) - \int_M (\mathrm{D}_A *\boldsymbol{E}|\xi)\,, \tag{11}$$

where $(.|.)$ denotes the scalar product in g.

In order to obtain the Second Noether Theorem we have to introduce the notion of localizable symmetries. Roughly speaking, an infinitesimal symmetry ξ is localizable if, for every pair κ, $\tilde{\kappa}$ of Cauchy data on disjoint Cauchy surfaces, there exists an infinitesimal symmetry $\tilde{\xi}$ such that $\xi_\mathcal{P}(\kappa) = \tilde{\xi}_\mathcal{P}(\kappa)$, and $\tilde{\xi}_\mathcal{P}(\tilde{\kappa}) = 0$, c.f. [6].

Theorem 3 (Second Noether Theorem) *The conserved momenta J_ξ corresponding to localizable infinitesimal symmetries ξ vanish identically on solutions of the field equations.*

In Yang-Mills theory localizable infinitesimal symmetries form a subalgebra of infinitesimal gauge symmetries. For Dirichlet boundary conditions localizable infinitesimal gauge symmetries vanish on the boundary. For the natural boundary condition all infinitesimal gauge transformations are localizable symmetries. Similarly, all infinitesimal gauge transformations are localizable symmetries if no boundary conditions are imposed. In any case the Second Noether Theorem gives rise to the constraint equation

$$\mathrm{D}_A *\boldsymbol{E} = 0\,, \tag{12}$$

which has to be satisfied by the Cauchy data of solutions of the field equations.

Taking into account Eqs. (11) and (12), we see that the values of the conserved charges on solutions of the field equations are given by the boundary data,

$$J_\xi(\kappa) = \int_{\partial M} (*\boldsymbol{E}|\xi)\,.$$

The algebra of infinitesimal gauge symmetries of a connection is isomorphic to a subalgebra of the structure algebra. Hence, for Dirichlet boundary conditions we have only a finite number of conserved charges which need not vanish identically. In the case of the natural boundary condition, or no boundary conditions, all conserved charges vanish identically on solutions of the field equations.

7 Structure of the Constraint Set

For a compact Cauchy surface without boundary the structure of singularitites of the solution set of the constraint equation, (12), was studied by Arms [17]. A generalization to zero levels of arbitrary equivariant momentum maps was studied by Arms, Marsden and Moncrief [18]. Under fairly general technical assumptions, which are satisfied by the Yang-Mills theory in absence of boundary conditions, the set of solution of the constraint equations is a stratified manifold. The strata are labeled by the dimension of the algebra of localizable infinitesimal symmetries of the Cauchy data. Cauchy data with no gauge symmetries form a submanifold of the space of all Cauchy data. For Cauchy data with gauge symmetries we have conical singularities.

Let $\xi \neq 0$ be an infinitesimal localizable gauge symmetry of Cauchy data $(\boldsymbol{A}, \boldsymbol{E})$. That is

$$D_{\boldsymbol{A}}\xi = 0 \tag{13}$$

and

$$[\xi, \boldsymbol{E}] = 0. \tag{14}$$

Equation (13) implies that ξ is uniquely determined by its restriction to the boundary. For Dirichlet boundary conditions localizable infinitesimal gauge symmetries vanish on the boundary. Hence, in the case of Dirichlet boundary conditions, infinitesimal localizable symmetries act on the space of Cauchy data without fixed points. If the results of [18] were applicable in this case, one could conclude that for Dirichlet boundary data the constraint equation (12) defines a submanifold of $\mathcal{P}$. However, the lack of an existence theorem for a Cauchy-Dirichlet problem suggests a possibilty of further constraints.

For the natural boundary condition, as well as for no boundary conditions, all infinitesimal gauge transformations are localizable infinitesimal symmetries. Hence the action of the Lie algebra of localizable infinitesimal symmetries in the space of Cauchy data has fixed points, and we can expect that the constraint set has singularities similar to those described in [18].

It is of interest to investigate more closely the stratification of $\mathcal{P}$ given by the stability algebras of Cauchy data in $\mathcal{P}$. Following [17] we illustrate this stratification in the case of Yang-Mills theory with structure group SU(2). Since infinitesimal gauge symmetries of a given connection form a Lie algebra h isomorphic to a subalgebra of the structure algebra, we have three possibilities: (i) $h = 0$ (no symmetries), (ii) $h = $ u(1), and (iii) $h = $ su(2).

The first case, $h = 0$, corresponds to points with no symmetry. It is a submanifold of $\mathcal{P}$ consisting of irreducible connections.

In the second case, $h = $ u(1), let $\xi \neq 0$, be an infinitesimal gauge symmetry of Cauchy data $(\boldsymbol{A}, \boldsymbol{E})$. It follows from (13) that the connection $\boldsymbol{A}$ is reducible to the U(1) subgroup of SU(2) generated by ξ. Equation (14) implies that $\boldsymbol{E}$ is parallel to ξ. Hence the Yang-Mills field with the Cauchy data $(\boldsymbol{A}, \boldsymbol{E})$ is gauge equivalent to an electromagnetic field. This property is propagated by the evolution equations.

Thus, the stratum (ii) with u(1) symmetry algebra consists of electromagnetic fields disguised as Yang-Mills fields.

In the third case, $h = \mathrm{su}(2)$, there are three linearly independent infinitesimal gauge symmetries satisfying (13) and (14). Hence, $E = F = 0$, and A is a flat connection. These properties are propagated by the evolution equations and the Yang-Mills field with these Cauchy data is trivial.

8 Concluding Remarks

It appears that the main difficulty of Yang-Mills theory in spatially bounded domains is the lack of existence and uniqueness theorems. Without the knowledge of a complete set of data ensuring the existence of solutions and their uniqueness (up to a gauge transformation), we cannot characterise the space of physical degrees of freedom (the reduced phase space). This makes it impossible to analyze a canonical structure of the system or its quantization.

It should be noted that this difficulty is not restricted to spatially bounded domains. The present existence theorems for Yang-Mills equations in Minkowski space are valid only for the Cauchy data in appropriate Sobolev spaces. The restriction to Sobolev spaces means that the charges and the topological charges have to vanish identically. It has been shown by Jaffe and Taubes [19], that there exist static solutions of monopole equations with a prescribed monopole number. Thus, the space of Cachy data admitting solutions of Yang-Mills-Higgs equations with a prescribed monopole number is not empty. However, we do not know how big it is. Moreover, we do not have a Hamiltonian theory in which the monopole number is a dynamical variable. Therefore, it is difficult to discuss classical or quantum interactions between states with different monopole numbers.

References

1.   D. Bohm: *Quantum Theory* (Prentice Hall, New York, 1952)
2.   M. MacCallum: "Quantum Cosmological Models", in *Quantum Gravity*, ed. by C. Isham, R. Penrose, and D.W. Sciama (Clarendon Press, Oxford, 1973)
3.   B. Lawruk, J. Śniatycki, W.M. Tulczyjew: "Special Symplectic Spaces", J. Diff. Eq. **17** 477-497 (1974)
4.   J. Kijowski and W. M. Tulczyjew, *A Symplectic Framework for Field Theories*, Springer Lecture Notes in Physics **107** (Springer, Berlin, 1979)
5.   E. Binz, J. Śniatycki: "Conservation Laws in Spacetimes with Boundary", Class. Q. Grav. **3** 1191–1197 (1986)
6.   E. Binz, J. Śniatycki, H. Fischer: *Geometry of Classical Fields* (North Holland, 1988)
7.   J. Kijowski: "On Energy Localization in Gauge Field Theories and Gravitation", lecture at 22nd Symposium on Mathematical Physics, Toruń, December 1989
8.   I. Segal: "The Cauchy Problem for the Yang-Mills Equations", J. Funct. Anal. **33** 175–194 (1979)

9. J. Ginibre, G. Velo: "The Cauchy Problem for Coupled Yang-Mills and Scalar Fields in the Temporal Gauge", Commun. Math. Phys. **82** 1–21 (1981)
10. D. Eardley, V. Moncrief: "The Global Existence of Yang-Mills-Higgs Fields in 4-Dimensional Minkowski space", Commun. Math. Phys. **83** 171–179 (1982)
11. I. Segal: "Non-Linear Semi-Groups", Ann. Math. **78** 339–364 (1963)
12. J. Śniatycki, K. Foltinek, G. Bolton: "Perturbation Approach to the Mixed Problem for Yang-Mills Equations" (unpublished)
13. Th. De Donder: "Theorie invariantive du calcul des variations", Bull. Acad. de. Belg. (1929)
14. J. Śniatycki: "On the Geometric Structure of Classical Field Theory in Lagrangian Formulation", Proc. Camb. Phil. Soc. **68** 475–483 (1970)
15. C. Günther: "The Polysymplectic Hamiltonian Formalism in Field Theory and the Calculus of Variations", J. Diff. Geom. **25** 23–53 (1987)
16. Th. Lepage: Acad. Roy. Belg. Bull. (Cl. Sci. V.) **22** 716 (1936)
17. J. Arms: "The Structure of the Solution Set for the Yang-Mills Equations", Math. Proc. Camb. Phil. Soc. **90** 361–372 (1981)
18. J. Arms: J. Marsden, V. Moncrief: "Symmetry and Bifurcations of Momentum Mappings", Commun. Math. Phys. **78** 455–478 (1981)
19. A. Jaffe, C. Taubes: *Vortices and Monopoles: Structure of Static Gauge Theories* (Birkhauser, Boston, 1980)

Parallel Transport of Phases

Armin Uhlmann

University Leipzig, Department of Physics

Abstract: General features of the concept of Berry's phase are reported and extended to parallel transport based on curves of density operators. Product integral representations and a natural connection are introduced.

1 Introduction

Parallel transport of phases is a natural structure in the fundamentals of Quantum Theory. It is my aim to describe some essentials of that structure according to Berry [1] and Simon [2], which is defined via transport conditions for vectors and phases along curves of pure states. A further purpose is to introduce the extension of these constructions to curves of more general states (i.e. mixtures) [3]. To do so is a problem of internal consistency: In Quantum Theory - and in contrast to Classical Statistical Mechanics - the question whether a state is a pure or a mixed one is decided by the set of observables and can, consequently, be changed by adding or neglecting observables (operators). The criteria for parallelity should be compatible with this feature. On the other hand, the case of pure states is basic and most important, and serves as a guide. See also [4].

The vectors of a Hilbert space $\mathcal{H}$ represent *pure* states if two of them can be distinguished by their expectation values provided they are linearly independent. To do so one needs enough *observables* acting as operators on $\mathcal{H}$. The simplest and also natural assumption for this is that potentially every self-adjoint operator is allowed to become an observable. It is however sufficient, and for technical reasons highly desirable, to use the bounded hermitian operators on $\mathcal{H}$, i.e. the hermitian elements of the algebra $\mathcal{B}(\mathcal{H})$ of all bounded operators acting on $\mathcal{H}$.

A vector ψ describes a state by the collection of its *expectation values*

$$A \mapsto \frac{<\psi, A\psi>}{<\psi, \psi>}$$

and for this reason two vectors describe the same state if and only if they are linearly dependent. Excluding the zero of $\mathcal{H}$ and identifying two linearly dependent vectors defines the *projective space*, $\mathbb{P}\mathcal{H}$, which labels uniquely the pure states. It can hence be considered as the *space of pure states*. $\mathbb{P}\mathcal{H}$ can be realized either

a) as the space of 1-dimensional linear subspaces of $\mathcal{H}$ – the first Grassmann manifold of $\mathcal{H}$, or
b) as the space of rays of $\mathcal{H}$ (a ray is 1-dimensional linear subspace with the exclusion of the zero of $\mathcal{H}$), or
c) as the space of the 1-dimensional projection operators, i.e. of the operators $P = P^2 = P^*$ which project $\mathcal{H}$ onto an 1-dimensional subspace.

Here always exclusively $\mathbb{P}\mathcal{H}$ is interpreted as the set of 1-dimensional projections. The merit in doing so is: The points of $\mathbb{P}\mathcal{H}$ appear as operators, and $\mathbb{P}\mathcal{H}$ is canonically imbedded into $\mathcal{B}(\mathcal{H})$ as a subset. As an example, the distance between two elements of $\mathbb{P}\mathcal{H}$ can be given by the operator norm $\| P_2 - P_1 \|$ of their difference. An inconvenience in using case c) above is in the double role the projections of rank one are playing: Such an operator represents as well a state as a genuine observable asking with which apriori probability this state is realized.

$\mathcal{H} - \{0\}$, the Hilbert space without its zero element, can be considered as a $\mathbb{C}^\times$-fibre bundle over $\mathbb{P}\mathcal{H}$. Because the norming of vectors is a topological trivial operation it is further useful to introduce the unit sphere

$$\mathbb{S}(\mathcal{H}) = \{\psi \in \mathcal{H} : <\psi, \psi> = 1\} \tag{1}$$

of $\mathcal{H}$ which is a S^1-bundle over $\mathbb{P}\mathcal{H}$.

Every Schrödinger equation

$$H(t)\psi = i\dot{\psi} \tag{2}$$

determines a (non-canonical) lift from $\mathbb{P}\mathcal{H}$ for the integral curves of

$$[H(t), P] = i\dot{P}. \tag{3}$$

Indeed, if $t \mapsto P_t$ is a solution of (3) and $P_0 = |\psi_0><\psi_0|$ then there is just one solution $t \mapsto \psi(t)$ of (2) with $\psi(0) = \psi_0$. Now $t \mapsto \psi(t)$ is clearly a lift of $t \mapsto P_t$ into $\mathcal{H} - \{0\}$. This lift sits in the subbundle (1) because of the conservation of the norm.

Replacing within (2)

$$H(t) \mapsto H_{new}(t) = H(t) - a(t)\mathbb{1}, \tag{4}$$

the new curve

$$H_{new}(t)\psi_{new} = i\dot{\psi}_{new}, \qquad \psi_{new}(0) = \psi_0$$

in $\mathcal{H} - \{0\}$ is again a lift of $t \mapsto P_t$ with

$$\psi_{new}(t) = \exp i \int_0^t a(t)dt \cdot \psi(t).$$

This shows that the lifting may produce rather arbitrary phases. Furthermore, (2) and (3) produce lifts only for solutions, a rather restricted class of curves in the space of pure states. This explains why the procedure above is *not* a canonical lifting procedure. A *canonical* or *natural* lifting procedure should be valid for all (sufficiently smooth) curves of $\mathbb{P}\mathcal{H}$, and the lifts should be uniquely determined by their basic curves up to its initial value.

The arbitrariness mentioned above can be avoided in going to the *adiabatic limit* [5] – provided this is possible. To do this one considers together with (2) the family of Schrödinger equations

$$H(t/T)\psi_T(t) = i\dot\psi_T(t), \qquad \psi_T(0) = \psi_0$$

with $T > 0$ and the corresponding family of equations (3) on $\mathbb{P}\mathcal{H}$ with solutions

$$t \mapsto P_{T,t} = |\psi_T(t) ><\psi_T(t)|.$$

One refers to *adiabatic convergence* if

$$\lim_{T\to\infty} P_{T,tT} = P_t^{\text{adi}} \tag{5}$$

is converging towards a new curve $t \to P_t^{\text{adi}}$ in $\mathbb{P}\mathcal{H}$.

If it is possible – after a suitable substitution (4) – to reach convergence of ψ_T towards a curve ψ^{adi} in the sense of

$$\text{w}-\lim_{T\to\infty} T\left(\psi_T(tT) - \psi^{\text{adi}}(t)\right) = 0, \tag{6}$$

then one may heuristically (i.e. up to the interchange of two limiting procedures) argue as follows:

$$\begin{aligned} <\psi^{\text{adi}}, \dot\psi^{\text{adi}}> &= \lim_{T\to\infty} <\psi^{\text{adi}}, \frac{\mathrm{d}}{\mathrm{d}t}\psi_T(tT)> \\ &= \lim -iT <\psi^{\text{adi}}, H(t)\psi_T(tT)> \\ &= \lim -iT <H(t)\psi^{\text{adi}}, \psi_T(tT)>. \end{aligned} \tag{7}$$

Because of (6) this can become reasonable only with $<H(t)\psi^{\text{adi}}, \psi^{\text{adi}}>= 0$ and vanishing right hand side.

The question whether convergence (5) and (6) takes place is difficult and only solved [6] using rather strong assumptions. However, in the cases one can prove adiabatic convergence it results in

$$<\psi^{\text{adi}}, \frac{\mathrm{d}}{\mathrm{d}t}\psi^{\text{adi}}>= 0 \quad \text{with} \quad <\psi^{\text{adi}}, \psi^{\text{adi}}>= 1. \tag{8}$$

It is perhaps better to consider (8) as a necessary condition for the convergence of (6). It forces the vanishing of the *dynamical phase* by requiring a suitable shift (4) before performing (6). It is thus a kind of renorming the hamiltonian in order that adiabatic convergence (5) in the state space can imply (6).

At this point we arrived at a *natural* or *canonical* lifting procedure which induces indeed a well known *parallel transport* in the bundle $S(\mathcal{H})$, respectively

$\mathcal{H} - \{0\}$. It is reasonable to 'forget' the adiabatic origin of (8) and to treat this transport condition as a concept in its own right. Let

$$s \mapsto P_s, \qquad 0 \le s \le 1,$$

be an arbitrary (but sufficiently regular) curve in $\mathbb{P}\mathcal{H}$. A lift

$$s \mapsto \psi(s) \quad \text{with} \quad P_s = |\psi(s)><\psi(s)| \tag{9}$$

is called *parallel* iff it fulfills

$$< \psi, \frac{\mathrm{d}}{\mathrm{d}s} \psi > = < \frac{\mathrm{d}}{\mathrm{d}s} \psi, \psi > . \tag{10}$$

However, $< \psi, \dot\psi >$ is purely imaginary for a curve (9) of constant norm, and (10) reduces to

$$< \psi, \frac{\mathrm{d}}{\mathrm{d}s} \psi > = 0. \tag{11}$$

Parallel lifts are integral curves of connection 1-forms. A good choice for them is

$$< \psi, \mathrm{d}\psi >$$

for the fibre bundle $\mathbb{S}(\mathcal{H})$ and

$$\frac{1}{2} \frac{< \psi, \mathrm{d}\psi > - < \mathrm{d}\psi, \psi >}{< \psi, \psi >} \tag{12}$$

for the larger bundle $\mathcal{H} - \{0\}$.

At this place I like to give a first account for an extension to curves of not necessarily pure states. Let the algebra of observables be a unital *-subalgebra $\mathcal{A}$, i.e. a subalgebra containing the identity map and with every operator its hermitian conjugate. Then two linearly independent vectors may not be distinguishable by the elements of $\mathcal{A}$, and the *vector states* of $\mathcal{A}$

$$\omega = \omega_\psi : \ A \mapsto \omega(A) := \frac{< \psi, A\psi >}{< \psi, \psi >}, \qquad A \in \mathcal{A}, \tag{13}$$

generate a foliation of $\mathcal{H} - \{0\}$. Two vectors belong to the same leaf of this foliation iff their vector states (13) coincide. The unitaries (and, in a certain way, the partial isometries) act on every leaf of the commutant $\mathcal{A}'$ of $\mathcal{A}$.

Given a curve

$$s \mapsto \omega_s, \qquad 0 \le s \le 1, \tag{14}$$

of vector states of $\mathcal{A}$ there are in general many essentially different lifts

$$s \mapsto \psi(s) \quad \text{with} \quad \omega_s = \omega_{\psi(s)} \tag{15}$$

into $\mathcal{H} - \{0\}$. It is an obviously meaningful question whether there is a natural criterion distinguishing certain of these lifts, a *transport condition* selecting up to the choice of the initial vector just one lift (15) of a given curve (14). Let me call such a transport condition a *natural parallel transport* where the word *natural* means that the transport depends on $\mathcal{H}$ and $\mathcal{A}$ only.

Such a natural parallel transport gives rise to a *holonomy problem*: A closed curve of vector states will generally not induce a closed parallel lift. If things work well, and (15) is a parallel lift of (14), then the linear functional

$$A \mapsto \nu(A) = <\psi(0), A\psi(1)>, \qquad A \in \mathcal{A}, \tag{16}$$

should depend *only* on the original curve (14). In particular, $\nu(\mathbb{1})$ then would generalize what is called Berry's phase factor (see next section).

An ansatz which will be sufficient for an important class of curves (14) is the following preliminary definition [7]: A lift (15) of (14) is parallel if it is of constant norm and fulfills

$$(\dot{\psi}, B\psi) = (\psi, B\dot{\psi}) \quad \text{for all} \quad B \in \mathcal{A}'. \tag{17}$$

This is, as will be shown later on, a reasonable set of conditions which are similar to the Berry - Simon one.

If the algebra of observables is $\mathcal{B}(\mathcal{H})$, as it was assumed at the beginning, its commutant consists of the multiples of the identity map only, and (17) means that the lift (15) in this case satisfies (10) resp. (11), i.e. the condition of Berry and Simon.

In the setting above it is without further assumptions unclear, which states of the algebra $\mathcal{A}$ can be given by vector states, and how to handle the other states. To circumvent this, a more satisfying way is performing extensions instead of reductions of states. First of all this is nothing but inverting the point of view: One starts with $\mathcal{A}$ and asks for unital embeddings of $\mathcal{A}$ into $\mathcal{B}(\mathcal{H})$ such that all or a reasonable part of the states of $\mathcal{A}$ become reductions of pure vector states of $\mathcal{B}(\mathcal{H})$. One has to ensure, however, that the final results do *not* depend on the choice of the embedding.

2 Parallel Transport

The parallel transport can be realized in rather different bundle spaces and I describe one which is embedded in $\mathcal{B}(\mathcal{H})$. Again I start with problems for pure states before switching to a slightly larger class.

An operator V is called a *partial isometry* iff VV^* and (consequently) V^*V are projection operators referred to as the *left* and *right support* of V, respectively.

Working with pure states one remains in the set of partial isometries of rank one. A partial isometry of rank one, V, can be written as

$$V = |\psi(1) >< \psi(0)|, \quad VV^* = P_1 \quad V^*V = P_0$$

with two normed vectors, and it may be interpreted as annihilating the 'in-state' $P_0 = |\psi(0) >< \psi(0)|$ and creating the 'out-state' $P_1 = |\psi(1) >< \psi(1)|$. Given the in- and out-states, this operation is fixed up to a phase factor because every $\mathbb{P}\mathcal{H}$-invariant for pairs of states depends on the transition probability

$$\text{tprob}(P_0, P_1) = |<\psi(0), \psi(1)>|^2 = \text{tr}(P_0 P_1).$$

This slight arbitrariness cannot be removed without introducing a new structural element.

This new structural element is a curve, $\mathbf{c}$, connecting smoothly P_0 and P_1:

$$\mathbf{c} : s \mapsto P_s, \quad 0 \le s \le 1, \tag{18}$$

With the aid of the following construction it is possible to fix the phase factor in dependence on $\mathbf{c}$. One takes subdivisions

$$1 > s_1 > s_2 > \ldots > s_m > 0 \tag{19}$$

of the parameter s of the curve and performs ([8], [9])

$$V = V(\mathbf{c}) := \lim P_1 P_{s_1} P_{s_2} \ldots P_{s_m} P_0, \tag{20}$$

where the limiting procedure is taken over finer and finer subdivisions (19). To calculate V one uses a *lifted* path

$$\mathbf{c}^{\text{lift}} : \quad s \mapsto \psi(s), \quad \text{with} \quad P_s = |\psi(s) >< \psi(s)| \tag{21}$$

of unit vectors with which (20) is converted into

$$V = |\psi(1) >< \psi(0)| \lim <\psi(1), \psi(s_1) >< \psi(s_1), \psi(s_2) > \ldots < \psi(s_m), \psi(0) > . \tag{22}$$

If (21) is twice differentiable one estimates by Taylor's theorem

$$|1 + (t - s) < \dot{\psi}(s), \psi(s) > - < \psi(s), \psi(t) > | \le (t - s)^2 \, \text{const.}, \tag{23}$$

where the constant is independent of s and t. One knows that (22) converges absolutely if

$$\lim \sum |<\psi(s_{k+1}), \psi(s_k) > -1| \tag{24}$$

is absolutely converging. But (23) guarantees that (24) converges absolutely towards

$$\int |<\dot{\psi}(s), \psi(s) > | \, ds.$$

The existence of (20) is now established.

It is convenient to require

$$<\psi(s), \dot{\psi}(s) >= 0 \tag{25}$$

before performing (22). At this place the parallelity condition appears as a technical device, and the result of (20) or (22) does *not* depend on it. With (25) the estimate (23) results in

$$V(\mathbf{c}) = |\psi(1) >< \psi(0)| \quad \text{if} \quad <\psi, \dot{\psi} >= 0. \tag{26}$$

For an *arbitrary* lift (20) it follows

$$V(\mathbf{c}) = |\psi(1) > < \psi(0)| \exp \int < d\psi, \psi >,$$

because its right hand side is compatible with (25) and invariant under gauge transformations

$$\psi(s) \mapsto \epsilon(s)\psi(s), \qquad |\epsilon(s)| = 1.$$

There is at most one non-zero eigenvalue of (26), and its value is Berry's phase factor

$$\text{Berry}(\mathbf{c}) = \exp \int < d\psi, \psi >= \text{tr}\, V(\mathbf{c}). \tag{27}$$

The modulus of (27) is not greater than one. It equals one iff $\mathbf{c}$ is closed, i.e. a loop.

Essential parts of what was and will be said in this section is true for projections and partial isometries of arbitrary finite rank. A first assertion is:

If (18) is a smooth curve of projection operators of rank k then (20) converges, and the result is a partial isometry $V(\mathbf{c})$ of rank k with left support P_1 and right support P_0.

It will further become evident that for these curves there is a completely invariant characterization of (16) by

$$\nu_{\mathbf{c}}(A) = \frac{1}{k}\text{tr}\,(V(\mathbf{c})A) \tag{28}$$

such that Berry's phase factor is

$$\text{Berry}(\mathbf{c}) = \frac{1}{k}\text{tr}\,(V(\mathbf{c})) = \nu_{\mathbf{c}}(1). \tag{29}$$

To prove (18) for projections of rank k one writes

$$P_s = \sum |\psi_j(s) > < \psi_j(s)|$$

and requires for the curve of orthonormal k-frames $\psi_1, \cdots, \psi_k$ the auxiliary condition

$$< \psi_j, \dot{\psi}_i >= 0 \quad \text{for all} \quad i, j. \tag{30}$$

To my knowledge (30) appeared first in an appendix of Fock's paper [10] as a condition that the phases of k-frames belonging to a degenerate eigenvalue of a time-dependent hamiltonian change as slowly as possible in the course of time. (30) is also known as defining a parallel transport in the fibre bundle of orthonormal k-frames (Stiefel manifolds). Extending Berry's anholonomy to curves of degenerate eigenstates and introducing the associated gauge theory is the idea of [11].

With (30) the right hand side of (20) decomposes into k independent product integrals (22). But their convergence to rank one projections is already established. Hence the assertion is proved.

For k fixed the mapping

$$\mathbf{c} \mapsto V(\mathbf{c}) \tag{31}$$

can be interpreted as a morphism from the groupoid of curves onto the groupoid of rank k partial isometries. The term *groupoid* indicates that two curves can be multiplicated if and only if the end of the first coincides with the beginning of the second. In the same spirit the multiplication of two partial isometries is allowed iff the right support of the first equals the left support of the second. It is now plain to see from (20)

$$V(\mathbf{c}_1\mathbf{c}_2) = V(\mathbf{c}_1)V(\mathbf{c}_2), \tag{32}$$

$$V(\mathbf{c}^{-1}) = V(\mathbf{c})^*. \tag{33}$$

Let me comment on (33) as follows. (20) implies that $V(\mathbf{c})$ does not depend on the way $\mathbf{c}$ is parametrized. But it depends on its orientation. Reversing the orientation gives $\mathbf{c}^{-1}$.

Using (20) one can get a differential equation for the morphism (31) of a curve (18) with varying endpoint. To this end one considers the curve

$$\mathbf{c}_s : t \mapsto P_t, \quad 0 \le t \le s \tag{34}$$

and the corresponding

$$V_s := V(\mathbf{c}_s) \tag{35}$$

to arrive at

$$\dot{V}_s = \dot{P}_s V_s . \tag{36}$$

(35) as defined by (34) and (20) is the unique solution of the differential equation (36) with initial value $V_0 = P_0$.

One can give the solutions of (36) a special format. At first an arbitrary (sufficiently regular) curve $s \mapsto V_s$ may be represented in the following way. One chooses orthonormal k-frames

$$s \mapsto \{\psi_1(s), \ldots \psi_k(s)\} \in V_s\mathcal{H} \quad \text{with} \quad <\psi_j, \dot{\psi}_i> = 0 \tag{37}$$

fulfilling the transport condition (30). Then, reminding $P = VV^*$, there is a *unique* second orthoframe

$$s \mapsto \{\tilde{\psi}_1(s), \ldots \tilde{\psi}_k(s)\} \in P_s\mathcal{H} = V_s^*\mathcal{H} \tag{38}$$

such that

$$V_s = \sum |\psi_j ><\tilde{\psi}_j|. \tag{39}$$

(39) is a solution of (36) if and only if the orthoframe (38) does *not* depend on s, provided (37) is valid.

The proof is a simple matter of calculation after inserting (39) into (36). In the same straightforward manner one proves:

The following three conditions on a curve $s \mapsto V_s$ are mutually equivalent:

$$\dot{V} = \dot{P}V, \qquad V^*\dot{V} = 0, \qquad V^*\dot{V} = \dot{V}^*V. \tag{40}$$

If one – and hence all – of these conditions are fulfilled, the curve $s \mapsto V_s$ is called a *parallel* lift of $s \mapsto P_s$ into the space of partial isometries of rank k. Equivalently

one may characterize such parallel lifts as being *integral curves* of the differential
1-forms

$$dV - dPV, \qquad V^*dV, \qquad \frac{1}{2}(V^*dV - dV^*V). \tag{41}$$

The last one is an antihermitian connection form. This requires a comment and I
denote for that purpose by $\mathcal{I}_k$ the space of partial isometries of rank k. For any k-
dimensional projection operator, P, the fibre $\mathcal{I}_k^P$ is the set of all V with $VV^* = P$.
Let

$$V \mapsto VU, \qquad V^*V \leq UU^* \tag{42}$$

be a map with partial isometry U depending on V. Then one gets from (41)

$$V^*dV - dV^*V \mapsto U(V^*dV - dV^*V)U^* + U^*dU - dU^*U. \tag{43}$$

However, the partial isometries do not constitute a group. To get a *gauge group*
one has to use the *unitary* transformations in (42). But then $-dU^*U = U^*dU$.
Hence the third expression of (41) is a connection form of the unitary group of $\mathcal{H}$.

It remains to say how all this could fit to the last part of Sect. 1. Of course $\mathcal{I}_k$
is not a Hilbert space but it is elegantly embedded in the Hilbert space of Hilbert
Schmidt operators

$$\mathcal{H}^{HS} = \{W \in \mathcal{B}(\mathcal{H}) : \operatorname{tr} WW^* < \infty\}, \qquad < W_1, W_2 >= \operatorname{tr} W_1^* W_2.$$

To that space one applies what has been said at the end of Sect. 1 where the
*-subalgebra $\mathcal{A}$ of $\mathcal{B}(\mathcal{H}^{HS})$ is identified with the set of mappings

$$\mathcal{A} = \{W \mapsto AW, \ A \in \mathcal{B}(\mathcal{H})\}.$$

A curve of projections of rank k can be understood as a substitute of a curve
of density operators on $\mathcal{H}$ of the form

$$s \mapsto \varrho_s := \frac{1}{k}P_s. \tag{44}$$

This curve will now be interpreted as the *reduction* of any curve

$$s \mapsto \frac{1}{\sqrt{k}}V_s \in \mathcal{H}^{HS}, \qquad V_s V_s^* = P_s. \tag{45}$$

In turn, every curve (45) *purifies* the curve of mixed states (44). Because one has
in the present setting

$$\mathcal{A}' = \{W \mapsto WA, \ A \in \mathcal{B}(\mathcal{H})\}, \tag{46}$$

it can be verified straightforwardly that (17) is equivalent to the last equation of
(40). Indeed this calculation can be done in more general terms using the fact that
every density operator ϱ of $\mathcal{H}$ can be purified by decompositions

$$\varrho = WW^*, \qquad W \in \mathcal{H}^{HS}. \tag{47}$$

Thus *every* curve of density operators of H

$$\mathbf{c} : s \mapsto \varrho_s \tag{48}$$

can be purified, i.e. lifted into a curve of pure vector states of the Hilbert space of Hilbert-Schmidt operators, or, what is the same, can be gained by reductions of pure states

$$c^{\mathrm{lift}} : s \mapsto W_s \in \mathcal{H}^{HS}, \quad \varrho_s = W_s W_s^* . \tag{49}$$

Because of (46) rewriting (17) results in

$$\mathrm{tr}\, \dot{W}^*(WA) = \mathrm{tr}\, W^*(\dot{W}A)$$

for all bounded operators A on $\mathcal{H}$. This can be valid only if

$$\dot{W}^* W = W^* \dot{W}, \tag{50}$$

and in this form (13) has been derived in [3]. It is therefore reasonable to call (49) a *parallel lift* or a *parallel purification* of (48) if (50) is valid.

To look at the set of unit vectors of $\mathcal{H}^{HS}$ as a fibre bundle with the unitary group of $\mathcal{H}$ and with the (non-singular) density operators as its base space has been stressed in [12]. See also [13] for problems of interpretation.

An ansatz

$$\dot{W}_s = G_s W_s \quad \text{with} \quad G_s = G_s^* \tag{51}$$

obviously solves (50). Differentiating (47) and replacing $\dot{W}$ by (51) immediately shows

$$\dot{\varrho} = G\varrho + \varrho G. \tag{52}$$

That method appeared in [12], [9]. It fits very well with (36) for curves of parallel isometries where $G = \dot{P}$.

3 The Minimal Length Property

The inequality

$$\frac{<\dot{\psi}, \dot{\psi}>}{<\psi, \psi>} \leq \frac{<\dot{\psi}, \dot{\psi}>}{<\psi, \psi>} - \frac{<\psi, \dot{\psi}><\dot{\psi}, \psi>}{<\psi, \psi>^2}$$

where the right hand side is the lifted projective metric of $\mathbb{P}\mathcal{H}$, shows that Berry's parallelity condition results from minimizing $<\dot{\psi}, \dot{\psi}>$. Hence parallel lifts can be considered as those of shortest length.

Before combining this with the previously discussed scheme a historical remark is in order. If a k-dimensional subspace (or its projection) moves smoothly through its Hilbert space, there are numerous comoving orthonormal bases. How can one avoid 'unnecessary' rotations of these k-frames? The answer given in [10] was to require

$$\int \mathrm{dt} \sum <\dot{\psi}_j, \dot{\psi}_j> = \text{Min} !$$

This simple variational problem implies as its necessary condition (its 'Euler equations')

$$< \psi_j, \dot{\psi}_k >= 0.$$

These ideas can easily be used to produce the parallelity conditions (17), (50), and similar ones. To prepare this let $\mathcal{H}^{\text{ext}}$ be the Hilbert space of an extended system and $\mathcal{A}$ a unital *-subalgebra of $\mathcal{B}(\mathcal{H}^{\text{ext}})$.

Remark. Up to the notation 'ext' things are as in the last part of Sect. 1. In Sect. 2 the role of $\mathcal{H}^{\text{ext}}$ is played by $\mathcal{H}^{\text{HS}}$.

A curve

$$\mathbf{c} : s \mapsto \omega_s \qquad \text{with} \quad 0 \le s \le 1 \tag{53}$$

of states of $\mathcal{A}$ can be *purified* by embedding $\mathcal{A}$ into $\mathcal{B}(\mathcal{H}^{\text{ext}})$ with large enough $\mathcal{H}^{\text{ext}}$ so that there exists a curve

$$\mathbf{c}^{\text{lift}} : s \mapsto \psi(s) \in \mathcal{H}^{\text{ext}} \tag{54}$$

with

$$\omega_s(A) =< \psi(s), A\psi(s) > \qquad \text{for all} \quad A \in \mathcal{A}. \tag{55}$$

(54) is clearly not fixed by (53) and the arbitrariness is the larger the bigger $\mathcal{A}'$, the commutant of $\mathcal{A}$ in $\mathcal{B}(\mathcal{H}^{\text{ext}})$, is. Indeed, every curve of partial isometries

$$s \mapsto U_s \in \mathcal{A}' \quad \text{with} \quad \parallel \psi(s) \parallel = \parallel U_s \psi(s) \parallel \tag{56}$$

gives a new purifying curve

$$s \mapsto \psi'(s) = U_s \psi(s). \tag{57}$$

The purification ambiguity can be diminished by the requirement

$$\int \sqrt{< \dot{\psi}, \dot{\psi} >} \mathrm{d}s = \text{Min} ! \quad \text{or} \quad \int < \dot{\psi}, \dot{\psi} > \mathrm{d}s = \text{Min} ! \tag{58}$$

where the extrema are taken on the set of all lifts (54) satisfying (55). For sufficiently regular curves this is locally equivalent to

$$< \dot{\psi}, \dot{\psi} >= \text{Min} ! \tag{59}$$

If (54) is an admissible curve and $B \in \mathcal{A}'$ then (56) with $U_s = \exp isB$ gives rise to another such curve. The assumption that (54) is already solving (59) or (58) will result in

$$0 \le < B\psi, B\psi > +i[< \dot{\psi}, B\psi > - < \psi, B\dot{\psi} >].$$

This set of inequalities can be valid for all B iff

$$< \dot{\psi}, B\psi >=< \psi, B\dot{\psi} > \qquad \text{for all} \quad B \in \mathcal{A}'. \tag{60}$$

(60) is proved above for hermitian B. But these operators span $\mathcal{A}'$ linearly.

Because of (55) one is working with unit vectors by definition. For curves within $\mathcal{H}^{\text{ext}} - \{0\}$ one either requires the constancy of the vector norms explicitly, or, with the same effect, demands

$$\frac{<\dot{\psi}, \dot{\psi}>}{<\psi, \psi>} = \text{Min !}$$

for parallelity of the lifts (54). However, the conditions (60) remain valid under arbitrary rescaling of the vector norms.

It is highly desirable to know for what curves (53) there exists a unique holonomy, i.e. a unique

$$\nu_{\mathbf{c}}(A) = <\psi(0), A\psi(1)>, \qquad A \in \mathcal{A}, \tag{61}$$

only depending on $\psi(0)$ and $\psi(1)$, the initial and final vectors of *any* parallel lift. This amounts to the s-independence of U_s if (54) and (57) both produce the minima of (59) or fulfill (60). See also [9] for this problem.

Here I circumvent this problem by trying to establish the correctness of (61) directly for the particular but important case

$$\mathcal{A} = \mathcal{B}(\mathcal{H}) \quad \text{and} \quad \mathcal{H}^{\text{ext}} = \mathcal{H}^{\text{HS}}$$

already introduced in Sect. 2. Let (53) be given as a curve of density operators on $\mathcal{H}$

$$\mathbf{c} : s \mapsto \varrho_s \quad \text{with} \quad 0 \le s \le 1 \tag{62}$$

and a lift

$$\mathbf{c}^{\text{lift}} : s \mapsto W_s, \qquad \varrho_s = W_s W_s^* \tag{63}$$

of *minimal* length. The Bures length [14] of (62) then is the Hilbert space length of (63). The method is to use a polygon approximation to the curve (63), and to express this in terms of the curve (62).

With the aid of the polar decomposition

$$W_j = \varrho_j^{\frac{1}{2}} U_j$$

one gets

$$W_1 W_0^* = \varrho_1^{\frac{1}{2}} U_1 U_0^* \varrho_0^{\frac{1}{2}}. \tag{64}$$

For parallel lifts this gives rise to the definitions

$$V(\mathbf{c}) = U_1 U_0^*, \qquad \nu_{\mathbf{c}}(A) = \operatorname{tr} W_0^* A W_1 \quad \text{with} \quad A \in \mathcal{B}(\mathcal{H}) \tag{65}$$

and the aim is to show independence of the chosen parallel lift. This will be done for faithful (non-singular) density operators only. For every subdivision

$$1 > s_1 > s_2 > \ldots > s_m > 0$$

there is the identity

$$\begin{aligned} U_1 U_0^* &= U_1 (U_{s_1}^* U_{s_1})(U_{s_2}^* U_{s_2}) \cdots (U_{s_m}^* U_{s_m}) U_0 \\ &= (U_1 U_{s_1}^*)(U_{s_1} U_{s_2}^*) \cdots (U_{s_m} U_0^*). \end{aligned} \tag{66}$$

The next step is approximating $U_s U_t^*$ for small $s-t$. Because the curve in question is of minimal length the approximation is done by replacing two neighboring W's by

$$\tilde{W}_s = \varrho_s^{\frac{1}{2}} V_s, \quad \tilde{W}_t = \varrho_t^{\frac{1}{2}} V_t \tag{67}$$

such that these two vectors have minimal distance. This is settled by the requirement [3]

$$\tilde{W}_s \tilde{W}_t^* = \varrho_s^{\frac{1}{2}} V_s V_t^* \varrho_t^{\frac{1}{2}} > 0. \tag{68}$$

In this and only in this case $< \tilde{W}_t, \tilde{W}_s >$ is positive and attains its maximal value for all decompositions (67). That maximal value is the root of the transition probability [15] between the two density operators ϱ_s and ϱ_t

$$\text{tprob}(\varrho_s, \varrho_t) = \left(\text{tr}\,(\varrho_t^{\frac{1}{2}} \varrho_s \varrho_t^{\frac{1}{2}})^{\frac{1}{2}}\right)^2.$$

A solution of (68) is obviously

$$V_s V_t^* = \varrho_s^{-\frac{1}{2}} \varrho_t^{-\frac{1}{2}} (\varrho_t^{\frac{1}{2}} \varrho_s \varrho_t^{\frac{1}{2}})^{\frac{1}{2}}$$

and the solution is unique, for otherwise one comes into conflict with the uniqueness of the polar decomposition. Writing now

$$X_{s,t} = \varrho_t^{-\frac{1}{2}} (\varrho_t^{\frac{1}{2}} \varrho_s \varrho_t^{\frac{1}{2}})^{\frac{1}{2}} \varrho_t^{-\frac{1}{2}}, \tag{69}$$

(66) can be approximated by

$$\varrho_1^{-\frac{1}{2}} X_{1,s_1} X_{s_1,s_2} \cdots X_{s_m,0}\, \varrho_0^{\frac{1}{2}}.$$

Hence

$$W_1 W_0^* = \lim X_{1,s_1} X_{s_1,s_2} \cdots X_{s_m,0}\, \varrho_0. \tag{70}$$

This indicates that the left hand side of the non-commutative product integral (70) is independent of the choice of the shortest lift of (62), and the same is true with (64) and (65). The aim, to show the correctness of the holonomy problem for parallel lifts for curves of non-singular density operators, has been reached.

It is worthwhile to rewrite (69) and (70) with the help of the non-commutative *geometric* (or *quadratic*) *mean* [16] which can be defined for two positive definite operators by [17]

$$A \# B := A^{\frac{1}{2}} (A^{-\frac{1}{2}} B A^{-\frac{1}{2}})^{\frac{1}{2}} A^{\frac{1}{2}}.$$

Then

$$X_{s,t} = \varrho_s^{\frac{1}{2}} \# \varrho_t^{-\frac{1}{2}}.$$

Inserting into (70) yields

$$W_1 W_0^* = \lim (\varrho_1^{\frac{1}{2}} \# \varrho_{s_1}^{-\frac{1}{2}})(\varrho_{s_1}^{\frac{1}{2}} \# \varrho_{s_2}^{-\frac{1}{2}}) \cdots (\varrho_{s_m}^{\frac{1}{2}} \# \varrho_0^{-\frac{1}{2}})\, \varrho_0.$$

Therefore the parallel transport can be described by

$$W_1 = V(\mathbf{c}) W_0 \quad \text{with} \quad V(\mathbf{c}) = \lim (\varrho_1^{\frac{1}{2}} \# \varrho_{s_1}^{-\frac{1}{2}})(\varrho_{s_1}^{\frac{1}{2}} \# \varrho_{s_2}^{-\frac{1}{2}}) \cdots (\varrho_{s_m}^{\frac{1}{2}} \# \varrho_0^{-\frac{1}{2}}).$$

A cross check ist now that

$$G := \lim_{\epsilon \to 0} \frac{X_{t+\epsilon, t} - \mathbb{1}}{\epsilon} \tag{71}$$

fulfills (52), i.e.

$$\dot{\varrho} = \varrho G + G \varrho. \tag{72}$$

The same arguments can be applied for curves of density operators of *constant support*. It should be possible to require only *constant rank* in order that the product integrals above and the ones discussed in Sect. 2 should appear as special cases. Presently the correct format of that (hypothetical) product integral is not known to me.

4 The Connection Form

To get parallel lifts of a curve of states one needs at first a suitable extension in order to represent the original curve as the reduction of a curve of pure states, or, what is the same, to allow for a purification. The arbitrariness of the lifting involved gives rise to a gauge group (or gauge groupoid). It is the aim of the following to show the existence of a natural *connection form* (or *gauge potential*) for the parallel transport already discussed. This can and will be done for the normal states of $\mathcal{B}(\mathcal{H})$. Such a state is given by a density operator ϱ of a Hilbert space $\mathcal{H}$ and described by their expectation values

$$\varrho: \quad A \mapsto \varrho(A) := \operatorname{tr} A\varrho.$$

To achieve purification it is sufficient to consider factor extensions. The most important one, the space of Hilbert-Schmidt operators, has already been considered. It is convenient to represent these extensions as spaces of Hilbert-Schmidt mappings of a Hilbert space $\mathcal{H}'$ into the given Hilbert space $\mathcal{H}$:

$$\mathcal{H}^{\text{ext}} = \mathcal{L}^2(\mathcal{H}', \mathcal{H}) \tag{73}$$

consisting of all mappings

$$W: \quad \mathcal{H}' \to \mathcal{H} \quad \text{with} \quad \operatorname{tr}_{\mathcal{H}'} W^*W = \operatorname{tr}_{\mathcal{H}} WW^* < \infty. \tag{74}$$

Here, as usual, W^* is a map from $\mathcal{H}$ into $\mathcal{H}'$ defined by

$$<\psi, W\psi'> = <W^*\psi, \psi'> \quad \text{for all} \quad \psi \in \mathcal{H}, \quad \psi' \in \mathcal{H}'$$

so that

$$W^* \in \mathcal{L}^2(\mathcal{H}, \mathcal{H}') \quad \text{iff} \quad W \in \mathcal{L}^2(\mathcal{H}', \mathcal{H}).$$

The scalar product of $\mathcal{B}(\mathcal{H}^{\text{ext}})$ reads

$$(W_1, W_2) := \operatorname{tr}_{\mathcal{H}'} W_1^* W_2 = \operatorname{tr}_{\mathcal{H}} W_2 W_1^*,$$

where $W_2 W_1^*$ or $W_1^* W_2$ is in $\mathcal{B}(\mathcal{H})$ or $\mathcal{B}(\mathcal{H}')$, respectively. One observes that (73) is nothing than $\mathcal{H}^{HS}$ if $\mathcal{H}' = \mathcal{H}$. Contact with previous notations is reached with

$$\mathcal{A} = \{W \to AW, A \in \mathcal{B}(\mathcal{H})\}, \qquad \mathcal{A}' = \{W \to WB, B \in \mathcal{B}(\mathcal{H}')\}. \qquad (75)$$

In this setting a state ϱ can be purified if and only if

$$\operatorname{rank} \varrho \leq \dim \mathcal{H}'.$$

The set of all states (density operators) which satisfy (75) can now be regarded as the base space of the bundle $\mathcal{H}^{ext} - \{0\}$ with the bundle projection

$$\pi : \qquad W \mapsto \varrho := WW^* / (W, W). \qquad (76)$$

The bundle group is the group of unitaries of $\mathcal{B}(\mathcal{H}')$ acting as

$$W \mapsto \check{U} = WU \quad \text{with} \quad U \in \mathcal{B}(\mathcal{H}'). \qquad (77)$$

The parallelity condition can now be written

$$(\dot{W}, WB) = (W, \dot{W}B) \quad \text{for all} \quad B \in \mathcal{B}(\mathcal{H}')$$

which results in (50) with vectors W of the form (74) out of (73). This can be reexpressed in the following way. For a curve of density operators

$$s \mapsto \varrho_s \qquad \text{with} \quad 0 \leq s \leq 1$$

one looks for purifying curves

$$s \mapsto W_s \in \mathcal{H}^{ext} = \mathcal{L}^2(\mathcal{H}', \mathcal{H})$$

annihilating the differential 1-form

$$W^* dW - (dW^*)W.$$

This is a form with values in $\mathcal{B}(\mathcal{H}')$ living on the space (73). However, it is *not* a connection form for the gauge transformations (77). To remedy that defect I introduce another differential 1-form $\mathbf{A}$ of a similar structure by [18]

$$W^* dW - (dW^*)W = W^* W \cdot \mathbf{A} + \mathbf{A} \cdot W^* W \qquad (78)$$

which vanishes exactly along parallel lifts, and which is a connection form for the transformations (77). If the support of W equals $\mathcal{H}'$ then (78) determines $\mathbf{A}$ uniquely. Otherwise one has to require additionally

$$< \psi', \mathbf{A}\psi' >= 0 \quad \text{for all} \quad \psi' \in \mathcal{H}' \quad \text{with} \quad W\psi' = 0. \qquad (79)$$

With (78) and (79) the differential form $\mathbf{A}$ is completely defined up to those tangential directions $\dot{W}$ for which there does not exist a solution of (78). These directions correspond to tangential directions at the boundary of the base space along which the rank of the density operator is changing.

Using uniqueness it is elementary to show

$$\mathbf{A} + \mathbf{A}^* = 0$$

and it is a matter of straightforward calculation that a regauging (77) results in

$$\mathbf{A} \mapsto \tilde{\mathbf{A}} =: U^* \mathbf{A} U + U^* \mathrm{d} U. \tag{80}$$

It is remarkable that (80) remains valid if one exchanges the auxiliary Hilbert space $\mathcal{H}'$ by another one, say $\mathcal{H}''$, and if U in (77) is an isometry from $\mathcal{H}''$ into $\mathcal{H}'$. Thus the connection forms living on different spaces (73) appear to be 'all the same up to gauge transformations'.

The introduced connection form respects further scale transformations which do not change (76): $\mathbf{A}$ remains *invariant* under scale transformations

$$W \mapsto \lambda W \tag{81}$$

where λ may arbitrarily vary with W. Hence $\mathbf{A}$ can be considered directly as a connection form defined on $\mathbb{P}\mathcal{H}^{\mathrm{ext}}$.

Remark. If $\mathcal{H}' = \mathcal{H}$ and finite dimensional, and if W^{-1} exists, then $\mathbf{A}$ remains unchanged if W is replaced by $(W^*)^{-1}$. In the base space that transformation becomes $\varrho \to (\varrho^{-1})/\mathrm{tr}(\varrho^{-1})$.

There is a further differential 1-form, $\mathbf{G}$, defined on $\mathcal{H}^{\mathrm{ext}}$ as given by (73) but with values in $\mathcal{B}(\mathcal{H})$ and invariant with respect to gauge transformations (77). It is implicitly defined by

$$\mathrm{d}(WW^*) = \mathbf{G}\, WW^* + WW^*\, \mathbf{G}. \tag{82}$$

This is supplemented by

$$< \psi, \mathbf{A}\psi > = 0 \quad \text{for all} \quad \psi \in \mathcal{H} \quad \text{with} \quad W^*\psi = 0$$

to take care of the null space of W. Again the definition (82) works up to certain directions in the tangent space along which the rank (or von Neumann dimension) of the density operator is diminishing. From the definition follows easily

$$\mathbf{G} = \mathbf{G}^*. \tag{83}$$

The differential form $\mathbf{G}$ reflects the operator G introduced at the end of Sect. 2, Equation (52), and also in Sect. 3, Equations (71) and (72). Namely, $G\mathrm{d}s$ is the pull back of $\mathbf{G}$ into the base space of density operators.

It follows from (51) that $\mathrm{d}W - \mathbf{G}W$ vanishes allong parallel lifts. Hence

$$\Theta := \mathrm{d}W - W\mathbf{A} - \mathbf{G}W$$

is vanishing along every parallel lift. On the other hand, the covariant $\mathbf{A}$-derivative DW transforms with (77) like

$$DW := \mathrm{d}W - W\mathbf{A} \mapsto DWU = (\mathrm{d}W - W\mathbf{A})U.$$

Thus, because $\mathbf{G}$ is a gauge invariant, Θ transforms covariantly with (77). Because every (smooth enough) lift can be gauged to become a parallel lift, Θ is vanishing for all lifts and has to be zero:

$$dW - WA = GW. \tag{84}$$

Having a connection form (a gauge potential) it is tempting to introduce its curvature 2-form

$$\mathbf{F} = d\mathbf{A} + \mathbf{A} \wedge \mathbf{A}.$$

Performing the exterior derivative of (84) one gets

$$W(d\mathbf{A} + \mathbf{A} \wedge \mathbf{A}) + (d\mathbf{G} - \mathbf{G} \wedge \mathbf{G})W = 0,$$

$$(d\mathbf{A} + \mathbf{A} \wedge \mathbf{A})W^* = W^*(d\mathbf{G} + \mathbf{G} \wedge \mathbf{G}).$$

A more explicit representation of $\mathbf{A}$ is possible by sandwiching (78) with eigenstates of W^*W. This, however, demands knowledge of the eigenvectors of an arbitrary hermitian trace class operator. With the exception of low dimensions, in particul two, this can scarcely be solved effectively. Another method, using the integral representation (for positive definite X)

$$Y = \int_0^\infty (\exp -sX)Z(\exp -sX)\,ds \quad \text{if} \quad XY + YX = Z$$

is also not easy for calculating, say, $\mathbf{F}$. Therefore, with the exception of pure states, up to now, a satisfactory geometrical interpretation of the gauge potential and the curvature remains to be given.

If $\dim \mathcal{H}' = 1$ then $\mathcal{H}^{\text{ext}}$ coincides with $\mathcal{H}$ and it follows directly from (78), see also (12),

$$\mathbf{A} = \frac{1}{2} \frac{<\psi, d\psi> - <d\psi, \psi>}{<\psi, \psi>},$$

$$\mathbf{F} = \frac{<d\psi, d\psi>}{<\psi, \psi>} - \frac{<\psi, d\psi> \wedge <d\psi, \psi>}{<\psi, \psi>^2}.$$

If W is proportional or equal to a partial isometry, V, see (41), then

$$\mathbf{A} = \frac{1}{2}(V^*dV - dV^*V).$$

An explicit expression for $\mathbf{A}$ in the case $\dim \mathcal{H} = 2$ has been given in [19].

*   *   *   *   *

A considerable fraction of the material presented is due to a manuscript version of a lecture given at the Arnold-Sommerfeld-Institut, Clausthal 1987, which extended a talk at 15th International Conference on Differential Geometric Methods in Theoretical Physics, Clausthal 1986 [7]. For the interest, help, and kind hospitality I am grateful to H.-D. Doebner and his Colleagues.

References

1.   M.V. Berry: Proc. Roy. Soc. London **A392** 45 (1984)
2.   B. Simon: Phys. Rev. Lett. **51** 2167 (1983)
3.   A. Uhlmann: Rep. Math. Phys. **24** 229 (1986)
4.   *Geometric Phases in Physics*, ed. by A.Shapere and F.Wilczek (World Scientific Publishing Co., Singapore, 1990);
     Anomalies, Phases, Defects, ed. by M. Bregola, G. Marmo, and G. Morandi (Bibliopolis, Naples, 1990);
     Topological Phases in Quantum Theory, ed. by B. Markowski and S.I.Vinitski (World Scientific Publishing Co., Singapore, 1989)
5.   M. Born: Z. Phys. **40** 167 (1926);
     M. Born, V. Fock: Z. Phys. **51** 165 (1928);
     M. Born, J. Oppenheimer: Ann. Phys. (Leipzig) **84** 457 (1927)
6.   T. Kato: J. Phys. Soc. Japan **5** 435 (1950)
7.   A. Uhlmann: "Parallel Transport and Holonomy Along Density Operators", in *Differential Geometric Methods in Theoretical Physics*, ed. by H.D. Doebner and J.D. Hennig (World Sci. Publ., Singapore, 1987; pp. 246–254)
8.   R.G. Littlejohn: Phys. Rev. Lett. **61** 2159 (1988);
     M. Berry: "Quantum Adiabatic Holonomy", in *Anomalies, Phases, Defects* (see [4])
9.   A. Uhlmann: Ann. Phys. (Leipzig) **46** 63 (1989)
10. V. Fock: Z. Phys. **49** 323 (1928)
11. F. Wilczek, A. Zee: Phys. Rev. Lett. **52** 2111 (1984)
12. L. Dabrowski, H. Grosse: "On Quantum Holonomy for Mixed States" (Wien, UWThPh-1988-36)
     L. Dabrowski, A. Jadcyk: "Quantum Statistical Holonomy" (Trieste, 155/88/FM)
13. L. Dabrowski: "A Superposition Principle for Mixed States ?" (Trieste, 156/88/FM)
14. D.J.C. Bures: Trans. Amer. Math. Soc. **135** 199 (1969)
15. A. Uhlmann: Rep. Math. Phys. **9** 273 (1976)
16. W. Pusz, L. Woronowicz: Rep. Math. Phys. **8** 159 (1975)
17. T. Ando: Linear Algebra Appl. **26** 203 (1979)
18. A. Uhlmann: "Gauge Field Governing Parallel Transport along Mixed States" (Wien, UWThPh-1990-25)
19. G. Rudolph: "A Connection Form Governing Parallel Transport Along 2 × 2-Density Matrices" (Leipzig-Wroclaw seminar, Leipzig 1990)

II

Classical and Quantum Field Theory

An Alternative Approach to the Quantization of Linear Relativistic Field Equations

Gerhard C. Hegerfeldt

Institut für Theoretische Physik, Universität Göttingen, Göttingen, Germany

Abstract: A procedure is outlined for quantizing linear wave equations as operators in a Hilbert space. It leads to the same quantum fields as the customary procedures like canonical quantization whenever the latter work. But our approach is applicable in more general situations such as the 3-component Oppenheimer equation for which canonical quantization was recently shown to violate relativity requirements. The proposal is analyzed for the general case and seen to always work for nonzero mass where it leads to relativistic quantum fields. Locality has to be checked separately. For zero mass an extra condition is needed. It appears that a linear equation for which our proposal does not work cannot be quantized at all in a Hilbert space in a relativistic way.

1 Introduction

Linear wave or field equations are of importance as they provide a basis for adding interactions between fields. Their quantization is therefore of intrinsic interest. There are essentially two approaches to this problem, and a third indirect one through group representations. The first two approaches are widely used in textbooks for the standard equations. One is based on the canonical formalism and imposes equal-time canonical commutation or anticommutation relations, the other constructs a Fock space from the one-particle space determined by the wave equation. The third uses representations of the Poincaré group as described by Wigner [1], and then constructs fields and field equations from the one-particle space of fixed mass; cf. e.g. [2] and [3]. This last approach yields fields and field equations only as a by-product and does not start from them. The Fock-space construction of the second approach depends on the prior knowledge of what the one-particle space should be and on an invariant scalar product in it. The first approach through the canonical formalism is closest to ordinary quantum mechanics but is not manifestly relativistic. After quantization one has to check the existence of a unitary representation of the Poincaré group, invariance of the vacuum, transformation properties of the fields, and so on. For the usual textbook equations the above approaches work equally well.

Recently, however, it was shown [4] that difficulties arise for the 3-component and 4-component Oppenheimer equations [5]. These equations do not immediately suggest an invariant scalar product, but they admit a canonical formalism. It was shown in [4] that the corresponding canonical quantization leads to quantum fields which although satisfying the original field equations do not admit a unitary representation of the Lorentz group under which they transform covariantly and which leaves the vacuum invariant. A necessary, and for fields with c-number commutators or anticommutators also necessary, condition is the relativistic invariance of the 2-point function. Just this condition was shown in [4] to be violated.

The question therefore arises of how to quantize linear field equations for which canonical quantization does not give a relativistic quantum theory and for which there is no obvious invariant scalar product. May be such equation cannot be quantized at all?

In this paper we propose an alternative procedure for quantizing linear relativistic field equation in a Hilbert space. This proposal works in all cases in which the older approaches worked, and it gives the same result. It also works for cases for which the former failed. Our proposal is based on the requirement that the 2-point function be relativistically invariant. This requirement is analyzed in the sequel and shown to lead to uniquely specified quantum fields if it can be satisfied.

As an illustration of our proposal we show explicitly and in detail how the 3-component Oppenheimer can and must be quantized to yield a local relativistic field with positive Hamiltonian. For nonzero mass our proposal is shown always to work. The Hamiltonian is always positive, but locality has to be checked individually. For general mass-zero equations we establish conditions under which our proposal leads to relativistic quantum fields. From our construction it appears that there may be linear wave equations which cannot be relativistically quantized with positive energy in a Hilbert space.

2 Preliminaries on Linear Wave Equations

Linear relativistic wave or field equations may be written as first order equations of the form

$$\{-iL^\alpha \partial_\alpha + m\}\, \psi = 0 \tag{1}$$

where $L^0, ..., L^3$ are matrices and m is a number (or possibly also a matrix [6], but this case will not be considered here). The field ψ transforms according to a (spinor or ordinary) representation of the Lorentz group,

$$\psi_\Lambda(x) = D(\Lambda)\psi(\Lambda^{-1}x). \tag{2}$$

For $m \neq 0$ the condition for relativistic invariance of (1) is [7]

$$D(\Lambda)^{-1} L^\alpha D(\Lambda) = \Lambda^\alpha{}_\beta L^\beta. \tag{3}$$

For $m = 0$ one has the weaker condition

$$V(\Lambda)^{-1}L^{\alpha}D(\Lambda) = \Lambda^{\alpha}{}_{\beta}L^{\beta} \tag{4}$$

where now the L^{α}'s may be rectangular matrices and where V is some additional representation of the Lorentz group [7].

The general solution of (1) can in principle be obtained by Fourier transform as a superposition of plane-wave solutions $\boldsymbol{u}(p)\exp\{-ip\cdot x\}$. The allowed momentum values p satisfy [7]

$$p^2 = m^2/\lambda_k^2 \equiv \mu_k^2 \tag{5}$$

where λ_i is any nonzero eigenvalue of L^0. We first consider the case for which there is only a single mass value μ so that $p^0 = \pm\sqrt{\boldsymbol{p}^2 + \mu^2}$. For $p^0 > 0$ one has to determine a maximal set of linear independent $\boldsymbol{u}$'s satisfying

$$(-iL^{\alpha}p_{\alpha} + m)\boldsymbol{u}_i^+(\boldsymbol{p}) = 0. \tag{6}$$

For the negative root one has a similar equation for $\boldsymbol{u}_j^-(\boldsymbol{p})$. Then the general solution of (1) can be written, with p^0 now defined as

$$p^0 = +\sqrt{\boldsymbol{p}^2 + \mu^2},$$

in the form

$$\boldsymbol{\psi}(x) = (2\pi)^{-3/2}\int \frac{d^3p}{p^0}\left\{\sum_i a_i(\boldsymbol{p})\boldsymbol{u}_i^+(\boldsymbol{p})e^{-ip\cdot x} + \sum_j b_j(\boldsymbol{p})^*\boldsymbol{u}_j^-(-\boldsymbol{p})e^{ip\cdot x}\right\} \tag{7}$$

$$\equiv \boldsymbol{\psi}^+(x) + \boldsymbol{\psi}^-(x)$$

In case of several mass values μ_k one would have an additional sum. We note that for $m = 0$ one can choose $\boldsymbol{u}_i^-(\boldsymbol{p}) = \boldsymbol{u}_i^+(-\boldsymbol{p})$.

The above $\boldsymbol{u}_i^+(\boldsymbol{p})$ have definite transformation properties which are crucial for our quantization approach. For $m \neq 0$ one obtains from (6)

$$D(\Lambda)[-iL^{\alpha}p_{\alpha} + m]D(\Lambda^{-1})D(\Lambda)\boldsymbol{u}_i^+(\boldsymbol{p}) = 0.$$

Using (3) this yields, with $p' = \Lambda p$,

$$(-iL^{\alpha}p'_{\alpha} + m)D(\Lambda)\boldsymbol{u}_i^+(\boldsymbol{p}) = 0 .$$

Hence $D(\Lambda)\boldsymbol{u}_i^+(\boldsymbol{p})$ must be a linear combination of $\boldsymbol{u}_i^+(\boldsymbol{p}')$'s, i.e., of the form

$$D(\Lambda)\boldsymbol{u}_i^+(\boldsymbol{p}) = \sum_l \kappa_{li}^{\Lambda}(p)\boldsymbol{u}_l^+(\underline{\Lambda p}), \tag{8}$$

where $\underline{\Lambda p} \equiv \boldsymbol{p}'$ is the 3-vector part of Λp. A similar relation holds for $\boldsymbol{u}^-(\boldsymbol{p})$.

A field $\psi(x)$ quantized in a Hilbert space and satisfying a linear field equation must, by general principles, have c-number commutators or anticommutators. The a^*, a, b^*, b will then obey relations of the form

$$[a_i(\boldsymbol{p}), a_j(\boldsymbol{p}')^*]_{\pm} = p^0 Z_{ij}(p)\delta^{(3)}(\boldsymbol{p} - \boldsymbol{p}') \tag{9}$$

with still unspecified $Z_{ij}(p)$, and similarly for b and b^*. The operators a and b have to annihilate the vacuum state if the energy is positive.

We now propose to determine the unknown functions $Z_{ij}(p)$ in such a way that the 2-point function of the field becomes relativistically invariant. This is certainly a necessary condition and will be shown to lead, if it can be satisfied, to a unique solution for Z_{ij} and thus to a unique quantization of the field. Before we study this proposal for the general case we will apply it to the 3-component Oppenheimer equation for which the usual canonical quantization leads to a violation of relativity requirements.

3 Quantization of the Oppenheimer Equation

In the usual notation (j_1, j_2) of representations of the Lorentz group [2] the field ψ of the 3-component Oppenheimer equation transforms according to the 3-dimensional complex representation $(0,1)$.

In this case the L^α's are 4×3-matrices given in terms of the 3×3-matrices S_k, the generators of rotations. One has

$$S_1 = \begin{pmatrix} 0 & 0 & 0 \\ 0 & 0 & -i \\ 0 & i & 0 \end{pmatrix}, \quad S_2 = \begin{pmatrix} 0 & 0 & i \\ 0 & 0 & 0 \\ -i & 0 & 0 \end{pmatrix}, \quad S_3 = \begin{pmatrix} 0 & -i & 0 \\ i & 0 & 0 \\ 0 & 0 & 0 \end{pmatrix} \tag{10}$$

and then

$$L^0 = \begin{pmatrix} 1 & 0 & 0 \\ 0 & 1 & 0 \\ 0 & 0 & 1 \\ 0 & 0 & 0 \end{pmatrix}, \quad L^1 = \begin{pmatrix} & S_1 & \\ 1 & 0 & 0 \end{pmatrix},$$

$$L^2 = \begin{pmatrix} & S_2 & \\ 0 & 1 & 0 \end{pmatrix}, \quad L^3 = \begin{pmatrix} & S_3 & \\ 0 & 0 & 1 \end{pmatrix}. \tag{11}$$

The equations

$$L^\alpha \, \partial_\alpha \, \psi = 0 \tag{12}$$

are then relativistic field equations. This follows either from the original paper of Oppenheimer [5] or from the general theory [7].

Equation (12) is equivalent to a set of two equations,

$$\dot{\psi} = -\boldsymbol{S} \cdot \nabla \psi \qquad \nabla \cdot \boldsymbol{\psi} = 0 . \tag{13}$$

In this form the relativistic invariance is not apparent, and especially the second (transversality) equation has led to some confusion because it is tempting to regard it as a subsidiary condition which might then give rise to difficulties with quantization. This viewpoint is misleading since neither equation in (13) is relativistically invariant on its own and since there is a completely consistent canonical formalism for (12) and the set in (13).

Equation (6) yields, for given $p = (p^0, \boldsymbol{p})$ with $p^0 = |\boldsymbol{p}|$, only one $\boldsymbol{u}^\pm(\boldsymbol{p})$ satisfying

$$\boldsymbol{S} \cdot \boldsymbol{p} \, \boldsymbol{u}^{\pm}(\boldsymbol{p}) = \pm \mid \boldsymbol{p} \mid \boldsymbol{u}^{\pm}(\boldsymbol{p}) \quad \text{and} \quad \boldsymbol{p} \cdot \boldsymbol{u}^{\pm}(\boldsymbol{p}) = 0 \, . \tag{14}$$

Note that one can choose $\boldsymbol{u}^{-}(\boldsymbol{p}) = \boldsymbol{u}^{+}(-\boldsymbol{p})$. As for the Dirac and Weyl equation we normalize $\boldsymbol{u}^{\pm}(\boldsymbol{p})$ to $\mid \boldsymbol{p} \mid$. By (7) the general solution now is

$$\psi(x) = (2\pi)^{-3/2} \int \frac{d^3 p}{p^0} \left\{ a(\boldsymbol{p}) \, \boldsymbol{u}^{+}(\boldsymbol{p}) \, e^{-ip \cdot x} + b(\boldsymbol{p})^{*} \boldsymbol{u}^{-}(-\boldsymbol{p}) \, e^{ip \cdot x} \right\}$$
$$\equiv \psi^{+}(x) + \psi^{-}(x) \, . \tag{15}$$

Canonical quantization of ψ leads to a 2-point function which is not Lorentz invariant [4]. We now show how our alternative proposal works. Imposing commutation relations for a and a^{*} as in (9), with still unspecified $Z(p)$, and defining

$$\psi^{+}(\boldsymbol{f}) \equiv \int d^4 x \, \boldsymbol{f}(x)^{*} \cdot \psi^{+}(x) \equiv \sum_i \int d^4 x \, f_i(x)^{*} \psi_i^{+}(x) \tag{16}$$

one obtains by (2)

$$\langle 0 | \psi_A^{+}(\boldsymbol{f}) \, \psi_A^{+}(\boldsymbol{g})^{*} \mid 0 \rangle = (2\pi)^{-3} \int d^4 x \, d^4 x' \, \frac{d^3 p}{p^0} \, \frac{d^3 p'}{p'^0} \, \boldsymbol{f}(x)^{*} \cdot D(\Lambda) \boldsymbol{u}^{+}(\boldsymbol{p})$$
$$e^{-i\Lambda p \cdot x} \left(D(\Lambda) \boldsymbol{u}^{+}(\boldsymbol{p}') \right)^{*} \cdot \boldsymbol{g}(x') \, e^{i\Lambda p' \cdot x'} \langle 0 | a(\boldsymbol{p}) a^{*}(\boldsymbol{p}') | 0 \rangle \, . \tag{17}$$

By (8) there is a function $\kappa^A(p)$ such that

$$D(\Lambda) \boldsymbol{u}^{+}(\boldsymbol{p}) = \kappa^A(p) \boldsymbol{u}^{+}(\underline{\Lambda p}) \, . \tag{18}$$

Either by direct calculation or by a group-theoretical argument [4] one can show that

$$|\kappa^A(p)|^2 = (\Lambda p)^0 / p^0 \, . \tag{19}$$

Since a annihilates the vacuum one has

$$< 0 \mid a(\boldsymbol{p}) a^{*}(\boldsymbol{p}') \mid 0 > = p^0 Z(p) \delta^{(3)}(\boldsymbol{p} - \boldsymbol{p}') \, . \tag{20}$$

Insertion into (17) and a change of variables yield

$$\langle 0 | \psi_A^{+}(\boldsymbol{f}) \, \psi_A^{+}(\boldsymbol{g})^{*} | 0 \rangle = 2\pi \int \frac{d^3 p}{p^0} \left| \kappa^A(\Lambda^{-1} p) \right|^2 Z(\Lambda^{-1} p) \tilde{\boldsymbol{f}}(p)^{*} \cdot \boldsymbol{u}^{+}(\boldsymbol{p}) \, \boldsymbol{u}^{+}(\boldsymbol{p})^{*} \cdot \tilde{\boldsymbol{g}}(p) \tag{21}$$

since $d^3 p / p^0$ is invariant. This will, by (19), be independent of Λ if and only if we choose

$$Z(p) = p^0 \tag{22}$$

or a multiple thereof so that

$$[a(\boldsymbol{p}), a^{*}(\boldsymbol{p}')] = (p^0)^2 \delta^{(3)}(\boldsymbol{p} - \boldsymbol{p}') \, . \tag{23}$$

The operators $\hat{a}(\boldsymbol{p})$ and $\hat{a}^{*}(\boldsymbol{p})$ defined by

$$\hat{a}(\boldsymbol{p}) = (p^0)^{-1/2} a(\boldsymbol{p}) \tag{24}$$

then satisfy canonical commutation relations in the usual relativistic normalization. We note, that up to now, anticommutation relations would have worked just as well.

For the negative-frequency part ψ^- the argument is completely analogous if one uses $\boldsymbol{u}^-(\boldsymbol{p}) = \boldsymbol{u}^+(-\boldsymbol{p})$. As a consequence b and b^* have to satisfy the analog of (23).

Putting $\boldsymbol{f}(x) = \delta^{(4)}(x - y')\boldsymbol{e}_i$ and $\boldsymbol{g}(x) = \delta^{(4)}(x - y)\boldsymbol{e}_j$ one finds from (22) and (23)

$$< 0 \mid \psi_i^+(x)\psi_j^+(y)^* \mid 0 >= (2\pi)^{-3} \int \frac{d^3p}{p^0} e^{-ip\cdot(x-y)} p^0 u_i^+(\boldsymbol{p})u_j^+(\boldsymbol{p})^* \,. \qquad (25)$$

Now, since $\boldsymbol{u}^+, \boldsymbol{u}^-$ and $\boldsymbol{p}$ are a basis and are eigenvectors of $\boldsymbol{S}\cdot\boldsymbol{p}$ for the eigenvalues $\mid \boldsymbol{p} \mid, - \mid \boldsymbol{p} \mid$, and 0, respectively, one easily finds that

$$p^0 u_i^+(\boldsymbol{p})u_j^+(\boldsymbol{p})^* = \frac{1}{2} \left\{ (p^0)^2 \delta_{ij} - p_i p_j + i\, \varepsilon_{ijk} p_k p^0 \right\} \,. \qquad (26)$$

Insertion into (25) gives

$$< 0|\psi_i^+(x)\psi_j^+(y)^*|0 >= (2\pi)^{-3} \int \frac{d^3p}{p^0} e^{-ip\cdot(x-y)} \frac{1}{2} \left\{ (p^0)^2 \delta_{ij} - p_i p_j + i\, \varepsilon_{ijk} p_k p^0 \right\} \,. \qquad (27)$$

Similarly one finds

$$< 0|\psi_i^-(x)^*\psi_j^-(y)|0 >= (2\pi)^{-3} \int \frac{d^3p}{p^0} e^{-ip\cdot(x-y)} \frac{1}{2} \left\{ (p^0)^2 \delta_{ij} - p_i p_j - i\, \varepsilon_{ijk} p_k p^0 \right\} \,. \qquad (28)$$

Since the right-hand sides of (27) and (28) can be written as a combination of time and space derivatives of $\Delta^+(x-y)$, the expressions are local. All relevant quantities can be obtained from (27) and (28), such as the field commutator $[\psi_i(x), \psi_j(y)^*]$ which is also local.

One can also determine the Hamiltonian, which must satisfy

$$\dot{\psi} = i[H, \psi] \,.$$

It is easily checked that H is unique up to an additive constant and that

$$H = \int \frac{d^3p}{p^0} \left\{ a^*(\boldsymbol{p})a(\boldsymbol{p}) + b^*(\boldsymbol{p})b(\boldsymbol{p}) \right\} \qquad (29)$$

does the trick. It is evidently positive.

In contrast to canonical quantization our approach has thus led to a quantized Oppenheimer field satisfying all axioms of relativistic field theory. From the construction it is apparent that the requirement of invariant 2-point functions fixes the quantized field uniquely, up to a constant factor.

Is the resulting field related to any example already known? Not surprisingly, it is. Defining field operators $\boldsymbol{E}$ and $\boldsymbol{B}$ by

$$\boldsymbol{E} = \text{Re}\,\psi\,, \qquad \boldsymbol{B} = \text{Im}\,\psi$$

one checks that they have the same 2-point function as the electric and magnetic field operators in the Coulomb gauge. Indeed, already classically $\mathrm{Re}\,\psi$ and $\mathrm{Im}\,\psi$ satisfy Maxwell's equations as checked by explicit calculation. Therefore, from the uniqueness mentioned above, this relationship carries over to the quantized fields.

4 The Quantization Proposal in the General Case

It suffices to consider the case of a single mass value in (5). The analog of (17) is

$$\langle 0|\psi_\Lambda^+(\boldsymbol{f})\,\psi_\Lambda^+(\boldsymbol{g})^*\,|\,0\rangle = (2\pi)^{-3} \int d^4x\ d^4x'\ \frac{d^3p}{p^0}\ \frac{d^3p'}{p'^0} \sum_{ij} \boldsymbol{f}(x)^* \cdot D(\Lambda)\boldsymbol{u}_i^+(\boldsymbol{p})$$

$$e^{-i\Lambda p \cdot x}\left(D(\Lambda)\boldsymbol{u}_j^+(\boldsymbol{p}')\right)^* \cdot \boldsymbol{g}(x')\ e^{i\Lambda p' \cdot x'} \langle 0\,|a_i(\boldsymbol{p})a_j^*(\boldsymbol{p}')|\,0\rangle\ . \quad (30)$$

By (8) and (9) this becomes

$$\langle 0\,|\,\psi_\Lambda^+(\boldsymbol{f})\,\psi_\Lambda^+(\boldsymbol{g})^*\,|\,0\rangle = 2\pi \int \frac{d^3p}{p^0}\ \tilde{\boldsymbol{f}}(\Lambda p)^* \cdot \sum_{ij\alpha\beta} \kappa_{\alpha i}^\Lambda(p)\boldsymbol{u}_\alpha^+(\underline{\Lambda p})$$

$$Z_{ij}(p)\left(\kappa_{\beta j}^\Lambda(p)\boldsymbol{u}_\beta^+(\underline{\Lambda p})\right)^* \cdot \tilde{\boldsymbol{g}}(\Lambda p)\ . \quad (31)$$

By a change of integration variable, $p \to p' = \Lambda p$, it is seen that this is independent of Λ for all $\boldsymbol{f}$ and $\boldsymbol{g}$ if and only if, in matrix notation,

$$\kappa^\Lambda(p)\ Z(p)\ \kappa^\Lambda(p)^* = Z(\Lambda p)\ . \quad (32)$$

In order to see if this can be achieved by a judicious choice of $Z_{ij}(p)$ we have to determine the general form of the matrix $\kappa^\Lambda(p)$ in (8). Applying first $D(\Lambda')$ and then $D(\Lambda)$ to $\boldsymbol{u}_i^+(\boldsymbol{p})$ and using

$$D(\Lambda)\ D(\Lambda') = D(\Lambda\Lambda')$$

one obtains from (8)

$$\kappa^\Lambda(\Lambda'p)\ \kappa^{\Lambda'}(p) = \kappa^{\Lambda\Lambda'}(p)\ . \quad (33)$$

To exploit this relation we pick a fixed $\hat{p}$ on the mass hyperboloid or, respectively, on the forward light-cone if the mass is 0, e.g. $\hat{p} = (\mu,0,0,0)$ or $(1,0,0,1)$, and for each p on this manifold we pick a Lorentz transformation $L(p)$ such that

$$L(p)\hat{p} = p\ . \quad (34)$$

Putting $\Lambda' = L(p)$ and replacing p by $\hat{p}$ in (33) we obtain

$$\kappa^\Lambda(p) = \kappa^{\Lambda L(p)}(\hat{p})\ \kappa^{L(p)}(\hat{p})^{-1}\ . \quad (35)$$

Now let S be an element of the stabilizer group or little group of $\hat{p}$, i.e.

$$S\hat{p} = \hat{p}\ .$$

Then (33) gives

$$\kappa^{\Lambda S}(\hat{p}) = \kappa^{\Lambda}(\hat{p})\kappa^{S}(\hat{p}) \tag{36}$$

so that, in particular, $\kappa^{S}(\hat{p})$ is a representation of the little group of $\hat{p}$. Now,

$$\Lambda L(p)\,\hat{p} = \Lambda p = L(\Lambda p)\,\hat{p},$$

and therefore

$$L(\Lambda p)^{-1}\Lambda L(p) \equiv S(\Lambda, p)$$

is in the little group of $\hat{p}$ so that

$$\Lambda L(p) = L(\Lambda p)S(\Lambda, p) . \tag{37}$$

From (35), (36), and (37) we thus obtain

$$\kappa^{\Lambda}(p) = \kappa^{L(\Lambda p)}(\hat{p})\,\kappa^{S(\Lambda,p)}(\hat{p})\,\kappa^{L(p)}(\hat{p})^{-1} . \tag{38}$$

Conversely, given a representation $S \rightarrow \kappa^{S}(\hat{p})$ of the little group of $\hat{p}$ and a mapping $p \rightarrow \kappa^{L(p)}(\hat{p})$ then $\kappa^{\Lambda}(p)$ defined by the right-hand side of (38) satisfies (33).

In case of *nonzero mass* the little group is $SO(3)$, and since $\kappa^{S}(\hat{p})$ is then a representation of a compact group we may assume it to be *unitary*, by a change of basis. In this case we can choose

$$Z(p) = \kappa^{L(p)}(\hat{p})\,\kappa^{L(p)}(\hat{p})^{*} . \tag{39}$$

Then, from (38)

$$\begin{aligned}\kappa^{\Lambda}(p)Z(p)\kappa^{\Lambda}(p)^{*} &= \kappa^{L(\Lambda p)}(\hat{p})\kappa^{L(\Lambda p)}(\hat{p})^{*} \\ &= Z(\Lambda p)\end{aligned} \tag{40}$$

and by the remark in connection with (32) this yields the invariance of the 2-point function in (31). Apart from a constant factor (39) appears to be the only possible choice for $Z(p)$. It is important to note that $Z(p)$ is a positive-definite matrix because otherwise the a_i^*'s could not be interpreted as creation operators. In particular, the operators $\hat{a}_i$ defined through

$$a_i(\boldsymbol{p}) = \sum_{\alpha} \kappa_{i\alpha}^{L(p)}(\hat{p})\,\hat{a}_{\alpha}(\boldsymbol{p}) \tag{41}$$

satisfy commutation or anticommutation relations in standard form.

For the negative frequency part of the argument is similar. Hence, for nonzero mass our quantization proposal yields a relativistic 2-point function. As for the 3-component Oppenheimer equation one shows that the Hamiltonian is positive.

For *zero mass* we return to (38). The little group is now isomorphic to the Euclidean group in 2 dimensions and hence non-compact. Its compact rotational part does not cause any problems, by the same argument as in the nonzero mass case. Its noncompact part, however, cannot be represented unitarily in finite dimension except trivially. Just this happens if the representation of the little group can be decomposed into one-dimensional representations or is already one-dimensional, as for the 3-component Oppenheimer equation. Thus it appears that for mass zero a relativistic quantization is possible only if, in the representation of the little group E_2, the translational part is represented trivially in $\kappa^S(\hat{p})$. In this case the Hamiltonian is also positive.

5 Conclusion

It has been proposed to quantize linear relativistic field equations by the requirement that their 2-point function be relativistically invariant. This proposal has been illustrated by the quantization of the 3-component Oppenheimer equation for which canonical quantization has recently been shown to fail relativity requirements. In contrast, our approach leads to a local relativistic field. The general case has then been analyzed. The approach always works for nonzero mass, while for zero mass an extra condition is needed. The resulting quantum fields are relativistic and have positive energy. Locality has to be checked in each case separately. From our construction it appears that our proposal leads to the same fields as other quantization methods if the latter work, but is more general than for example canonical quantization. It also appears that equations for which our approach does not work cannot be quantized at all in a Hilbert space.

Part of this work was done while the author was at the Institute for Advanced Study, Princeton, New Jersey. The research was partially supported by the Monell Foundation.

References

1   E.P. Wigner: Ann. of Math. **40** 149 (1939)
2   E. Weinberg: Phys. Rev. **B133** 1318 (1964); **B134** 882 (1964)
3   R. Omnès: *Introduction to Particle Physics* (Wiley-Interscience, New York, 1971)
4   G.C. Hegerfeldt, M.K. Hinders, B.A. Rhodes, G. v. H. Sandri:
      Institute for Advanced Study preprint (IASSNS-HEP-90/28)
5   J.R. Oppenheimer: Phys. Rev. **38** 725 (1931)
6   A.S. Wightman: in *Invariant Wave Equations*, Lecture Notes in Physics **73**,
      ed. by G. Velo and A.S. Wightman (Springer, Berlin, 1978)
7   I.M. Gelfand, R.A. Minlos, Z. Ya, Shapiro: *Representations of the Rotation and Lorentz Groups and Their Applications* (MacMillan, New York, 1963)

A Lattice Approximation of the Dirac Equation

Jerzy Kijowski, Artur Thielmann

Institute for Theoretical Physics, Polish Academy of Sciences
Al. Lotników 32/46, 02-668 Warsaw, Poland

Abstract: A different point of view on discretisation of the classical theory of the Dirac equation is given. Canonical structure of the model is given, the Cauchy problem is formulated and solved, fermion doubling is discussed and a solution via time conserved constraints is proposed.

1 Introduction

The aim of the present paper is to construct a lattice approximation of the Dirac equation, which preserves as much as possible of the specific properties of the continuous Dirac theory, such as canonical structure, the structure of the space of initial data, etc. The formulation is meant as a starting point for quantization. Various versions of discrete approximations of spinor field theories have already been proposed (see e.g. [1–2]). The main disease of these approximations lies in doubling of the number of solutions with respect to the original Dirac equation. This means that the lattice version of the Dirac equation becomes rather a second order than a first order equation. In general there is no satisfactory lattice description of first order partial differential equations. Our idea is to interpret Dirac theory as a combination of two ingredients: 1) a second order dynamical equation and 2) a constraint imposed on initial data (see Sect. 2). We show that a similar idea can be realized on the level of the lattice approximation of the theory. This way we obtain the theory without any fermion doubling.

The theory is constructed on a four-dimensional, Minkowski space hypercubic lattice Λ with sites at the points $\mathbb{R}^4 \ni x = \delta \cdot n$, where δ denotes the lattice constant and $n = (n^0, n^1, n^2, n^3) \in \mathbb{Z}^4$ is a point in $\mathbb{R}^4$ with integer coordinates. To represent momenta canonically conjugate to field variables we will also need the geometrically dual lattice Λ^*, i.e., the lattice whose sites are precisely the centers of the four-dimensional hypercubes of Λ. Finally, the complete version of the theory will be given on the full composed lattice $\Lambda \cup \Lambda^*$. Sites of Λ will be denoted by $x, \ldots$, links $(x, x + \hat{\mu}), \ldots$ or $(x; \hat{\mu}), \ldots$, where $\hat{\mu}$ is a vector of length δ and direction of the oriented μ-th axis.

2 Continuous Theory

2.1 Lagrangian Formulation

The classical theory of a complex bispinor field $\psi(x)$ can be derived from the lagrangian

$$\mathcal{L} = \frac{i}{2}(\overline{\psi}\gamma^{\mu}\partial_{\mu}\psi - \partial_{\mu}\overline{\psi}\gamma^{\mu}\psi) - m\overline{\psi}\psi , \qquad (1)$$

which implies the following second-type constraint

$$p^{\mu} = \frac{\delta\mathcal{L}}{\delta\partial_{\mu}\overline{\psi}} = -\frac{i}{2}\gamma^{\mu}\psi \qquad (2)$$

relating the canonical momentum with the field variable. The field equation derived from the lagrangian (1) is the Dirac equation

$$\frac{\delta\mathcal{L}}{\delta\overline{\psi}} - \partial_{\mu}\frac{\delta\mathcal{L}}{\delta\partial_{\mu}\overline{\psi}} = \frac{i}{2}\gamma^{\mu}\partial_{\mu}\psi - m\psi - \partial_{\mu}(-\frac{i}{2}\gamma^{\mu}\psi) = i\gamma^{\mu}\partial_{\mu}\psi - m\psi = 0 . \qquad (3)$$

2.2 Hamiltonian Formulation

Now we pass to the hamiltonian formulation of the theory. It is based on the hamiltonian obtained from the above lagrangian via the standard Legendre transformation

$$\mathcal{H} = \overline{p^{0}}\partial_{0}\psi + \partial_{0}\overline{\psi}p^{0} - \mathcal{L} = -\frac{i}{2}(\overline{\psi}\gamma^{k}\partial_{k}\psi - \partial_{k}\overline{\psi}\gamma^{k}\psi) + m\overline{\psi}\psi . \qquad (4)$$

The hamiltonian is defined on the phase space $\mathcal{P}$ consisting of initial data (p^{0}, ψ) which fulfill the constraint equation (2). Following [8] we parameterize the phase space by the "real" q and the "imaginary" p parts of the bispinor ψ (in the sense of the charge conjugation). For the sake of simplicity we use the Majorana representation, where the charge conjugation coincides with the complex conjugation. This means that both q and p are four-dimensional real bispinors. Substituting $\psi = q + ip$ to the hamiltonian (4) we obtain

$$\mathcal{H} = -2p^{T}\gamma^{0}\gamma^{k}\partial_{k}q - 2imp^{T}\gamma^{0}q . \qquad (5)$$

The canonical structure of the phase space of the theory is given by the standard symplectic form

$$\omega = d\overline{p}^{0} \wedge d\psi - d\overline{\psi} \wedge dp^{0}$$

which reduces on the subspace defined by the constraint (2) to

$$\omega = id\overline{\psi}\gamma^{0} \wedge d\psi .$$

Then in terms of the unconstrained variables q and p one gets

$$\omega = 2\mathrm{d}p^T \wedge \mathrm{d}q \ .$$

This means that $2p$ is now the momentum canonically conjugate to q. Therefore, the field dynamics is governed by the Hamilton equations

$$\dot{q} = \frac{1}{2}\frac{\delta \mathcal{H}}{\delta p^T}$$
$$\dot{p} = -\frac{1}{2}\frac{\delta \mathcal{H}}{\delta q^T} \ . \tag{6}$$

Using the specific form (5) of the hamiltonian the reader may easily check that these are precisely the real and the imaginary parts of the Dirac equation

$$i\gamma^\mu \partial_\mu q = mq$$
$$i\gamma^\mu \partial_\mu p = mp \ . \tag{7}$$

2.3 Complexification of the Phase Space

Consider now the complexification $\widetilde{\mathcal{P}}$ of the above phase space and define on it the hamiltonian by the formula

$$\widetilde{\mathcal{H}} = -\overline{p}\gamma^k \partial_k q - im\overline{p}q - \partial_k \overline{q}\gamma^k p + im\overline{q}p \ , \tag{8}$$

where q and p are no longer real. Obviously, the hamiltonian $\widetilde{\mathcal{H}}$ coincides with $\mathcal{H}$ on the subspace $\mathcal{P} \subset \widetilde{\mathcal{P}}$. Define also the canonical structure in $\widetilde{\mathcal{P}}$ by the following symplectic form

$$\Omega = \mathrm{d}\overline{p}\gamma^0 \wedge \mathrm{d}q - \mathrm{d}\overline{q}\gamma^0 \wedge \mathrm{d}p \ ,$$

which on $\mathcal{P}$ coincides obviously with ω. The hamiltonian field equations of this theory take the form

$$\dot{q} = \frac{\delta \mathcal{H}}{\delta \overline{p}}$$
$$\dot{p} = -\frac{\delta \mathcal{H}}{\delta \overline{q}} \tag{9}$$

coinciding practically with (7), except that now both q and p are complex. The theory is thus the doubling of the original Dirac theory. Limiting ourselves to the subspace of real data

$$q = q^*$$
$$p = p^* \tag{10}$$

at one instant of time, we will remain forever in this subspace, since it is invariant with respect to the dynamics. This way we reproduce the original theory as a sub-theory corresponding to the invariant subspace $\mathcal{P}$ in the phase space $\widetilde{\mathcal{P}}$ of the doubled Dirac theory. As we show later, similar construction can also be performed in the lattice version of the theory.

3 Primary Lattice Approximation

We approximate the action of the Dirac theory, defined by (1), on Λ by the following expression

$$
\mathcal{A} = \frac{i}{2}\delta^4 \sum_{(x,\hat{\mu})} \left\{ \left[\frac{\overline{\psi}(x+\hat{\mu}) + \overline{\psi}(x)}{2} \right] \gamma^\mu \left[\frac{\psi(x+\hat{\mu}) - \psi(x)}{\delta} \right] + \right.
$$
$$
\left. - \left[\frac{\overline{\psi}(x+\hat{\mu}) - \overline{\psi}(x)}{\delta} \right] \gamma^\mu \left[\frac{\psi(x+\hat{\mu}) + \psi(x)}{2} \right] \right\} +
$$
$$
- m\delta^4 \sum_x \overline{\psi}(x)\psi(x) .
\tag{11}
$$

In the above formula the derivative related to a lattice link is approximated by the simple difference term and the value of the field on the link by the arithmetic mean. We could approximate ψ by the value at one of the link's endpoints, but this would differ from our formula only by a rearrangement of terms in $\mathcal{A}$ (in continuous theory this corresponds to adding a boundary integral to the action). Such a change does not influence the field equations. The field equations resulting from the action (11) are

$$
0 = \frac{\partial \mathcal{A}}{\partial \overline{\psi}(x)} = \frac{i\delta^3}{2} \sum_\mu \gamma^\mu [\psi(x+\hat{\mu}) - \psi(x-\hat{\mu})] - m\delta^4\, \psi(x)
\tag{12}
$$

or rewritten in a different form

$$
\psi(x+\hat{\mu}) = \psi(x-\hat{\mu}) - \sum_k \gamma^0 \gamma^k [\psi(x+\hat{k}) - \psi(x-\hat{k})] - 2i\delta \cdot m\gamma^0 \psi(x) .
\tag{13}
$$

In the limit $\delta \to 0$ they tend to the Dirac equation. Unfortunately they combine the variables ψ at three consecutive space-like hypersurfaces, that of $x - \hat{0}$, that of x itself and that of $x + \hat{0}$ ($\hat{0}$ is a vector of the length δ and the direction of the oriented 0-th axis). As a dynamical equation (13) is therefore of the second order. To find a unique solution of this equation we have to specify initial values of ψ at two consecutive instants of time or, in other words, the value of the function and of its time derivative. There are therefore twice as many initial conditions (and consequently twice as many solutions) as we would expect for the Dirac equation. For this reason the primary lattice description is not satisfactory and we will look for a more detailed, canonical structure of the lattice model.

4 Dual Lattice Theory

4.1 Construction of the Action

According to the symplectic philosophy of field theories (see e.g., [5–6]) we describe field dynamics as a superposition of local dynamics related to each spacetime cell separately. Within a single lattice cell the dynamics is expressed in terms of the local configurations and momenta. The composition of the local dynamics corresponding to neighboring cells follows from the matching condition on their common wall. The condition states that both the configurations and momenta on the wall are equal.

To construct the lattice version of the phase space $\widetilde{\mathcal{P}}$ we have to represent infinitely many degrees of freedom of the field in a spacetime cell by a finite number of configurations and momenta. Whereas the configurations were described on the primary lattice, the momenta need the use of the dual lattice. This is due to the fact that geometrically the momenta are vector densities or differential three-forms in spacetime i.e., objects which have to be integrated over three-dimensional domains (e.g., walls of lattice cells). It is natural therefore to represent configurations with their sample values on primary lattice elements and momenta as averaged over the corresponding dual lattice elements (see [5]).

We add therefore the configurations $\psi(x + \frac{\mu}{2})$ at the centers of links $x + \frac{\mu}{2}$ to the description of the field configuration at the primary lattice sites $\psi(x)$. The points $x + \frac{\mu}{2}$ are also centers of the three-dimensional walls of the dual lattice cells. To find momenta conjugate to both types of variables (configurations at the primary lattice sites and at the centers of links), we want to rewrite the action (11) of the theory in terms of these variables. We would like to have the action in the form

$$\mathcal{A} = \delta^4 \sum_x \mathcal{L}_x , \qquad (14)$$

with local lagrangians $\mathcal{L}_x$ depending on configurations contained in the lattice cell K_x dual to the site x only, i.e. on $\psi(x)$ and on $\psi(x + \frac{\mu}{2})$ at all centers of the links outgoing from x.

To keep the dynamics on the primary lattice unchanged, we want our new action to reduce to the old one when the new variables $\psi(x + \frac{\mu}{2})$ are eliminated. By the elimination we mean the variation with respect to $\psi(x + \frac{\mu}{2})$ and the substitution of the stationary value to the action. The equation obtained by the above variation relates the value of ψ at the center of a link to the values at its endpoints and may therefore be called "interpolation equation". The reduction requirement guarantees that variation of the new action with respect to the primary lattice sites' configurations will reproduce the previous approximation of the Dirac equation.

Assuming that the action is quadratic with respect to the variables $\psi(x)$ and $\psi(x + \frac{\mu}{2})$ connecting only nearest neighbors and assuming some symmetry conditions we have the solution unique up to a change of sign of the imaginary unit i. We start from a general expression for a quadratic form of the variables $\psi(x)$ and $\psi(x + \frac{\mu}{2})$ connecting only nearest neighbors

$$\mathcal{L}_x = \sum_\mu \left\{ a\,\overline{\psi}(x)\gamma^\mu\psi(x + \tfrac{\dot{\mu}}{2}) + b\,\overline{\psi}(x)\gamma^\mu\psi(x - \tfrac{\dot{\mu}}{2}) + \right.$$

$$+ c\overline{\psi}(x + \tfrac{\dot{\mu}}{2})\gamma^\mu\psi(x) + d\overline{\psi}(x - \tfrac{\dot{\mu}}{2})\gamma^\mu\psi(x) +$$

$$\left. + f\overline{\psi}(x)\gamma^\mu\psi(x) + e\overline{\psi}(x - \tfrac{\dot{\mu}}{2})\gamma^\mu\psi(x - \tfrac{\dot{\mu}}{2}) + e\overline{\psi}(x + \tfrac{\dot{\mu}}{2})\gamma^\mu\psi(x + \tfrac{\dot{\mu}}{2}) \right\} +$$

$$- m\overline{\psi}(x)\psi(x) \ . \tag{15}$$

We decided the last two terms under the summation sign to have the same coefficient e, since they simply add up in the sum (14) of neighboring cells' actions. First, we want the action to be real, so we have

$$a^* = c$$
$$b^* = d$$
$$e^* = e$$
$$f^* = f \ .$$

Variation of (15) with respect to $\overline{\psi}(x + \tfrac{\dot{\mu}}{2})$ gives the interpolation equation

$$-2e\psi(x + \tfrac{\dot{\mu}}{2}) = c\psi(x) + d\psi(x + \hat{\mu}) \ .$$

Eliminating all $\psi(x + \tfrac{\dot{\mu}}{2})$ from the action we get

$$-2e\mathcal{A} = \delta^4 \sum_x \sum_\mu \left\{ ad\,\overline{\psi}(x)\gamma^\mu\psi(x + \hat{\mu}) + bc\,\overline{\psi}(x + \hat{\mu})\gamma^\mu\psi(x) + \right.$$

$$\left. + [ac + bd - 2ef]\,\overline{\psi}(x)\gamma^\mu\psi(x) \right\} +$$

$$+ 2em\delta^4 \sum_x \overline{\psi}(x)\psi(x) \ . \tag{16}$$

Comparing this to the primary lattice formula (11) we get the following system of equations for coefficients

$$a^*a + b^*b = 2ef \ , \tag{17a}$$

$$ab^* = -\frac{i}{\delta} \ , \tag{e(17b)}$$

$$a^*b = \frac{i}{\delta e} \ , \tag{17c}$$

where the third equation is the complex conjugate of the second one (since e is real). Because the terms in the action containing a and b differ only by link orientation, we assume that these coefficients have the same modulus. Decompose a and b into the moduli and phase factors $a = r\exp(i\varphi)$ and $b = r\exp(i\chi)$, and substitute to (17b):

$$r\exp(i\varphi) \cdot r\exp(-i\chi) = -\frac{i}{\delta}e \ . \tag{18}$$

Taking moduli of both sides of (18) we obtain $|e| = \delta r^2$. Comparing the phase factors instead we get $b = \text{sgne} \cdot ia$. From (17a) we have now $f = \text{sgne} \cdot \frac{1}{\delta}$. We substitute all the coefficients to the lagrangian (15) and calculate the momentum. To obtain the momentum with a proper continuum limit we have to compare the variation of the action related to a single cell $\delta^4 \cdot \mathcal{L}_x$ with respect to the boundary configuration and the momentum integrated over the corresponding wall

$$\frac{\partial(\delta^4 \mathcal{L}_x)}{\partial \overline{\psi}(x + \frac{\mu}{2})} =: \delta^3 \cdot p(x + \frac{\mu}{2}) \, .$$

This way

$$p(x + \tfrac{\mu}{2}) := \delta \cdot \frac{\partial \mathcal{L}_x}{\partial \overline{\psi}(x + \frac{\mu}{2})} = \delta r \exp(-i\varphi)\gamma^\mu \psi(x) + \text{sgne} \cdot \delta^2 r^2 \gamma^\mu \psi(x + \tfrac{\mu}{2}) \, .$$

Comparing the result with the continuum expression (2) we see that r has to behave like $\frac{1}{\delta}$ when $\delta \to 0$, in order to provide the correct continuum limit of our theory. The simplest choice is $\delta r = R$. Then in the limit we obtain

$$p = [R \exp(-i\varphi) + \text{sgne} \cdot \mathrm{I\!R}^2]\gamma^\mu \psi$$

and therefore the expression in the square brackets has to be equal to $-\frac{i}{2}$. The real part gives

$$R = -\text{sgne} \cdot \cos \varphi$$

whereas the imaginary part

$$R \sin \varphi = \frac{1}{2} \, .$$

Performing the same operations with the opposite wall momentum $p(x - \frac{\mu}{2})$ we get

$$R = \sin \varphi$$

and

$$-\text{sgne} \cdot R \cos \varphi = \frac{1}{2} \, .$$

Since R is positive we have the unique solution for $R = \frac{1}{\sqrt{2}}$ and $\exp(i\varphi) = \frac{1+i}{\sqrt{2}}$.

The only choice which is still left is the sign of e. It is easy to check that a change of the sign corresponds to the change of i to $-i$ everywhere. We choose e to be negative. Finally we have

$$a = d = \frac{1+i}{2\delta}$$

$$b = c = \frac{1-i}{2\delta}$$

$$e = -\frac{1}{2\delta}$$

$$f = -\frac{1}{\delta} \, .$$

This way the dual lattice action becomes $\mathcal{A} = \delta^4 \sum_x \mathcal{L}_x$ with

$$
\mathcal{L}_x = \frac{i}{2\delta} \sum_\mu \left\{ (1-i)\overline{\psi}(x)\gamma^\mu \psi(x + \tfrac{\hat{\mu}}{2}) - (1+i)\overline{\psi}(x)\gamma^\mu \psi(x - \tfrac{\hat{\mu}}{2}) + \right.
$$

$$
- (1+i)\overline{\psi}(x + \tfrac{\hat{\mu}}{2})\gamma^\mu \psi(x) + (1-i)\overline{\psi}(x - \tfrac{\hat{\mu}}{2})\gamma^\mu \psi(x) +
$$

$$
\left. + i\overline{\psi}(x - \tfrac{\hat{\mu}}{2})\gamma^\mu \psi(x - \tfrac{\hat{\mu}}{2}) + 2i\overline{\psi}(x)\gamma^\mu \psi(x) + i\overline{\psi}(x + \tfrac{\hat{\mu}}{2})\gamma^\mu \psi(x + \tfrac{\hat{\mu}}{2}) \right\} +
$$

$$
- m\overline{\psi}(x)\psi(x) . \quad (19)
$$

4.2 Lagrangian Formulation

Variation of the constructed action over $\psi(x + \tfrac{\hat{\mu}}{2})$ gives the interpolation equation

$$
0 = \frac{\delta \mathcal{A}}{\delta \overline{\psi}(x + \tfrac{\hat{\mu}}{2})} = \frac{i\delta^3}{2}\gamma^\mu \left[-(1+i)\psi(x) + (1-i)\psi(x + \hat{\mu}) + 2i\psi(x + \tfrac{\hat{\mu}}{2}) \right] =
$$

$$
= \delta^3 \gamma^\mu \left[\frac{1-i}{2}\psi(x) + \frac{1+i}{2}\psi(x + \hat{\mu}) - \psi(x + \tfrac{\hat{\mu}}{2}) \right] , \quad (20)
$$

which reproduces exactly the original action (11) when substituted to (19). Now, the momentum canonically conjugate to $\psi(x + \tfrac{\hat{\mu}}{2})$ corresponds to the three-dimensional wall of an elementary lattice cell and is given by variation of the single-cell lagrangian

$$
p(x + \tfrac{\hat{\mu}}{2}) = \delta \cdot \frac{\partial \mathcal{L}_x}{\partial \psi(x + \tfrac{\hat{\mu}}{2})} = -\frac{i}{2}\gamma^\mu [(1+i)\psi(x) - i\psi(x + \tfrac{\hat{\mu}}{2})] =
$$

$$
= -\frac{i}{2}\gamma^\mu \left[\psi(x) - i\frac{\delta}{2}\frac{\psi(x + \tfrac{\hat{\mu}}{2}) - \psi(x)}{\frac{\delta}{2}} \right] . \quad (21)
$$

It is easy to see that the continuum limit of (21) provides the standard second-type constraint (2), because the additional term in the square brackets corresponds to the derivative of ψ multiplied by $\frac{\delta}{2}$, so it tends to zero in the continuum limit. Comparing the two expressions (20) and (21) one can notice that the interpolation equation is equivalent to the requirement that the momenta corresponding to the neighboring cells are equal, i.e.

$$
p(x + \tfrac{\hat{\mu}}{2}) + p(y - \tfrac{\hat{\mu}}{2}) = 0 \quad (22)
$$

for $y = x + \hat{\mu}$.

The variation of (19) with respect to $\psi(x)$ gives us the field equation

$$
\sum_\mu \gamma^\mu \left[\frac{1+i}{2\delta}\psi(x + \tfrac{\hat{\mu}}{2}) + \frac{1-i}{2\delta}\psi(x - \tfrac{\hat{\mu}}{2}) - \frac{1}{\delta}\psi(x) \right] - m\psi(x) = 0 . \quad (23)
$$

The above expression tends to the ordinary Dirac equation (3) when $\delta \to 0$. To prove this, let us eliminate the configurations at the centers of links using the interpolation formula (20)

$$\sum_{\mu} \gamma^{\mu} \left\{ \frac{1+i}{2\delta} \left[\frac{1-i}{2} \psi(x) + \frac{1+i}{2} \psi(x+\hat{\mu}) \right] + \right.$$

$$\left. + \frac{1-i}{2\delta} \left[\frac{1-i}{2} \psi(x-\hat{\mu}) + \frac{1+i}{2} \psi(x) \right] - \frac{1}{\delta} \psi(x) \right\} - m\psi(x) = 0 \ .$$

This way we obtain the primary lattice expression (12).

4.3 Hamiltonian Formulation

We will describe now the time evolution of the Cauchy data. Our dynamic equation (23) is a second-order difference equation, so the Cauchy data consist of configurations <u>and</u> momenta on the initial hyperplane composed from the fully space-like three-dimensional walls of the elementary cells of the lattice Λ^*. Using equations (20) – (23) we compute Cauchy data on the later hyperplane Σ_1 through the data on the earlier hyperplane Σ_0.

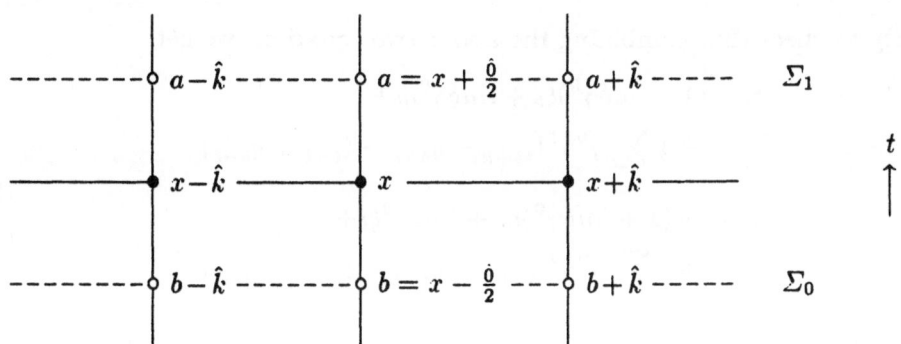

Fig. 1.

Denoting the points as on Fig. 1 and denoting $\pi := 2\gamma^0 p$ we have

$$\psi_a = -im\delta\gamma^0 \psi_b + i(1+m\delta\gamma^0)\pi_b +$$

$$- \frac{1}{2} \sum_k \gamma^0 \gamma^k (\psi_{b+k} - \pi_{b+k} - \psi_{b-k} + \pi_{b-k}) \tag{24a}$$

$$\pi_a = -i(1-m\delta\gamma^0)\psi_b - im\delta\gamma^0 \pi_b +$$

$$+ \frac{1}{2} \sum_k \gamma^0 \gamma^k (\psi_{b+k} - \pi_{b+k} - \psi_{b-k} + \pi_{b-k}) \ . \tag{24b}$$

We introduce new variables

$$\xi(x + \tfrac{\dot{\mu}}{2}) = \psi(x + \tfrac{\dot{\mu}}{2}) + i\,\pi(x + \tfrac{\dot{\mu}}{2})$$
$$\eta(x + \tfrac{\dot{\mu}}{2}) = i\,\psi(x + \tfrac{\dot{\mu}}{2}) + \pi(x + \tfrac{\dot{\mu}}{2})$$

(25)

Let us notice that vanishing of η is by definition equivalent to the constraint (2) in the continuous theory. To understand the meaning of the new parameters we express them in terms of the primary lattice variables using (20) and (21). We get

$$\xi(x + \tfrac{\dot{\mu}}{2}) = \psi(x) + \psi(x + \hat{\mu})$$
$$\eta(x + \tfrac{\dot{\mu}}{2}) = \psi(x) - \psi(x + \hat{\mu}) \, .$$

(26)

As we expected, the second equation assures vanishing of η in the continuum limit. We can now rewrite the Dirac equation (23) in the following form

$$\xi_a - \xi_b + \tfrac{1}{2}\sum_k \gamma^0\gamma^k(\xi_{a+k} - \xi_{a-k} + \xi_{b+k} - \xi_{b-k}) +$$
$$-\eta_a - \eta_b + \tfrac{1}{2}\sum_k \gamma^0\gamma^k(\eta_{a+k} - \eta_{a-k} - \eta_{b+k} + \eta_{b-k}) =$$
$$= -im\delta\gamma^0(\xi_a + \xi_b + \eta_a - \eta_b)$$

(27)

and the interpolation condition as

$$\xi_a + \eta_a = \xi_b - \eta_b \, .$$

(28)

It is easy to check that combining the above two equations we get

$$\xi_a = (1 - im\delta\gamma^0)\xi_b + im\delta\gamma^0\eta_b +$$
$$- \tfrac{1}{2}\sum_k \gamma^0\gamma^k[\xi_{b+k} - \eta_{b+k} - \xi_{b-k} + \eta_{b-k}]$$
$$\eta_a = -(1 + im\delta\gamma^0)\eta_b + im\delta\gamma^0\xi_b +$$
$$+ \tfrac{1}{2}\sum_k \gamma^0\gamma^k[\xi_{b+k} - \eta_{b+k} - \xi_{b-k} + \eta_{b-k}] \, .$$

(29)

Observe that the same result can be obtained directly rewriting (24) or (13) in terms of ξ and η. In the limit when $\delta \to 0$ the second equation implies vanishing of η, whereas the first one becomes the Dirac equation for ξ. These are two equations for the complete complex bispinors.

4.4 Momentum Picture

It is convenient for our purposes to rewrite the lattice field dynamics (29) in terms of the Fourier transformations of the fields. In the present paper we assume that the lattice is infinite. In this case the space dual to the three-dimensional lattice (on which we defined the Cauchy data) is the three-dimensional torus T_δ^3 parameterized by three angles

$$\phi_1, \phi_2, \phi_3 \in \left] -\frac{\pi}{\delta}, \frac{\pi}{\delta} \right] \, .$$

Every function on the lattice can be written in the standard form of the Fourier transformation

$$\xi_b = \int_{T_\delta^3} d^3\phi \, e^{ib\cdot\phi} \alpha(\phi) \tag{30a}$$

$$\eta_b = \int_{T_\delta^3} d^3\phi \, e^{ib\cdot\phi} \beta(\phi) \, , \tag{30b}$$

where $b = ((n^0 - \frac{1}{2})\delta, n^1\delta, n^2\delta, n^3\delta)$ for a fixed n^0 and $n^k \in Z, k = 1, 2, 3$.

In the case of a finite lattice we would have $1 \leq n^k \leq N^k$. The dual space is the three-dimensional cyclic lattice parameterized by the angles ϕ_k such that for every $k = 1, 2, 3$ we have $\exp(iN^k\phi_k) = 1$. The integrals over the torus coordinates would be substituted with finite sums. Otherwise all the formulae will remain unchanged.

For the infinite lattice let us denote

$$\sigma = \sum_k \gamma^k \sin(\delta \cdot \phi_k) \, . \tag{31}$$

The evolution is given by the following operator (we omit here the obvious dependence of the variables α and β on the angles ϕ)

$$\binom{\alpha}{\beta}(t+\delta) = -i\left(\begin{array}{cc} 1 - i\gamma^0(\sigma + m\delta) & i\gamma^0(\sigma + m\delta) \\ i\gamma^0(\sigma + m\delta) & -1 - i\gamma^0(\sigma + m\delta) \end{array} \right)\binom{\alpha}{\beta}(t) \, . \tag{32}$$

Introduce the following operator

$$\Gamma = i\gamma^0(\sigma + m\delta) \, . \tag{33}$$

Its square is proportional to the unit matrix with the coefficient

$$\Gamma^2 = -(m\delta)^2 + \sigma^2 < 0 \, .$$

The evolution operator simplifies now to

$$\binom{\alpha}{\beta}(t+\delta) = \left(\begin{array}{cc} 1 - \Gamma & \Gamma \\ \Gamma & -1 - \Gamma \end{array} \right)\binom{\alpha}{\beta}(t) \, . \tag{34}$$

5 Constraint Equation

As we have pointed out in the previous section, the Dirac equation in the lattice version is a second-order difference equation. The Cauchy data for such a theory consist both of configurations and canonical momenta. Therefore, we have twice as many solutions as we would expect for a first-order equation (that we have in the standard continuum case). This manifests the fermion doubling problem in our model.

The situation is similar as in the case of the complexified phase space we described in the Sect. 2.3, where the initial data were composed of two independent complex bispinor fields. Only solutions fulfilling the constraint equation (10) were physically admissible. The constraint (10) is conserved by the dynamics (9). Once fulfilled by initial data, it will always be valid during the evolution. Our idea to remove the fermion doubling consists in reproducing this mechanism on the lattice level. We look for a subspace $\mathcal{P}$ of the phase space $\widetilde{\mathcal{P}}$ having the following two properties:

1) it is parameterized by a single complex bispinor field (corresponding to the two real fields q and p in the continuum case),
2) it is conserved by the dynamics (29).

Observe that the dynamics (29) is given by real matrices, so the real and the imaginary parts of ξ and η do not interact during the evolution. Therefore a constraint

$$\xi = \xi^*$$
$$\eta = \eta^*$$

or a constraint

$$\xi = -\xi^*$$
$$\eta = -\eta^*$$

or their arbitrary linear combinations are conserved. Due to (26) this would correspond in the continuum limit to the constraint $\mathrm{Im}\psi = 0$ or $\mathrm{Re}\psi = 0$. Such a theory is not a complete continuous Dirac theory and therefore the idea to define $\mathcal{P}$ via constraint of this type has to be abandoned.

Local constraints mixing ξ and η at one point are not conserved, so we have to look for another possibility. Our idea is to mimic the second-type constraint (2) of the continuum theory. Assume that the constraint is given by a linear operator A:

$$\mathcal{P} := \left\{ \begin{pmatrix} \alpha \\ \beta \end{pmatrix} \middle| \beta = A\alpha \right\} .$$

Assuming that the constraint $\mathcal{P}$ is conserved by the dynamics (34) we obtain the following condition on A

$$\begin{pmatrix} \alpha \\ A\alpha \end{pmatrix}(t+\delta) = \begin{pmatrix} 1-\Gamma & \Gamma \\ \Gamma & -1-\Gamma \end{pmatrix}\begin{pmatrix} \alpha \\ A\alpha \end{pmatrix}(t) ,$$

or equivalently

$$A(1-\Gamma) + A\Gamma A = \Gamma - (1+\Gamma)A . \tag{35}$$

The simplest solution of this equation is obtained assuming $A\Gamma = \Gamma A$. In this case (35) implies

$$A = \frac{\sqrt{1+\Gamma^2}-1}{\Gamma^2} \cdot \Gamma . \tag{36}$$

Rewriting the (36) in terms of the original fields ξ and η we obtain the following, nonlocal constraint for the Cauchy data

$$\eta(b_0) = i \sum_{b \in \Sigma_0} \int_{T_\delta^3} d^3\phi \; e^{i(b_0-b)\cdot\phi} \frac{1-\sqrt{1-(m\delta)^2+\sigma^2}}{(m\delta)^2-\sigma^2} \cdot \gamma^0(\sigma+m\delta)\xi(b)$$

with

$$\sigma^2 = - \sum_k \sin^2(\delta \cdot \phi_k) \ .$$

This constraint defines a linear subspace of the Cauchy data conserved by the time evolution. The solutions fulfilling (36) on the initial hyperplane will remain in the same subspace forever, so the number of solutions of the lattice Dirac equation is reduced to the expected value, i.e. the doubling is canceled.

To show that (36) provides the constraint (2) in the continuum limit, observe that the fraction in the integrand tends to $-\frac{1}{2}$ when $\delta \to 0$. Since the term $\sigma + m\delta$ vanishes as δ, the constraint assures vanishing of η. This is equivalent (see (25)) to the classical constraint (2).

Finally, the theory based on the phase space $\mathcal{P}$ and the evolution (29) is the desired lattice version of the Dirac theory, without fermion doubling and with the proper continuum limit.

References

1.   K.G. Wilson: Phys. Rev. **D10** 2445 (1974)
2.   L. Susskind: Phys. Rev. **D16** 3031 (1977)
3.   J.M. Rabin: Nucl. Phys. **B201** 315 (1982)
4.   P. Rossi, U. Wolff, D. Zwanziger: Phys. Rev. **D30** 2233 (1984)
5.   J. Kijowski, W.M. Tulczyjew: *A Symplectic Framework for Field Theories*, Lecture Notes in Physics **107** (Springer-Verlag, Berlin, 1979)
6.   J. Kijowski, G. Rudolph: Rep. Math. Phys. **21** 309 (1985)
7.   J. Kijowski, G. Rudolph: Lett. Math. Phys. **15** 119 (1988)
8.   A. Jakubiec: Lett. Math. Phys. **9** 171 (1985)

Some Hidden Aspects of Hidden Symmetry

L. O'Raifeartaigh

Dublin Institute for Advanced Studies, 10 Burlington Road, Dublin 4, Ireland

Abstract: It is shown that the Goldstone theorem is actually a special case of the Noether theorem in the presence of spontaneous symmetry breakdown, and is thus immediately valid for quantized as well as classical fields. The situation when gauge fields are introduced is discussed, emphasis being placed on some points that are not often discussed in the literature such as the compatibility of the Higgs mechanism and the Elitzur theorem and the extent to which the vacuum configuration is determined by the choice of gauge.

1 Introduction

Although spontaneously-broken gauge theories have been in existence for more than two decades [1], and their experimental relevance has been put beyond all doubt by the theories of (low-temperature) superconductivity [2] and the electroweak interactions [3], new aspects of the theories continue to surface even at the most fundamental and elementary levels. Typical examples are the discovery of the non-trivial topological content of these theories (the existence of monopoles [4], instantons [5] and the Gribov effect [6], for example) and the emergence of paradoxes such as Elitzur's theorem [7], which states that, strictly speaking, gauge symmetries are not spontaneously broken. This means, of course, that, for gauge theories, the name 'spontaneously broken symmetry' is particularly inappropriate, and should be replaced by some alternative such as 'hidden symmetry'. The purpose of the present note is to discuss some of the aspects of hidden gauge symmetries that still arise, with a view to simplifying and clarifying them. Such a discussion would seem to me to be appropriate in a birthday Festschrift for Professor Doebner since much of his outstanding scientific endeavour has been devoted to simplifying and clarifying the subtler aspects of various theories.

The first, and perhaps most interesting, point to be considered is the intimate connection between the Noether [8] and Goldstone [9] theorems. In the literature these theorems are usually treated as separate and unrelated, but it is shown in Sect. 2 that, far from their being unrelated, the Goldstone theorem can actually be derived as a special version of the Noether theorem in the case of a spontaneous breakdown. Some advantages that result from deriving the Goldstone theorem in this way are that it becomes evident at once that the theorem is valid not merely for the minima of the energy (or the Euclidean action) but for any stationary points, and that it does not matter whether the fields are quantized or not [10, 11]. All that is actually required is that the condensation point be stationary, which comes out as the absence of 'tadpoles' in the quantized case.

The above discussion is, of course, valid only in the absence of gauge fields, and a question that has to be discussed is the modifications that occur when gauge fields are coupled into the system. The Noether derivation of the Goldstone theorem turns out to be very convenient for this purpose. It is first shown (Sect. 3) that the procedure of minimizing the energy (or the Euclidean action) is a gauge covariant one in the sense that the vacuum configuration lies on a gauge-invariant orbit. This forms the basis for the Elitzur result, which is essentially the statement that in quantum field theory the functional integration [11] averages over this orbit and so preserves the gauge symmetry.

The most characteristic feature of spontaneously broken gauge theories is, of course, the Higgs mechanism [1, 3], through which the Goldstone fields disappear and gauge fields acquire masses. However, since the Goldstone fields disappear only in a particular gauge (the so-called physical one) and the mass-generation is usually computed for a particular vacuum configuration (the constant one in the topologically trivial case) *without* averaging over the orbit, some question arises as to the compatibility of the Higgs mechanism and the Elitzur theorem. This question is discussed for the two parts of the Higgs mechanism, namely the vanishing of the Goldstone fields and the generation of the gauge-field masses, in Sects. 4 and 5 respectively. In Sect. 4 the Noether derivation of the Goldstone theorem is used to show that, although the Goldstone fields vanish only in a particular gauge, the Goldstone theorem fails in all gauges, and in Sect. 5, it is shown that the generation of the gauge-field masses is a completely gauge-invariant phenomenon and is thus insensitive to the Elitzur averaging. In both these sections the discussion is not only gauge invariant but also independent of the topology of the underlying configuration space.

Finally, in Sect. 6 we consider a point which arises out of the discussion of Sects. 4 and 5, but is not often discussed, namely the compatibility of the general gauge choice with the choice of vacuum configuration. We show that, generically, the choice of gauge for general configurations determines the vacuum configuration, but that this is not always the case, and that, in particular, the physical gauge leaves the choice of vacuum configuration completely free.

2 Noether Version of Goldstone Theorem

We begin by recalling the Noether theorem. Let $L(\eta, \partial_\mu \eta)$, where ∂_μ means $\partial/\partial x_\mu$, be the Lagrangian density for any set of fields $\eta(x)$ and let the Euler-Lagrange field equations for L be written in the form

$$\partial_\mu \pi_\mu = \frac{\partial L}{\partial \eta} , \quad \text{where} \quad \pi_\mu = \frac{\partial L}{\partial(\partial_\mu \eta)} . \tag{1}$$

Now suppose that the fields transform linearly with respect to some x-independent (rigid) continuous group G i.e.

$$\eta(x) \rightarrow e^{ia \cdot \sigma} \eta(x) , \tag{2}$$

where a_α and σ_α for $\alpha = 1,2...\dim G$, are the (rigid) group parameters and generators respectively. The representation generated by the σs need not, of course, be irreducible, and will certainly not be irreducible if the set of fields $\eta(x)$ includes fields of different spin. Noether's theorem then states that, if one defines the currents

$$j_\mu^\alpha(x) = \sum_\eta \pi_\mu(x) \sigma^\alpha \eta(x) , \tag{3}$$

then, as a consequence of the field equations, one has

$$\partial_\mu j_\mu^\alpha = \left(\frac{\delta L(\phi, \partial_\mu \phi)}{\delta a_\alpha} \right)_{a=o} , \tag{4}$$

and hence if the Lagrangian density is invariant with respect to the group transformations (2) the currents $j_\mu^\alpha(x)$ are conserved.

In order to derive the Goldstone theorem from this result let us denote the scalar fields in the set $\eta(x)$ by $\phi(x)$ and assume for simplicity that the Lagrangian is of the standard form

$$L(\phi, \partial_\mu \phi, \eta) = \frac{1}{2}(\partial_\mu \phi)^2 + \Lambda(\phi, \eta) , \tag{5}$$

where η now denotes all the other fields except the scalar fields and the term $\Lambda(\phi, \eta)$ contains no derivatives of ϕ. Then the Noether currents take the form

$$j_\mu^\alpha(x) = (\partial_\mu \phi(x))\sigma^\alpha \phi(x) + j_\mu^\alpha(\eta(x)) , \tag{6}$$

where $j_\mu^\alpha(\eta(x))$ denotes the part coming from the fields other than the scalar fields, and may contain the scalar fields but not their derivatives.

Suppose now that the scalar field undergoes a spontaneous symmetry breakdown i.e. the field takes the form

$$\phi(x) = \phi^o + \theta(x) , \quad \text{where} \quad \phi^o \neq 0 , \tag{7}$$

and where ϕ^o, which we shall call the condensate, is a constant stationary point of the action. Then clearly the Noether currents may be written as

$$j_\mu^\alpha = (\partial_\mu \theta(x)) \sigma^\alpha \phi^o + j_\mu^\alpha(\theta(x)) + j_\mu^\alpha(\eta(x)) \,, \tag{8}$$

and the Noether theorem may be written as

$$\partial_\mu j_\mu^\alpha = \partial^2 (\theta(x) \sigma^\alpha \phi^o) + \partial_\mu j_\mu^\alpha(\theta(x)) + \partial_\mu j_\mu^\alpha(\eta(x)) = 0 \,. \tag{9}$$

But since the constant field ϕ^o is supposed to be a stationary point of the action the currents $j_\mu(\theta)$ and $j_\mu(\eta)$ in this equation are bilinears in the fields. Hence equation (9) may be written as

$$\partial^2 (\theta(x) \sigma^\alpha \phi^o) = \text{bilinear in the fields} \,. \tag{10}$$

But this shows that the fields $(\theta(x) \sigma^\alpha \phi^o)$ are massless, which is precisely the Goldstone theorem. Of course, if the condensate ϕ^o is only a stationary point but not a minimum of the energy the word massless is not quite appropriate, since, strictly speaking, masses are defined only at the minima. But the statement that the field equations for the Goldstone fields $(\theta(x) \sigma^\alpha \phi^o)$ contain no linear terms may then be regarded as the generalization of the usual Goldstone statement for arbitrary stationary points. The most interesting point concerning the above derivation, however, is that it makes no explicit distinction between classical and quantized fields. The Noether theorem holds for both kinds of fields, so the only place in which the distinction enters is in the condensate. In the classical case the assumption that the condensate ϕ^o is stationary requires no further consideration after the initial choice, but in the quantized case it requires more careful consideration. In particular, in perturbation theory it requires that the condensate remain stationary in each order of perturbation i.e. it requires that the vacuum be corrected at each order of perturbation so that no tadpole graphs appear. But this is the only distinction between the classical and quantum cases.

It should, perhaps, be emphasized that the question as to whether a spontaneous breakdown actually occurs i.e. whether the field $\phi(x)$ can actually condense to a non-zero constant ϕ^o, lies outside the scope of the above discussion, which is concerned only with the *consequences* of there being a spontaneous breakdown. In particular the well-known statements [12] to the effect that there can be no spontaneous symmetry breakdown in low dimensions are not in contradiction with the above discussion but complementary to it.

3 The Effect of Gauge Fields

The assumption that the kinetic part of the Lagrangian for the scalar fields is of the form shown in equation (5) eliminates the possibility of the scalar fields being coupled to gauge fields and in this section we wish to consider the modifications that arise if a gauge-field coupling is introduced. In that case the standard Lagrangian density becomes

$$L(\phi, A_\mu) = \frac{1}{4} \text{tr} F^2 + \frac{1}{2} (D\phi)^2 + \Lambda(\phi) \,, \tag{11}$$

where F is the gauge field, D_μ is the covariant derivative, and, for simplicity, all other fields such as fermion fields have been omitted, so that Λ is actually a potential for ϕ. The field equation for ϕ is still (1) since this was derived for general Lagrangians, but since $\pi_\mu = D_\mu \phi$ is covariant, it is clear that the two sides of (1) are not separately covariant. It is convenient to make them covariant by adding a term $(A_\mu^\alpha \sigma_\alpha \pi_\mu)$ to each side, in which case the equation takes the form

$$D_\mu \pi_\mu = \frac{\partial \Lambda}{\partial \phi} , \tag{12}$$

each side of which is manifestly covariant. The expression (3) for the Noether currents also remains the same, and is covariant, but, by using the field equation (11) one sees that the divergence equation becomes the covariant one

$$D_\mu j_\mu^\alpha = \left(\frac{\delta L}{\delta a_\alpha} \right)_{a=o} . \tag{13}$$

Thus if L is group invariant the covariant divergence vanishes. This is the covariant version of Noether's theorem.

To consider the question of the Goldstone theorem one must now consider the question of a spontaneous symmetry breakdown. The important point to note is that, whereas in the non-gauge case the breakdown is given by equation (7), where the condensate ϕ^o is *constant*, in the gauge case it is given by equation (7) where the condensate ϕ^o is only *covariantly* constant. To see this, we note that the conditions for a minimum of the energy (or the Euclidean action) for the Lagrangian density (11), denoted by superscript zero, are

$$F^o = 0, \quad D^o \phi^o(x) = 0, \quad \text{and} \quad \Lambda^o = \text{minimum} , \tag{14}$$

the first of which implies that the gauge field is pure gauge and the second two imply that $\phi(x)$ lies on gauge invariant group orbit i.e.

$$A_\mu^o = (g^o)^{-1} \partial_\mu g^o \quad \text{and} \quad \phi^o(x) = U(g^o(x)) \phi(0) , \tag{15}$$

where $x = 0$ is some arbitrary origin in x-space. (Of course, if the x-space is topologically non-trivial $g^o(x)$ may become singular so that (15) holds only in coordinate patches). From (15) one sees that there is no need for $\phi^0(x)$ to be constant. Indeed it is well-known in monopole and instanton theory that if the x-space is topologically non-trivial and the topological charge is non-zero $\phi^0(x)$ *cannot* be constant. On the other hand it is clear that $\phi^0(x)$ lies on a gauge-invariant group orbit and this is the origin of the Elitzur theorem, which is based on the fact that in quantum theory the functional integral averages over the orbit and is thus gauge-invariant. It states essentially that, while the condensate $\phi^o(x)$ in (15) is not zero, the *average* of the condensate with respect to the functional integral (which is equivalent to the vacuum expectation of the field in the canonically quantized version of the theory) is zero. Thus $\phi^o(x) \neq 0$ but $< \phi^o(x) >= 0$.

On the other hand, in the usual treatments of spontaneously broken gauge theories one chooses as vacuum configuration a definite point on the orbit (e.g.

ϕ=constant and $A_\mu = 0$ in the topologically trivial case) and this breaks the gauge symmetry. Since this procedure produces the characteristic features of the theory such as the disappearance of the Goldstone fields and the generation of masses for the gauge fields the question then arises as to how these effects can be compatible with the Elitzur result. These are the questions that will be discussed in the following sections.

4 Gauge Analogue of the Noether-Goldstone Theorem

To obtain the gauge analogue of the Noether-Goldstone theorem we note that, since the Noether currents in the gauge case are of the form

$$j_\mu^\alpha(\phi(x)) = (D_\mu \phi(x))\sigma^\alpha \phi(x) , \tag{16}$$

in the case of a spontaneous breakdown they become

$$j_\mu^\alpha(\phi) = (D_\mu \phi^o(x))\sigma^\alpha \phi^o(x) + (D_\mu \phi^o(x))\sigma^\alpha \theta(x) + (D_\mu \theta(x))\sigma^\alpha \phi^o(x) + j_\mu^\alpha(\theta) . \tag{17}$$

But since the condensate $\phi^o(x)$ is gauge covariant with respect to the vacuum gauge-field A_μ^o we may write

$$D_\mu \phi^o(x) = (D_\mu - D_\mu^o)\phi^o(x) = \Delta A_\mu(x)\phi^o(x) \quad \text{where} \quad \Delta A_\mu = A_\mu - A_\mu^o . \tag{18}$$

and hence

$$\begin{aligned} j_\mu^\alpha(\phi(x)) = \Delta A_\mu^\beta(x)(\sigma_\beta \phi^o(x), \sigma^\alpha \phi^o(x)) + D_\mu(\theta(x), \sigma^\alpha \phi^o(x)) \\ + 2\Delta A_\mu^\beta(x)(\sigma_\beta \phi^o(x), \sigma^\alpha \theta(x)) + j_\mu^\alpha(\theta(x)) \end{aligned} , \tag{19}$$

where the inner product is in the group-representation space, and is used explicitly here and henceforth in order to clarify the notation. Note that ΔA_μ, being the difference of two connections, is a vector field. Applying the Noether theorem in the covariant form (13) one then obtains as the gauged version of the Goldstone theorem (10) the result

$$D_\mu j_\mu^\alpha(\phi(x)) = D^2(\theta(x), \sigma^\alpha \phi^o(x)) + D_\mu[\Delta A_\mu^\beta(\sigma_\beta \phi^o(x), \sigma^\alpha \phi^o(x))] + \text{bilinears} = 0 . \tag{20}$$

Thus even in the gauge theory there is a Goldstone theorem. However, it is only a formal theorem in the sense that it no longer implies the existence of massless fields. To see this, let us write (20) in the non-covariant form

$$\partial^2(\theta(x), \sigma_\alpha \phi^o(x)) + \partial_\mu[\Delta A_\mu^\beta(\sigma_\beta \phi^o(x), \sigma_\alpha \phi^o(x))] = \text{bilinears} . \tag{21}$$

From this equation one cannot conclude that the Goldstone field $(\theta(x), \sigma_\alpha \phi^o(x))$ is massless, but only that there is a relationship between its d'Alembertian and the divergence of the vector field $\Delta A_\mu^\alpha(x)$. Thus, in the sense that it predicts the existence of massless fields the Goldstone theorem fails in the presence of gauge fields. This is true in any gauge, but, of course, it is well-known that there exists a gauge, namely the physical gauge, in which it fails even in the stronger sense that the Goldstone fields actually vanish. Since the proof of this is usually only given for the constant vacuum configuration i.e. for $\phi^o(x) = $ constant, it may be worthwhile reproducing it here in a form that is applicable for any vacuum configuration: Consider the functions $f(x, a) = (\theta(x), U(a)\phi^o(x))$, where a^α are the group parameters and $U(a)$ the group-representation to which the scalar fields belong. For each value of x this is a continuous (even analytic) function of the a's and since the range of the a's is compact (assuming the group is compact) it is a function with at least one stationary point. Denoting any stationary point (the minimum, say) for each fixed x by $a_s^\alpha(x)$ we have

$$\left(\frac{\partial f(x,a)}{\partial a_\alpha}\right)_{a=a_s} = v_\beta^\alpha(a_s)(\theta(x), U(a_s)\sigma^\beta\phi^o(x)) = 0 \,, \tag{22}$$

where the $v_\beta^\alpha(a_s)$ is the group velocity matrix. Since this matrix is invertible from the general theory of Lie groups one sees that

$$(\theta(x), U(a_s(x))\sigma_\beta\phi^o(x)) = 0 \,, \tag{23}$$

and hence if we make the gauge transformation

$$\theta(x) \to \theta_g(x), \quad \text{where} \quad \theta_g(x) = U^{-1}(a_s(x))\theta(x) \,. \tag{24}$$

we have

$$(\theta_g(x), \sigma_\alpha\phi^o(x)) = 0 \,. \tag{25}$$

But this shows that the Goldstone fields, which are just the $\sigma_\alpha\phi^o(x)$ components of the scalar fields, vanish in the gauge $\theta = \theta_g$.

5 Covariant Mass Generation

In this section we wish to show that the mass-generation part of the Higgs mechanism also is a gauge-invariant phenomenon. For this we consider the form that the kinetic term for the scalar fields in the Lagrangian density (11) takes in the case of a spontaneous symmetry breakdown. It is easy to see that the form is

$$L(\phi(x)) = \frac{1}{2}(D_\mu\phi^o(x), D_\mu\phi^o(x)) + (D_\mu\phi^o(x), D_\mu\theta(x)) + L(\theta(x)) \,, \tag{26}$$

which, on using (18), reduces to

$$L(\phi(x)) = \triangle A_\mu^\alpha \triangle A_\mu^\beta(\sigma_\alpha\phi^o(x), \sigma_\beta\phi^o(x)) + \triangle A_\mu^\alpha(\sigma_\alpha, D_\mu\theta(x)) + L(\theta(x)) \,. \tag{27}$$

If we recall that $\triangle A_\mu$ is a vector rather than a connection we see that each term in (27) is separately gauge-invariant. On the other hand, the leading term on the right-hand-side is a mass-term for this vector field with mass-matrix $M_{\alpha\beta}(x)$ given by $(\sigma_\alpha\phi^o(x), \sigma_\beta\phi^o(x))$. Furthermore, since, by definition, we have

$M(x) = U_{\text{adj}}^{-1}(g(x))M(0)U_{\text{adj}}(g(x))$, where $U_{\text{adj}}(g)$ denotes the elements of the adjoint representation, the eigenvalues of $M(x)$ are independent of x (even in the topologically non-trivial case) and are therefore genuine physical masses. This generation of physical masses constitutes the covariant version of the Higgs mechanism. It shows that a spontaneously broken potential for the scalar fields will produce masses for the vector fields $\triangle A_\mu$ which are gauge-invariant, and therefore quite independent of the Elitzur averaging. Since, from (10), the directions $\sigma_\alpha \phi^o(x)$ are the Goldstone directions, it is, as usual, just the gauge fields in the Goldstone directions that acquire masses. Note that the mass-generation is not only gauge-invariant but independent of the topology of the x-space and is thus valid even for cases (such as monopole configurations) in which the x-dependence of the scalar condensates $\phi^o(x)$ cannot be gauged away.

6 Gauge-Fixing and Vacuum Configurations

A remarkable feature of the vanishing of the Goldstone fields as described in Sect. 4 is that in the case of the physical gauge it seems to be possible to fix the gauge for general configurations and at the same time leave the vacuum configuration arbitrary. To understand why this is remarkable, and is not to be expected for general gauge choices, let us consider the general relationship between the choice of gauge and the choice of vacuum configuration (restricting ourselves, for simplicity, to the topologically trivial case). First we note that, in general, gauge fixing imposes dim G conditions on the fields. On the other hand, the vacuum conditions (15) express all the fields in terms of dim G gauge functions, namely the parameters $a^o(x)$ in $g^o(x)=\exp(a^o(x)_\alpha \sigma^\alpha)$. Hence a complete gauge fixing would be expected to fix the functions $g^o(x)$ uniquely. For example, for scalar QED with one real gauge field $A_\mu(x)$ and one complex scalar field $\phi(x)$ the usual gauge chosen is the real gauge, in which the imaginary part of $\phi(x)$ is set equal to zero. Since the scalar-field part of the vacuum conditions in this case are

$$\phi^o(x) = \exp(ia^o(x))\phi^o(0) , \tag{28}$$

we see that the real gauge forces the vacuum configuration to be

$$a^o(x) = \text{constant} \quad \Rightarrow \quad \phi^o(x) = \text{constant} \quad \text{and} \quad A_\mu^o(x) = 0 . \tag{29}$$

Thus in this case the vacuum configuration is completely fixed by the general gauge fixing, and is the conventional, constant, configuration. Perhaps a more interesting set of gauges (valid for general groups) are the gauges

$$\partial_\mu A_\mu^\alpha(x) = f^\alpha(\phi(x)) , \tag{30}$$

where, for example, the functions $f^\alpha(\phi)$ are chosen as

$$f^\alpha(\phi) = 0, \quad f^\alpha(\phi) = (\phi^o, \sigma^\alpha \phi), \quad \text{and} \quad f^\alpha(\phi) = \phi^\alpha, \quad (\phi \, \epsilon \, \text{adjoint}) , \tag{31}$$

respectively, the first and second being the Landau and 't Hooft gauges, which have the property that they eliminate the bilinear term $\Delta A_\mu^\alpha(\sigma_\alpha, \partial_\mu \theta(x))$ in (27). It is easy to see that for these three gauges the conditions for the vacuum gauge functions $g^o(x)$ are

$$\partial_\mu(g^o(x)^{-1}\partial_\mu g^o(x)) = 0, \quad 0, \quad \text{and} \quad \phi^o(x), \tag{32}$$

respectively, where in the second case we have used the fact that in the adjoint representation the generators are anti-symmetric and in the third case we have, of course, the relation $\phi^o(x) = (g^o(x))^{-1}\phi^o(0)g^o(x)$. It is clear that in all three cases the vacuum gauge function $g^o(x)$ is determined uniquely up to a function which is a solution of (32) with zero right hand side (and which we shall call a quasi-harmonic function since it reduces to a harmonic function in the abelian case). The lack of complete determination of the vacuum in these cases is, of course, simply due to the fact that the original gauge fixing is itself complete only up to quasi-harmonic functions. From (32) we see that in both the Landau and 't Hooft cases the vacuum is forced to be a trivial configuration, i.e. either the constant vacuum configuration or a quasi-harmonic equivalent. But for the third gauge chosen in (31) one sees by inspection that the vacuum configuration *cannot* be the trivial one or a quasi-harmonic equivalent. This is the reason that it was chosen as an example, and is probably one of the reasons that such gauges are not usually considered in the literature!

Let us now consider the physical gauge in the light of these examples. The condition for the physical gauge is (25). Now, in order to keep the independence of all the scalar fields explicit one must consider them to be real fields, in which case the generators σ are anti-symmetric. On setting $\phi(x)$ equal to $\phi^o(x)$ for the physical gauge and using the anti-symmetry of the σ's one sees that the gauge condition (25) is *automatically* satisfied. Thus in the case of the physical gauge the general gauge condition imposes no condition on the $g^o(x)$ and hence no condition on the choice of vacuum configuration (except of course that $g^o(x)$ lie on the group orbit determined by the potential). From the discussion of the other gauges it is clear that in this respect the physical gauge is quite exceptional.

References

1.   See for example:
      L. Faddeev, A. Slavnov: *Gauge Fields* (Benjamin/Cummings, New York, 1980);
      T-P. Cheng, L-F. Li: *Gauge Theory of Elementary Particle Physics*
      (Clarendon Press, Oxford, 1984);
      C. Quigg: *Gauge Theories of Strong, Weak and Electromagnetic Interactions*
      (Benjamin/Cummings, New York, 1983);
      C. Lai, R. Mohapatra: *Gauge Theories of the Fundamental Interactions*
      (World Scientific, Singapore, 1981)
2.   A. Fetter, J.Walecka: *Quantum Theory of Many Particle Systems* (McGraw-Hill, New York, 1971)

3.   D. Bailin: *Weak Interactions* (Hilger, Bristol, 1982);
        E. Commins: *Weak Interactions* (McGraw-Hill, New York, 1973);
        C. Lai: *Gauge Theory of the Weak and Electromagnetic Interaction*
        (World Scientific, Singapore, 1981)
4.   N. Craigie, P. Goddard, W. Nahm: *Monopoles in Quantum Field Theory*
        (World Scientific, Singapore, 1982)
5.   S. Coleman: *Aspects of Symmetry* (Cambridge University Press, 1985)
6.   P. Ramond: *Field Theory* (Benjamin/Cummings, New York, 1981, page 296);
        V. Gribov: Nucl. Phys. **B139** 1 (1978)
7.   S. Elitzur: Phys. Rev. **D12** 3978 (1975)
8.   E. Noether: Gott. Nachr., 235 (1918)
9.   J. Goldstone: Nuovo Cimento **19** 154 (1961)
10.   J. Goldstone: A. Salam, S. Weinberg, Phys. Rev. **127** 965 (1962)
11.   D. Amit: *Field Theory, Renormalization Group and Critical Phenomena*
        (McGraw-Hill, New York, 1978)
12.   N. Mermin, H. Wagner: Phys. Rev. Lett. **17** 1133 (1966) ;
        S. Coleman: Comm. Math. Phys. **31** 259 (1973);
        S-K. Ma, R. Rajaraman: Phys. Rev. **D11** 1701 (1975)

A Baryon Standard Model for Electroweak and Strong Interactions

R. Rączka*

International School for Advanced Studies (SISSA), Trieste, Italy
and
Institute for Nuclear Studies, Warsaw, Poland

Abstract: A new model for electroweak and strong interactions based on observed particles only is proposed. The strong sector of this model is determined by the spontaneously broken SU(6) gauge group, which naturally arises from analysis of properties of observed vector mesons. In the low energy region the model shares properties of the Vector Dominance Model whereas in the high energy region the strong sector of the model is asymptotically free. The model gives approximately the same results as the Standard Model in the electroweak sector and provides a natural framework for a reliable description of lepton-hadron and hadron-hadron interactions.

1 Introduction

We propose a new model for description of weak, electromagnetic and strong interactions. Our motivation for introduction of this model is based on two facts:

1° The conventional standard model (SM) predicts in lepton-lepton processes the observed quantities like cross-sections, decay widths, vector meson masses, branching ratios etc. with impressive precision [1]. However in many lepton-hadron or hadron-hadron processes the SM predictions based on the structure of strong QCD sector seem to be in evident contradiction with many experimental data [2,3]. These results strongly suggest that perhaps the strong QCD sector of SM should be replaced by some other strong sector [4].

2° We are convicted that Nature for some reasons likes spontaneously broken gauge field theory models (SBGFTM) and works with observed particles only.

The difficulties of the strong sector of the conventional SM are summarized in Sect. 2. Hence I will clarify here Item 2. We know that SBGFTM based on the

* Supported in part by the contract RPBP 01.2 of the Institute of Theoretical Physics of Polish Academy of Sciences.

SU(2)$_L$×U(1) gauge group describes the electroweak lepton interactions with a remarkable accuracy. [1] One may expect therefore that some SBGFTM may also describe the strong sector of electroweak theory: if this would be the case then the observed distinct 1$^-$-vector mesons should form the adjoint representation of the requested gauge group. The inspection of the recent Review of Particle Properties shows up that the observed distinct vector meson multiplets with $J^P = 1^-$ have indeed the same quantum numbers as the vector meson multiplets in the 35-adjoint vector meson representation of the SU(6) gauge group [4]. This remarkable fact inspired us to propose an effective SBGFTM based on the SU(6) gauge group for strong interactions [4]. In this model it is natural to assume that fundamental fermions are baryons which belong to the lowest six dimensional fundamental representation of the SU(6) group i.e.

$$\Psi = (\psi_p, \psi_n, \psi_\Lambda, \psi_{\Lambda_c}, \psi_{\Lambda_b}, \psi_{\Lambda_t}) \tag{1}$$

where $\psi_p, \psi_n, \psi_\Lambda, \psi_{\Lambda_c}, \psi_{\Lambda_b}$, and ψ_{Λ_t} are the fermion field with the same quantum numbers as the physical proton, neutron, Λ, Λ_c-particle with charm, Λ_b-particle with beauty and Λ_t particle with top, respectively. It is noteworthy that the first five baryons in (1) were found experimentally. We show in Sect. 4 that the anomaly of the baryon sextet representation (1) cancels out the lepton anomaly of the electroweak sector. Thus the representation (1) is the simplest and the most natural baryon representation which gives the anomaly free model for electroweak and strong interactions.

From the point of view of the electroweak interactions it is natural to split out the sextet (1) into three baryon doublets

$$b_1 = \begin{bmatrix} p \\ n \end{bmatrix}, \quad b_2 = \begin{bmatrix} \Lambda_c \\ \Lambda \end{bmatrix}, \quad b_3 = \begin{bmatrix} \Lambda_t \\ \Lambda_b \end{bmatrix}. \tag{2}$$

We show in Sect. 3 that these doublets can be used instead of quark doublets

$$q_1 = \begin{bmatrix} u \\ d \end{bmatrix}, \quad q_2 = \begin{bmatrix} c \\ s \end{bmatrix}, \quad q_3 = \begin{bmatrix} t \\ b \end{bmatrix} \tag{3}$$

for the construction of a new version of SM which we shall call the Baryon Standard Model (BSM).

In Sect. 2 we analyze in detail the difficulties of conventional SM and QCD in a description of exclusive lepton-hadron and hadron-hadron interactions. Next in Sect. 3 we derive the form of the total lagrangian for BSM and discuss its implications. We prove in Sect. 4 that the proposed BSM is anomaly free. We present in Sect. 5 the analysis of proton-proton interactions at high energy. The simplicity of the BSM framework in a description of this most fundamental process of strong interactions is especially evident. Next we give in Sect. 6 the preliminary analysis of annihilation of polarized electrons on polarized positrons into proton-antiproton pair in the BSM. The obtained cross section shows a complex dependence on polarization of initial particles and provides therefore acute cross check for the proposed model. In Sect. 7 we analyze asymptotic freedom in BSM and a series of concrete processes which are under investigation in this model. Finally in Sect. 8 we discuss some general properties of BSM.

2 Difficulties with the QCD Strong Sector of the Standard Model

The conventional SM predictions for lepton-lepton processes agree with the experimental data with impressive precision [1]. However in the exclusive lepton-hadron or hadron-hadron processes the predictive power of SM is very limited. The main difficulties in calculations of predictions for these processes are connected with the fact that the QCD strong sector of SM is formulated not in the language of observed hadrons but in the language of unobserved quarks and gluons. For instance for the simplest two body lepton-hadron processes like

$$e^+e^- \rightarrow \begin{cases} \pi^+\pi^- \\ K^+K^- \\ \rho^+\rho^- \\ \bar{p}p \end{cases}$$

or for elastic scattering

$$e^-p \rightarrow e^-p$$

the SM is unable to provide a reliable predictions.

In fact any scattering amplitude $M_{\ell;h}$ for exclusive production of hadrons in lepton interaction is given as the integrated product

$$M_{\ell;h} = \int \mathrm{d}x M_{\ell;qg}(x) M_{q,g;h}(x) \tag{4}$$

where x represent the (continuous and discrete) quantum numbers of a quarks-gluons system, $M_{\ell;qg}$ represents the transition amplitude from the initial leptons into a quarks-gluons system and $M_{qg;h}$ represents the transition amplitude from a given quarks-gluons system to the final system of hadrons. The amplitude $M_{\ell;qg}$ can be calculated in SM up to any order in perturbation theory; however the hadronization amplitude $M_{qg;h}$ is completely unknown in the framework of SM, since hadrons do not appear in SM lagrangian. In order to overcome this fundamental difficulty one introduces various methods like factorization of $M_{\ell;h}$ amplitude [5], models of hadronization [6], models for hadrons as composites of quarks and leptons [7] etc.. However any of these methods introduces some ad hoc model dependent assumptions outside of SM framework and make many important predictions for exclusive processes model dependent [3]. For instance the Factorization Theorem for exclusive processes states that any hadronic amplitude for the hadron process $h_1 + h_2 \rightarrow h_3 + h_4$ can be represented in the form

$$M_{h_1 h_2; h_3 h_4} = \int dx \varPhi_{h_3}^*(x) \varPhi_{h_4}^*(x) T_H(x, Q^2, \theta_{\mathrm{cms}}) \varPhi_{h_2}(x) \varPhi_{h_1}(x)$$

where T_H is the hard scattering quark amplitude and $\varPhi_{h_i}$ is the distribution amplitude for the valence quarks in each initial and final hadron state [5 iii)]. Since the composition of hadron in terms of quarks and gluons is unknown in most applications one represents – by definition – every hadron in T_H-amplitude by a system of colinear valence quarks [3,5 iii)]. Clearly this definition is arbitrary and most probably incorrect: in fact the recent EMC experiment strongly indicates that proton must consist of quarks and gluons (see e.g. [8] and item v) of Sect. 2.1). Consequently most of QCD predictions for exclusive processes which use this hadron representation contain an error of unknown value; hence the scientific value of these QCD predictions for exclusive processeses is unclear.

Another example of strong dependence of QCD predictions on a chosen model for hadronization process is provided by the works which try to determine the value of the QCD scale parameter Λ_{QCD}: if one uses the Lund fragmentation model one obtains [9]

$$\Lambda_{\mathrm{QCD}} = (325 \pm 20)\mathrm{MeV};$$

on the other hand the Ali et al. Monte Carlo method including the perturbative calculations of order α_s^2 gives [10]

$$\Lambda_{\mathrm{QCD}} = (144 \pm 16)\mathrm{MeV}.$$

The similar large difference in quantitative predictions occurs for many processes with hadrons [3]. There are presently several experimental and theoretical results indicating that the strong QCD sector of SM based on unobserved quarks and gluons is not proper for a description of lepton-hadron and hadron-hadron interactions. We enumerate below the most important difficulties.

2.1 QCD Spin Crisis

The cross sections for scattering of polarized particles on a polarized target have a rich dependence on energy, polarizations of incident particles, polar and azimuthal angles, composition of final particles etc. [2,3]. Every given model of elementary particle interactions gives specific predictions for polarized particle scattering [3]. Hence the experimental results for polarized particle scattering provide the most severe available cross checks for validity of proposed models.

We recall for instance that the Serpukhov experiment

$$p + p_\uparrow \rightarrow p + p$$

for scattering of protons on polarized target protons carried out around 1975 demonstrated that experimental data contradict the Regge pole model predictions [11]. This experimental fact was decisive for a rejection of the Regge pole model and a Reggeon Field Theory as a reliable framework for a description of high energy scattering. One might expect therefore that analysis of polarization phenomena might provide the decisive information on a structure and properties of the strong sector in SM.

Presently all calculations of polarization effects in high energy collisions carried out in QCD or SM framework predict that polarization effects will decrease if energy increases [3]. However the experimental data demonstrate that spin and polarization effects play the significant role in high energy collisions contrary to QCD and SM predictions. This is known presently as the QCD spin crisis. We refer the interested reader to the recent Symposium Proceedings devoted to these problems [12]; here we mention the recent most interesting experimental results which illustrate the importance of spin and polarization phenomena in high energy physics.

i) *Inclusive Λ polarization*

It was found that when a Λ-particle is produced inclusively at large $p_\perp^2$ in the process $pp \to \Lambda X$ it has a large polarization [13]. It is remarkable that this polarization seems totally independent of energy from KEK energy of 12GeV [14] up to ISR energies equivalent to 2000GeV [15]. This striking experimental results puts serious questions on all QCD and SM predictions which predict that spin effects will disappear at higher energies [3].

ii) *High-$p_\perp^2$ polarized proton-proton scattering*

The several experiments on scattering of polarized proton beams on polarized targets were carried out by A.D. Krisch's group [2,16]. They considered the violent pp elastic collisions at large $p_\perp^2$. They found the large and unexpected spin effects. In particular they considered the behavior of the two-spin A_{nn} quantity given by the formula

$$A_{nn} = \frac{\mathrm{d}\sigma(\uparrow\uparrow) - \mathrm{d}\sigma(\uparrow\downarrow)}{\mathrm{d}\sigma(\uparrow\uparrow) + \mathrm{d}\sigma(\uparrow\downarrow)}.$$

They found for instance that for $\theta_{\mathrm{cms}} = 90^0$ the A_{nn} originally decreases from about $0,6$ for $p_{\mathrm{Lab}} = 3\,\mathrm{GeV}/c$ to about $0,1$ for $p_{\mathrm{Lab}} = 6\,\mathrm{GeV}/c$ in agreement with QCD predictions; however if energy increases then A_{nn} unexpectedly sharply increases to about $0,6$ for $p_{\mathrm{Lab}} = 11,75\,\mathrm{GeV}/c$. Notice that $A_{nn} = 0,6$ corresponds to $\mathrm{d}\sigma(\uparrow\uparrow)/\mathrm{d}\sigma(\uparrow\downarrow) = 4$; these results indicate the strong spin dependence of hadron interaction dynamics in hard region, contrary to QCD predictions [17,5 iii),3].

Similarly for larger energies for $p_{\mathrm{Lab}} = 18,5\,\mathrm{GeV}/c$ one finds strong dependence of cross-sections and A_{nn} on proton polarizations [2].

In case of exclusive scattering of unpolarized protons on polarized targets e.g. $p + p_\uparrow \to p + p$ the important role is played by the one-spin quantity A, (so called analyzing power) which is defined by the formula

$$A = \frac{\mathrm{d}\sigma(\uparrow) - \mathrm{d}\sigma(\downarrow)}{\mathrm{d}\sigma(\uparrow) + \mathrm{d}\sigma(\downarrow)}.$$

Analysis of helicity amplitudes in QCD for this process shows up that in the hard region A should be zero [5 iii),3]; on the other hand the experimental results show that A is quite large [18].

All these results put a serious question mark on the validity of the QCD framework [3].

iii) *Asymmetry in binary inelastic hadron reactions*

The problem of spin dependence was also investigated in several inelastic binary reactions of the type $h_1 + h_{2\uparrow} \rightarrow h_3 + X$ (e.g. $\pi^- + p_\uparrow \rightarrow \pi^0 + X$) [19]. It was verified that the asymmetry A in this processes for high $p_\perp$-values was large contrary to QCD prediction which states that A should be zero [3,5 iii)].

iv) *Parity violation in* σ_{Tot}

It was found in proton-proton scattering at energies of 45MeV and 800MeV that there is the parity violation of about one part in 10^7 [20], which is consistent with the weak contribution in SM to proton-proton scattering. However the similar experiment carried out at $6\,\text{GeV}/c$ gave the violation $(2, 65 \pm 0, 6 \pm 0, 35)10^{-6}$ [21]; this significant effect of parity violation in strong processes is difficult to reconcile with our understanding of the size of such weak interaction effects.

v) *QCD spin crisis of proton*

The original determination of proton structure functions $F_i(x, Q^2)\, i = 1, 2$ from deep lepton-proton inelastic scattering performed in SLAC provided the evidence on the existence of partons inside protons [22]. However – as it was stressed by Altarelli [23] – this idyllic picture was spoiled by the recent new, high precision EMC experiment on polarized lepton-nucleus scattering [8,24]. Using their and earlier results the EMC group concluded that the integral of polarized $g_1(x, Q^2)$ proton structure function has the value

$$\int_0^1 g_1(x, Q^2 = -10, 7\,\text{GeV}^2)\, \mathrm{d}x = 0, 126 \pm 0, 1 \pm 0, 015.$$

The above integral is connected with sum of the fractions Δq_f of the quark spins [25].

$$\int_0^1 g_1(x)\, \mathrm{d}x = \sum_f e_f^2 \Delta q_f.$$

Using this, Bjorken sum rules and flavor symmetry Mandula succeeded to determine Δq_f and obtained [25]

$$\sum_f \Delta q_f = 0, 04 \pm 0, 16.$$

This result leads to the shocking conclusion that quarks inside proton contribute at most 20% to the proton spin.

In order to avoid evident contradiction it was suggested in several works that there might be a significant contribution from gluons for each Δq_f [26]–[28]. This suggestion was numerically tested using the lattice gauge theory calculations [25]. It was found that gluons spin contribution Δg to the quark spin fraction Δq_f is consistent with 0; hence there remains the fundamental difficulty connected with the fact that quarks and gluons contribute at most 20% to the proton spin.

The other examples of QCD and SM spin crisis were presented during the recent Trieste Symposium [12].

All these results on interactions of polarized particles indicate that at present QCD and SM have serious problems in a reliable explanation of many experimental data on polarized particle interactions which seems to contradict perturbative predictions of these models [3].

2.2 Predictions of Unobserved Processes in Perturbation Theory

The quark fields enter into the SM lagrangian in the same manner as lepton fields; they have the same form of interactions with photon, $W^\pm$, Z_0-meson and Higgs particle as leptons. Consequently for many unobserved processes from a given lepton system to leptons, quarks and gluons the SM gives in perturbation theory nonzero probability amplitude. For instance the lowest order cross-section for a production of a quark pair $\bar{q}_i q_i$, $i = u, d, s, c, b$ or t in e^+e^- collisions has in the SM the form [29]:

$$\frac{d\sigma}{d\Omega} = \frac{9\alpha^2}{4s}\sqrt{1 - 4\mu_i}X(s,\theta)$$

where

$$X = G_1(s)(1 + \cos^2\theta) + 4\mu_i G_2(s)\sin^2\theta + 2\sqrt{1 - 4\mu_i}G_3(s)\cos\theta$$

$$s = (p_+ + p_-)^2, \quad \mu_i = \frac{m_i^2}{s}, \quad \theta = \angle(e^-, q_i)$$

and $G_i(s)$ are functions of SM coupling constants and the Z_0-propagator [29] (see also Sect. 6). We see that the cross-section for $\bar{u}u$ production in the high energy limit is 9-times bigger than that for $\mu^+\mu^-$-production. The inspection of the one-loop expression for this process given in [29] shows up that the cross section for the $e^+e^- \rightarrow \bar{u}u$ process with one-loop radiative corrections is also significant. Since higher orders of perturbation theory cannot reduce to zero one-loop cross section it is evident that in perturbation theory the cross section for the $e^+e^- \rightarrow \bar{u}u$ process will be nonzero. This strongly indicates that in SM the predicted cross-section for this unphysical process is nonzero.

Similarly SM predicts in perturbation theory nonzero cross sections for the processes

$$\mu^+\mu^- \rightarrow \bar{q}_i q_i; \quad e^+e^- \rightarrow g_i g_i$$

$$\gamma\gamma \rightarrow \bar{q}_i q_i; \quad e^+e^- \rightarrow n\bar{q}nq$$

$$W^+W^- \rightarrow \bar{q}_i q_i; \quad e^+e^- \rightarrow e^+e^- + n\bar{q}_\ell q_\ell + mg_i g_i$$

etc. which are also not observed.

One usually gives a hand waving argument that the free quarks or gluons will not appear in final states because of confinement; however the SM does not provide – up to now – any concrete calculational scheme which would forbid appearance of quarks or gluons in final states. The fact that SM predicts in perturbation theory

nonzero cross sections for unobserved processes with strongly interacting particles
makes for us the strong QCD sector of SM suspicious.

It could be in principle that SM predicts the non zero cross-sections for unphysical processes in perturbation theory but these cross sections calculated nonperturbatively would be zero. Although such phenomenon is against our entire
experience in Quantum Electrodynamics and theory of electroweak lepton interactions one cannot a priori reject this possibility.

In order to test this possibility we proposed to calculate the truncated amplitude

$$M^{\mathrm{Tr}}_{e^+e^-;\overline{u}u}(x,y;z,\xi) = <\overline{\psi}_{e^+}(x)\psi_{e^-}(y)\overline{\psi}_{\overline{u}}(z)\psi_u(\xi)>_{\mathrm{Tr}}$$

for the $e^+e^- \to \overline{u}u$ process in lattice gauge theory [30]. If this amplitude is nonzero
then the cross-section for the $e^+e^- \to \overline{u}u$ process will be also nonzero. Thus
the conventional SM would predict passage from the observed e_+e_--state to an
unobserved $\overline{u}u$-state and should be therefore abandoned.

Actually we proposed in [30] a simpler cross check of the strong sector in SM
based on QCD calculations only; consider the process $S \to \overline{u}u$ of a decay of the
scalar S-meson into a quark-antiquark pair. This process has non zero probability
amplitude in perturbation theory. Representing the S-meson as a $\overline{q}_i q_i$ system one
obtains that the decay amplitude in QCD is given by the following truncated
correlation function [30]

$$D^{\mathrm{Tr}}_{s;\overline{u}u}(x,y,z) = <:(\overline{q}_i q_i):(x)\overline{u}(y)u(z)>_{\mathrm{Tr}} .$$

If this amplitude in lattice gauge theory will be non-trivial then the unphysical decay $S \to \overline{u}u$ will be allowed in QCD framework! This then will present a significant
challenge for the QCD strong sector of the SM.

2.3 Physical Mass of Gluons and Quarks

Most of SM predictions are based on perturbation series in QCD – which according
to t'Hooft – might be even non-asymptotic [31]; hence the value of perturbative
predictions is unclear. In order to overcome these difficulties many people are
carrying out presently nonpertubative calculations based on the lattice gauge field
theory [32]. They calculated in this framework several important quantities like
baryon and meson masses, weak matrix elements, first and second order phase
transitions, etc. with a considerable success [32]. However in order to test validity
of the SM one should verify in my opinion nonperturbatively the fundamental
assumptions of the model like e.g. quark and gluon confinement. One can do this
by calculating for instance the lowest mass in the gluon or quark propagators or
the probability for the simplest unobserved processes discussed in item 2.2. If we
find that gluon and quark propagators have the definite nonzero lowest mass in
mass spectrum then we should see these particles in asymptotic states according
to general principles of quantum field theory.

The perturbative one-loop calculation of the gluon propagator showed that the
gluon has nonzero mass pole [33]: the two-loop calculation showed that the gluon

propagator is infrared divergent so higher order perturbation theory is inconclusive [34]. The first nonperturbative calculation of the gluon propagator in lattice gauge theory carried out by Mandula and Ogilvie demonstrated that the gluon has the physical mass around $600 MeV$ [35]. The subsequent calculations carried out by these authors confirmed this result [36].

The existence of a physical nonzero mass of the gluon creates a serious conceptual problem in QCD: if mass is nonzero then confinement does not hold and in physical reactions one should see free gluons: since we do not observe free gluons the validity of the QCD model is questionable.

The preliminary calculations of the quark mass spectrum in lattice gauge theory show that quarks have definite physical masses [37].

It would be interesting to repeat the Mandula-Ogilvie calculations of lowest gluon mass using bigger lattices as well as to calculate in the lattice gauge theory the lowest mass of the quark propagator.

The above analysis indicates that there exists a series of experimental and theoretical results which show that using the present form of QCD strong SM sector we have essential difficulties in describing many exclusive and inclusive physical processes. This encourage us to propose a form of SM based on observed particles only.

3 The Baryon Standard Model

We shall derive now a form of the total lagrangian for electroweak and strong interactions based on observed leptons and baryons. Since the GWS model is very successful in a description of electroweak lepton interactions we shall take the lepton sector in GWS form. Hence the lepton and Higgs sector will contain the following field multiplets:

$$\ell_{1L} = \begin{pmatrix} \nu_e \\ e_L^- \end{pmatrix}, \quad \ell_{2L} = \begin{pmatrix} \nu_\mu \\ \mu_L^- \end{pmatrix}, \quad \ell_{3L} = \begin{pmatrix} \nu_\tau \\ \tau_L^- \end{pmatrix},$$

$$\ell_{1R} = e_R^-, \quad \ell_{2R} = \mu_R^-, \quad \ell_{3R} = \tau_R^-, \tag{5}$$

$$\varphi = \begin{pmatrix} \varphi_+ \\ \varphi_0 \end{pmatrix}, \quad \varphi^c = \hat{c}\varphi^* = \begin{pmatrix} \varphi_0^* \\ -\varphi_- \end{pmatrix}, \quad \hat{c} = i\sigma_2, \quad \varphi_- = \varphi_+^*.$$

Since the gauge group $G = SU(2)_L \times U(1)$, the covariant derivative has the form

$$iD_\mu = i\partial_\mu + g'(Y/2)B_\mu + g(\tau_a/2)W_\mu^a \tag{6}$$

where g' is the coupling constant for the singlet B_μ-gauge field, g is the coupling constant for the triplet of the W_μ^a gauge field and Y is the hypercharge equal to $2 < Q >$, where $< Q >$ is the average charge of a given fermion or boson multiplet.

We now determine the baryon multiplet structure in this model. The baryon sextet (1) – similarly like in quark standard model – should interact with the Higgs doublet (5): hence by virtue of (6) it seems natural to introduce the following baryon multiplet structure

$$b'_{1L} = \begin{pmatrix} p'_L \\ n'_L \end{pmatrix}, \quad b'_{2L} = \begin{pmatrix} \Lambda'_{cL} \\ \Lambda'_L \end{pmatrix}, \quad b'_{3L} = \begin{pmatrix} \Lambda'_{tL} \\ \Lambda'_{bL} \end{pmatrix},$$

$$b'_{1R} = p'_R, b'_{2R} = \Lambda'_{cR}, b'_{3R} = \Lambda'_{tR}, \tag{7}$$

$$b'_{4R} = n'_R, b'_{5R} = \Lambda'_R, b'_{6R} = \Lambda'_{bR}.$$

According to the Cabibbo idea the leptons see a rotated baryon world [38]. Hence the baryon multiplets (7) do not yet correspond to the multiplets of the physical baryons $p, n, \Lambda, \Lambda_c, \Lambda_b$ and Λ_t, respectively.

The total lagrangian for the electroweak part of the BSM may be written in the form:

$$L = L_V + L_F + L_H + L_Y \tag{8}$$

where

$$L_V = -\frac{1}{4}(W^a_{\mu\nu} + \varepsilon^{abc}W^b_\mu W^c_\nu)^2 - \frac{1}{4}B^2_{\mu\nu} \tag{9}$$

$$W^a_{\mu\nu} = \partial_\mu W^a_\nu - \partial_\nu W^a_\mu, \quad B_{\mu\nu} = \partial_\mu B_\nu - \partial_\nu B_\mu$$

$$L_F = \sum_{j=1}^{3} \left\{ \left[\bar{\ell}_{jL}\gamma^\mu(i\partial_\mu - \frac{g'}{2}B_\mu + \frac{g}{2}\tau_a W^a_\mu)\ell_{jL} \right] \right.$$

$$+ \bar{\ell}_{jR}\gamma^\mu(i\partial_\mu - g'B_\mu)\ell_{jR} + \left[\bar{b}'_{jL}\gamma^\mu(i\partial_\mu + \frac{g'}{2}B_\mu + \frac{g}{2}\tau_a W^a_\mu)b'_{jL} \right] \tag{10}$$

$$\left. + \left[\bar{b}'_{jR}\gamma^\mu(i\partial_\mu + g'B_\mu)b'_{jR} \right] + \left[\bar{b}'_{j+3R}\gamma^\mu i\partial_\mu b'_{j+3R} \right] \right\}$$

$$L_H = \left[(i\partial^\mu + \frac{g'}{2}B^\mu + \frac{g}{2}\tau_a W^{a\mu})\varphi \right]^* \left[(i\partial_\mu + \frac{g'}{2}B_\mu + \frac{g}{2}\tau_a W^a_\mu)\varphi \right] - V(2\varphi^*\varphi) \tag{11}$$

with

$$V(2\varphi^*\varphi) = -\mu_0^2\varphi^*\varphi + \lambda(\varphi^*\varphi)^2, \quad \mu_0^2 > 0, \quad \lambda > 0 \tag{12}$$

and

$$L_Y = L_Y^\ell + L_Y^b$$

with

$$L_Y^\ell = -G_Y\bar{\ell}_{iR}(\varphi^*\ell_{iL}) + \text{h.c.} \tag{13}$$

and

$$L_Y^b = -\sum_{i,j=1}^{3} \left\{ F^{ij}\bar{b}'_{i+3R}(\varphi^*b'_{jL}) + G^{ij}\bar{b}'_{iR}(\varphi^{C*}b'_{jL}) \right\} + \text{h.c.} \tag{14}$$

We shall now reduce a general baryon Yukawa interaction lagrangian to the conventional form with physical baryons.

If the matrices $\hat{F} = \{F^{ij}\}$ and $\hat{G} = \{G^{ij}\}$ are hermitian then L_Y^b conserves CP-parity: otherwise the CP-parity is violated. Set now φ, $\hat{F}$ and $\hat{G}$ in the form

$$\varphi(x) = \frac{1}{\sqrt{2}}(v + \phi(x)), \quad \widehat{F} = \widehat{F}'\widehat{V}_F, \quad \widehat{G} = \widehat{G}'\widehat{V}_G \tag{15}$$

where $\widehat{F}'$ and $\widehat{G}'$ are hermitian whereas $\widehat{V}_F$ and $\widehat{V}_G$ are unitary. Set

$$b_L^{1'} = \begin{pmatrix} p_L' \\ \Lambda_{cL}' \\ \Lambda_{tL}' \end{pmatrix}, \quad b_L^{2'} = \begin{pmatrix} n_L' \\ \Lambda_L' \\ \Lambda_{bL}' \end{pmatrix}, \quad b_R^{1'} = \begin{pmatrix} p_R' \\ \Lambda_{cR}' \\ \Lambda_{tR}' \end{pmatrix}, \quad b_R^{2'} = \begin{pmatrix} n_R' \\ \Lambda_R' \\ \Lambda_{bR}' \end{pmatrix},$$

and

$$b_L^{1''} = \widehat{V}_F b_L^{1'}, \quad b_L^{2''} = \widehat{V}_G b_L^{2'}.$$

Then from (14) we obtain

$$L_Y^b = -2^{-1/2}(v + \phi)[\bar{b}_R^{2'}\widehat{G}'b_L^{2''} + \bar{b}_R^{1'}\widehat{F}'b_L^{1''} + \text{h.c.}]. \tag{16}$$

Setting $\widehat{m}_{b1} = 2^{-1/2}v\widehat{F}'$, $\widehat{m}_{b2} = 2^{-1/2}v\widehat{G}'$ and using the relations $\bar{b}_R^{1'}\widehat{F}'b_L^{1''} + \bar{b}_L^{1''}\widehat{F}'b_R^{1'} \equiv \bar{b}^{1'}\widehat{F}'b^{1'}$, where $b^{1'} = b_R^{1'} + b_L^{1''}$, $\bar{b}_R^{2'}\widehat{G}'b_L^{2''} + \bar{b}_L^{2''}\widehat{G}'b_R^{2'} = \bar{b}^{2'}\widehat{G}'b^{2'}$, where $b^{2'} = b_R^{2'} + b_L^{2''}$ we may write (16) in the form

$$L_Y^b = -(1 + \phi/v)[\bar{b}^{1'}\widehat{m}_{b1}b^{1'} + \bar{b}^{2'}\widehat{m}_{b2}b^{2'}].$$

Let $\widehat{B}^1$ and $\widehat{B}^2$ be unitary matrices which diagonalize the hermitian mass matrices $\widehat{m}_{b1}$ and $\widehat{m}_{b2}$ respectively, i.e.

$$\widehat{B}^{1*}\widehat{m}_{b1}\widehat{B}^1 = \begin{pmatrix} m_p & & 0 \\ & m_{\Lambda_c} & \\ 0 & & m_{\Lambda_t} \end{pmatrix}, \quad \widehat{B}^{2*}\widehat{m}_{b2}\widehat{B}^2 = \begin{pmatrix} m_n & & 0 \\ & m_\Lambda & \\ 0 & & m_{\Lambda_b} \end{pmatrix}.$$

Then setting $b^{1'} = \widehat{B}^1 b^1, b^{2'} = \widehat{B}^2 b^2$ we obtain

$$L_Y^b = -(1 + \phi/v) \sum_{i=1}^{3} [m_{b1}^i \bar{b}_i^1 b_i^1 + m_{b2}^i \bar{b}_i^2 b_i^2]. \tag{17}$$

Clearly the kinetic part is invariant with respect to $\widehat{B}^1$ and $\widehat{B}^2$ transformations, i.e.

$$\bar{b}^{1'}\gamma^\mu \partial_\mu b^{1'} = \bar{b}^1 \gamma^\mu \partial_\mu b^1, \quad \bar{b}^{2'}\gamma^\mu \partial_\mu b^{2'} = \bar{b}^2 \gamma^\mu \partial_\mu b^2.$$

Hence only the final fields b_i^1 and b_i^2, $i = 1, 2, 3$ have the meaning as the fields for the physical proton, Λ_c, Λ_t, neutron Λ, and Λ_b, respectively.

It follows from the form (8)–(14) of the total lagrangian that the action of matrices $\widehat{B}^1$ and $\widehat{B}^2$ will cancel out in all terms in which baryons form the neutral currents composed of $b_i^{1'}$ or $b_j^{2'}$ fields only. However, in charge currents e.g. of the form

$$J_\mu^{(-)} = (\bar{b}_{jL}^{2'}\gamma_\mu b_{jL}^{1'}) \tag{18}$$

the action of $\widehat{B}^1$ and $\widehat{B}^2$ transformation will not cancel out and it will lead to the current

$$J_\mu^{(-)} = \bar{b}_{jL}^2 \hat{K}^* \gamma_\mu b_{jL}^1 \tag{19}$$

with $\hat{K}$ the 3×3 Kobayashi-Maskawa-like matrix of the form

$$\hat{K} = \hat{B}^{1*} \hat{B}^2 . \tag{20}$$

Carrying out the conventional analysis in the spontaneous broken model given by (8) we obtain the total interaction lagrangian of the final theory in the form

$$L = L_W' + L_H' + L_F' + L_Y' \tag{21}$$

where

$$L_W' = -i\frac{g}{2}[(W_\mu^+ W_\nu^- - W_\nu^+ W_\mu^-)W_{\mu\nu}^3 + 2(W_{\mu\nu}^- W_\mu^+ - W_{\mu\nu}^+ W_\mu^-)W_\nu^3)]$$

$$-g^2[W_\mu^+ W_\mu^-(W_\lambda^3)^2 - (W_\lambda^+ W_\lambda^3)(W_\mu^- W_\mu^3) + \frac{1}{2}(W_\mu^+ W_\nu^-)^2 + \frac{1}{2}(W_\mu^+ W_\mu^-)^2] \tag{22}$$

with

$$W_\mu^3 = \sin\theta_W A_\mu + \cos\theta_W Z_\mu \tag{23}$$

$$L_H' = -(\lambda v\phi^3 + \frac{\lambda}{4}\phi^4) + (v\phi + \frac{\phi^2}{2})(\frac{\tilde{g}^2}{4}Z_\mu^2 + \frac{g^2}{2}W_\mu^+ W_\mu^-) - \mu_0^2/4\lambda \tag{24}$$

with

$$\tilde{g} = \frac{e}{\sin\theta_W \cos\theta_W} , \quad e = g' \cos\theta_W \tag{25}$$

$$L_F' = \sum_{i=1}^{6} \bar{\psi}_i \gamma^\mu \left\{ eQA_\mu + \frac{e(1-\gamma_5)}{2\sqrt{2}\sin\theta_W}[W_\mu^+ T^+ + W_\mu^- T^-] \right.$$

$$\left. + \frac{e}{\sin\theta_W \cos\theta_W}[\frac{1}{2}(1-\gamma_5)T_3 - \sin^2\theta_W Q]Z_\mu \right\} \psi_i \tag{26}$$

where

$$\psi_1 = \begin{pmatrix} \nu_e \\ e^- \end{pmatrix}, \quad \psi_2 = \begin{pmatrix} \nu_\mu \\ \mu^- \end{pmatrix}, \quad \psi_3 = \begin{pmatrix} \nu_\tau \\ \tau^- \end{pmatrix},$$

$$\psi_4 = \begin{pmatrix} p \\ n' \end{pmatrix}, \quad \psi_5 = \begin{pmatrix} \Lambda_c \\ \Lambda' \end{pmatrix}, \quad \psi_6 = \begin{pmatrix} \Lambda_t \\ \Lambda_b' \end{pmatrix} \tag{27}$$

with

$$\begin{vmatrix} n' \\ \Lambda' \\ \Lambda_b' \end{vmatrix} = \hat{K} \begin{vmatrix} n \\ \Lambda \\ \Lambda_b \end{vmatrix} .$$

The matrices $T^\pm$ and T_3 in (26) represent the weak isospin generators and Q represents the electric charge operator in unit of proton charge; in particular

$$T_3 = \begin{cases} \dfrac{1}{2} & \text{for} \quad \nu_e, \nu_\mu \nu_\tau, p, \Lambda_c, \Lambda_t \\[2mm] -\dfrac{1}{2} & \text{for} \quad e^-, \mu^-, \tau^-, n, \Lambda, \Lambda_b. \end{cases}$$

Finally

$$L'_Y = -\phi/v \left\{ m_N[(\overline{p}p) + (\overline{n}n)] + m_\Lambda(\overline{\Lambda}\Lambda) + m_{\Lambda_c}(\overline{\Lambda}_c \Lambda_c) \right.$$
$$\left. + m_{\Lambda_b}(\overline{\Lambda}_b \Lambda_b) + m_{\Lambda_t}(\overline{\Lambda}_t \Lambda_t) + m_e(\overline{e}e) + m_\mu(\overline{\mu}\mu) + m_\tau(\overline{\tau}\tau) \right\} \tag{28}$$

Comparing formulas (21)–(28) with the corresponding formulas in SM (see e.g. [39]) we see that three baryon doublets (2) play the same role as three quark doublets (3). However contrary to the conventional SM which predicts an unobserved process with nonzero probability, the present model predicts the observed processes only.

The starting Lagrangian (8) corresponds to a renormalizable gauge field theory [39]. The final Lagrangian (21) was obtained from (8) by the mechanism of spontaneous symmetry breaking. As we shall show in Sect. 4 the lepton and the baryon anomalies in the theory defined by the final Lagrangian (21) cancels out: consequently according to [39] the Lagrangian (21) leads also to a renormalizable theory.

One obtains the full Lagrangian of BSM adding to (21) the Lagrangian L_S of the strong sector: the form of this Lagrangian was given in [4].

4 Anomaly Cancellation

It is remarkable that the model (8) is anomaly free. In fact as follows from [40] the anomaly vanishes if

$$D^{abc}_{(R)} = \mathrm{Tr}\left[T^a\{T^b, T^c\}\right] = 0 \tag{29}$$

where R is a representation of $G = \mathrm{SU}_L(2) \times \mathrm{U}(1)$.

As is well-known for any $\mathrm{SU}(N)$ representation we have

$$D^{abc}(R) = d^{abc} A(R) \tag{30}$$

where d^{abc} is the totally symmetric invariant d-tensor of $\mathrm{SU}(N)$ and $A(R)$ is a scalar depending on R. It follows from (30) that if $D^{abc} = 0$ for a particular set of indices for which $d^{abc} \neq 0$ then $A(R) = 0$ and anomaly is absent. In our case the lepton and baryon representations are given by (5) and (7), respectively. If in the fermion triangle diagrams there are three W's then $D^{abc} = 0$ since $\mathrm{SU}(2)$ is anomaly free [40]. If there are one W and two B's then since for the direct product of groups $G = G^{(1)} \times G^{(2)}$

$$\mathrm{Tr}_R(T^{(1)}T^{(2)}) = \mathrm{Tr}_{R_1} T^{(1)} \times \mathrm{Tr}_{R_2} T^{(2)} \tag{31}$$

and

$$\mathrm{Tr}_R T^a = 0 \tag{32}$$

for $\mathrm{SU}(2)$ the anomaly also vanishes. Finally if there are two W's and one B then one may replace the generator $Y/2$ associated with B by $(Q - T_3)$. Then the term with T_3 vanishes and we remain with two cases corresponding to $W^+ W^- \gamma$ and $W_0 W_0 \gamma$ vertices proportional to

$$D^{+-\gamma} = \mathrm{Tr}[(T^+T^-)Q] \tag{33}$$

$$D^{33\gamma} = \mathrm{Tr}(T_3^2 Q). \tag{34}$$

Since $T^+T^- = 2(T_1^2 + T_2^2)$ the quantity (33) reduces to (34). Consequently our model is anomaly free if

$$A_{\mathrm{SU_L(2) \times U(1)}} = \mathrm{Tr}(T_3^2 Q) = 0.$$

From (5) we get

$$A^{\mathrm{leptons}} = -\frac{3}{4}.$$

Similarly from (7) we get

$$A^{\mathrm{baryons}} = \frac{1}{4}(Q_p + Q_{\Lambda_c} + Q_{\Lambda_t}) = \frac{3}{4}.$$

Hence our model is anomaly free.

We note that an anomaly of a given lepton generation is canceled out by the opposite anomaly of the corresponding baryon generation. Thus the condition of the anomaly absence does not determine the generation number.

Let us note that in case of the conventional standard model the quark anomaly has the form [40]

$$A_q = \frac{N_c}{12}.$$

This formula implies that there must be three colors for each quark in order to have anomaly cancellation. We see that in BSM we have achieved the anomaly cancellation without any color degrees of freedom.

5 High Energy Behavior of Proton-Proton Cross Sections

The simplicity of BSM framework is best seen in case of proton-proton or proton-antiproton interactions. If proton is represented as a system of three quarks then pp scattering is representing the six body problem: if proton is represented by three quarks and gluons – as it is strongly suggested by the recent EMC experiment [8] – then the proton-proton scattering process is representing in QCD a very many body problem. It is not surprising therefore that we have not practically unambiguous significant results in QCD or SM for low or high energy cross sections for pp or $\bar{p}p$ scattering [3].

Contrary, in the BSM framework the pp or $\bar{p}p$ process represents the two body process for analysis of which we have many calculational methods like leading logarithm approximation (LLA) method [41], pomeron theory approach [42] and others.

The problem of proton-proton and proton-antiproton scattering was considered in details in the framework of strong sector of BSM in a series of papers [4,43,44]. We present here only the main aspects of this analysis and the final results.

The scattering amplitude for elastic baryon-baryon scattering can be written in LLA in the form [41,4]

$$M_{ab;a'b'}(s,t) = \Gamma_{aa'}M^{(0)}(s,t)\Gamma_{bb'} + \Gamma^c_{aa'}M^{(35)}(s,t)\Gamma^c_{bb'}$$

where $M^{(0)}(s,t)$ and $M^{(35)}(s,t)$ are scattering amplitudes with scalar (pomeron) or 35-plet SU(6) quantum numbers in the t-channel, respectively. The vertices $\Gamma_{aa'}$ and $\Gamma^c_{aa'}$ have the form

$$\Gamma_{aa'} = cg\delta_{aa'}\delta_{\lambda_a\lambda_{a'}}; \quad \Gamma^c_{aa'} = \sqrt{2}gt^c_{aa'}\delta_{\lambda_a\lambda_{a'}}$$

where λ_a is the helicity of the baryon, a, t^c are the generators of the adjoint representation of SU(6), and c is a constant.

Let $M^{(X)}_\omega(t)$, $X = 0$ or 35, $\omega = J - 1$ be the partial wave amplitude connected with the amplitude $M^{(X)}(s,t)$; it can be shown that

$$M^{(X)}_\omega(t) = M^{(X)}_\omega(k, q - k)|_{k^2 = (q-k)^2 = m^2}$$

where $M^{(X)}_\omega(k, q - k)$ is the off shell Bethe-Salpeter amplitude satisfying the following integral equation [41]

$$\{\omega - \alpha(k^2) - \alpha[(q - k)^2] + 2\}M^{(X)}_\omega(k, q - k) =$$

$$= \frac{\omega}{A_X} + \frac{3g^2}{(2\pi)^3}\int d^2k' K^{(X)}_q(k, k')M^{(X)}_\omega(k', q - k') \tag{35}$$

where $\alpha(k^2)$ is the Regge trajectory function given by the formula ($k^2 = -t$) [4]

$$\alpha(t) = 1 - \frac{3g^2}{8\pi^2 m^2}(-t + m^2)(\gamma^2 - 1)^{-1/2}log[\gamma + (\gamma^2 - 1)^{1/2}] \tag{36}$$

with

$$\gamma(t) = \frac{1}{2m^2}(-t + 2m^2).$$

The kernel $K^{(X)}_q(k, k')$ is given in the closed form as the ratio of polynomials in scalar products of k, k' and q vectors.

It is remarkable that the solution $M^{(35)}_\omega(k, q - k)$ of (35) is k-independent and can be given in the closed form ($t = -q^2$)

$$M^{(35)}_\omega(k, q - k) = \frac{\tilde{c}^{(35)}\omega}{(t - m^2)(\omega + 1 - \alpha(t))}.$$

This gives for large s

$$\text{Im}\, M^{(35)}_\omega(s,t) = \frac{c^{(35)}}{t - m^2}(\frac{s}{s_0})^{\alpha(t)} + \quad \text{lower order terms.}$$

As follows from (36) in s-channel $\alpha(t) < 0$ for $t \leq 0$; hence for large s the amplitude $M^{(35)}(s,t)$ gives a negligible contribution to $\sigma^{pp}_T(s)$.

In case of the Pomeron amplitude one shows by analytic [41] or numerical [44] methods that the solution which determines the asymptotic in s behavior of $M^{(0)}(s,t)$ has the form

$$M_\omega^{(0)}(t) = \omega C_0(t) + \omega C_1(t)(\omega - \omega_0)^{1/2} \qquad (37)$$

where

$$\omega_0 \cong \pi^{-2} g^2 6 \log 2$$

and $C_0(t)$ and $C_1(t)$ are certain functions which in the limit when vector masses go to zero can be explicitly calculated. Using (36) one obtains

$$\operatorname{Im} M^{(0)}(s,0) = c^{(0)} \left(\frac{s}{s_0}\right)^{1+\omega_0} \left[\log\left(\frac{s}{s_0}\right)\right]^{-3/2} \sum_{n=0}^{\infty} c_n \left(\log \frac{s}{s_0}\right)^{-n} \qquad (38)$$

$$\operatorname{Re} M^{(0)}(s,0) = \widetilde{c}^{(0)} \left(\frac{s}{s_0}\right)^{1+\omega_0} \sum_{n=0}^{\infty} d_n \left(\log \frac{s}{s_0}\right)^{-n}$$

where the constants c_n and d_n are known [43]. Using (38) and the optical theorem one obtains:

$$\sigma_{Tot}^{pp}(s) = c^{(0)} \left(\frac{s}{s_0}\right)^{\omega_0} \left(\log \frac{s}{s_0}\right)^{-3/2} \sum_{n=0}^{\infty} c_n \left(\log\left(\frac{s}{s_0}\right)\right)^{-n}. \qquad (39)$$

The proton-antiproton elastic and total cross sections can be analyzed in a similar manner [4]. In fact in this case the scattering amplitude will have the form

$$\widetilde{M}_{ab;a'b'}(s,t) = \Gamma_{aa'} \widetilde{M}^{(0)}(s,t) \Gamma_{bb'} + \widetilde{\Gamma}_{aa'}^c \widetilde{M}^{(35)}(s,t) \widetilde{\Gamma}_{bb'}^c$$

with

$$\widetilde{\Gamma}_{bb'}^c = -g\sqrt{2} t_{b'b}^c \delta_{\lambda_b \lambda_{b'}}.$$

One shows that the partial wave amplitudes $\widetilde{M}_\omega^{(X)}(t)$ associated with $\widetilde{M}^{(0)}(s,t)$ and $\widetilde{M}^{(35)}(s,t)$ amplitudes respectively in LLA satisfy also (35). Thus as in the previous case the pomeron exchange amplitude $\widetilde{M}^{(0)}(s,t)$ will dominate the asymptotic behavior for $s \to \infty$: hence by (38) the total proton-antiproton cross section $\sigma_T^{\bar{p}p}(s)$ will be given also by the formula (39). Consequently in the present formalism for $s \to \infty$ we get

$$\sigma_T^{pp}(s) = \sigma_T^{\bar{p}p}(s).$$

We see therefore that in BSM in the considered approximation we have reproduced the celebrated Pomeranchuk theorem [45].

We present in Fig. 1 the energy dependence of $\sigma_T^{\bar{p}p}(s)$ for proton-antiproton scattering [4].

We see that the predicted total cross section at energies below 18GeV originally decreases, then reaches the minimum at energy around 18GeV and next increases up to around 130mb for Fly's Eye experiment at cosmic energy around $\sqrt{s} = 30\,000$GeV [46]. The obtained characteristic curve for $\sigma_T^{\bar{p}p}(s)$ which agrees with

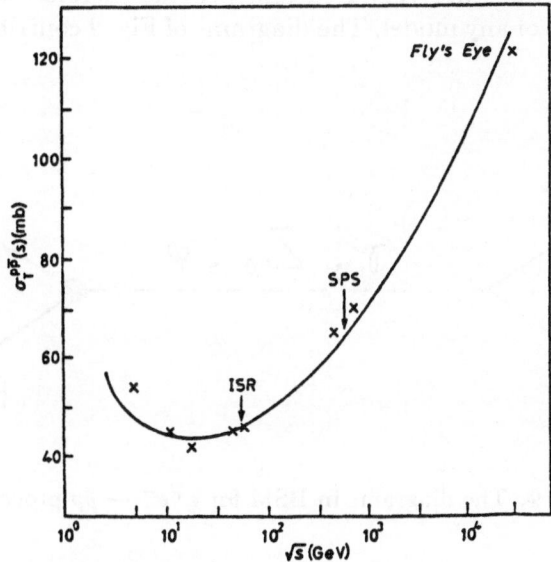

Fig. 1. The predicted total cross section $\sigma_T^{\bar{p}p}(s)$ for $\bar{p}p$ scattering: the crosses represent experimental points.

experimental results in the energy interval from 10 GeV to 30 000 GeV seems to us really remarkable: we do not know any other model which would give first the decrease of the $\sigma_T^{\bar{p}p}(s)$ and next increase if energy increases in agreement with experimental data.

We plan presently to carry out the analysis of scattering of polarized protons on polarized protons in the framework of BSM in order to verify if BSM gives a desired description of striking experimental effects observed in scattering of polarized protons [2]. In view of planned NEPTUN-A experiments in Serpukhov on scattering of protons at energy 400 GeV and 3 TeV on polarized target protons the existence of theoretical predictions for high $p_\perp^2$ polarized $p-p$ cross sections would be highly desirable [2,3].

6 Annihilation of Polarized Electron and Positron into a Proton-Antiproton Pair

We calculate now in the present model the differential cross section for annihilation of arbitrarily polarized electrons and positrons into proton-antiproton pairs. The calculation of this quantity in the conventional SM is practically impossible, since a reliable representation of the physical proton or antiproton in terms of quarks and gluons is unknown [3,8,25].

We consider the scattering of polarized leptons since polarization effects in differential cross sections are very rich and provide therefore a very acute cross check for predictions of any model. The diagrams of Fig. 2 contribute to the process $e^+e^- \to \bar{p}p$.

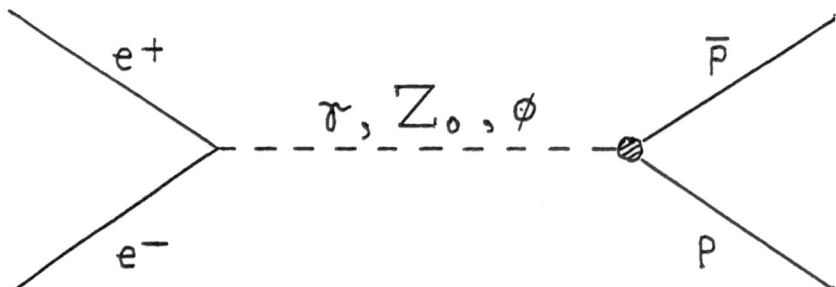

Fig. 2. The diagrams in BSM for $e^+e^- \to \bar{p}p$ process

The dashed circle at boson $\bar{p}p$ vertex represents the electromagnetic and weak nuclear form-factors $\Gamma_\mu(Q^2)$.

It was shown in [29] that in $e^+e^- \to \bar{f}f$ processes the Higgs exchange diagram due to the smallest of the electron mass gives a negligible contribution with respect to photon and Z exchange diagrams. Hence we shall disregard in our analysis the Higgs exchange diagram.

Let $P_L^\pm$ denotes a longitudinal and let $P_T^\pm$ denotes a transverse polarization of $e^\pm$. Let θ be the polar and ϕ the azimuthal angle between e^- and proton. Let m be the proton mass, $\mu_p = m^2/s$ and $s = (p_- + p_+)^2$. In order to show clearly the richness and complexity of polarization effects we first calculate the cross section using for diagrams in Fig. 2 the Born approximation.
Using the results of [29] for $e^+e^- \to \bar{f}f$ process we obtain

$$\frac{d\sigma}{d\Omega} = \frac{\alpha^2}{4s}\sqrt{1-4\mu_p}[(1 - P_L^+ P_L^-)X_U + (P_L^+ - P_L^-)X_L + P_T^+ P_T^- X_T] \qquad (40)$$

where

$$\begin{aligned}
X_U &= G_1(1 + \cos^2\theta) + 4\mu_p G_2 \sin^2\theta + \sqrt{1-4\mu_p}\,G_3 2\cos\theta \\
X_L &= H_1(1 + \cos^2\theta) + 4\mu_p H_2 \sin^2\theta + \sqrt{1-4\mu_p}\,H_3 2\cos\theta \qquad (41) \\
X_T &= (1 - 4\mu_p)(F_1 \cos 2\varphi + F_2 \sin 2\varphi)\sin^2\theta.
\end{aligned}$$

In the lowest order approximation the Z_0 meson propagator has the form:

$$\Pi(s) = \frac{s}{s - M_Z^2 + iM_Z\Gamma_Z}. \qquad (42)$$

The function G_i, H_j and F_k in (41) in the lowest order have the form: ($\Pi \equiv X + iY$)

$$G_1 = 1 - 2v_e v_p X + (v_e^2 + a_e^2)(v_p^2 + a_p^2 - 4\mu_p a_p)|\Pi|^2$$

$$G_2 = 1 - 2v_e v_p X + (v_e^2 + a_e^2)v_p^2|\Pi|^2$$

$$G_3 = -2a_e a_p X + 4v_e a_e v_p a_p|\Pi|^2$$

$$H_1 = -2a_e v_p X + 2v_e a_e(v_p^2 + a_p^2 - 4\mu_p a_p^2)|\Pi|^2$$

$$H_2 = -2a_e v_p X + 2v_e a_e v_p^2|\Pi|^2 \qquad (43)$$

$$H_3 = -2v_e a_p X + 2(v_e^2 + a_e^2)v_p a_p|\Pi|^2$$

$$F_1 = 1 - 2v_e v_p X + (v_e^2 - a_e^2)(v_p^2 + a_p^2)|\Pi|^2$$

$$F_2 = 2v_e a_e(v_p^2 + a_p^2)Y$$

where

$$v_e = \frac{-1 + 4\sin 2\theta_W}{2\sin 2\theta_W}, \qquad a_e = -\frac{1}{2\sin 2\theta W} \qquad (44)$$

$$v_p = -v_e; \quad a_p = -a_e.$$

The formulas (40)–(43) show that the dependence of differential cross sections on e^+ and e^- polarizations is highly nontrivial; since in a storage ring the transverse polarization is a natural one we shall carry out the analysis of $d\sigma/d\Omega$ for this case.

It follows from (41) that $d\sigma/d\Omega$ for a fixed s will have a rather characteristic structure in θ and φ variables. In Fig. 3 we first present the dependence of $d\sigma/d\Omega(s_0, \theta_0, \varphi)$ function on azimuthal angle φ for $\sqrt{s_0} = 90$GeV which correspond to the LEP energy obtained in CERN experiment and for $\theta_0 = \pi/10$.

The numerical analysis shows that for $\sqrt{s_0} = 90$ the function $F_1(s)$ in (41) is very small in comparison with $F_2(s)$: hence the φ-dependence has $\sin 2\varphi$-type shape. If energy is decreased say to the TRISTAN energy $\sqrt{s_0} = 55$GeV then $F_2(s)$ is small with respect to $F_1(s)$: consequently φ-dependence has $\cos 2\varphi$-type shape. In the region around $\sqrt{s_0} = 78.17$GeV the functions $F_1(s)$ and $F_2(s)$ are comparable: this gives a characteristic phase shift presented in Fig. 4.

It is remarkable that the passage from $\sin 2\varphi$-type shape to $\cos 2\varphi$-type shape is very sudden within 0.1GeV interval around 78.17GeV.

One might expect that the s-dependence of $d\sigma/d\Omega$ will be most interesting around the pole $s = M_{Z_0}^2$ of Z_0-propagator. Hence we plot in Fig. 5 Tthe $d\sigma/d\Omega(s, \theta_0, \varphi_0)$ as a function of the variable $\sqrt{s}$ for fixed θ_0 and φ_0.

We see the expected peak in the cross section at the Z_0-pole mass energy.

As follows from formula (41) the dependence on the polar angle depends on relative values of G_1, G_2 and G_3 functions and can be quite interesting. We give in Fig. 6 the dependence of the cross section on the polar angle for LEP energy $\sqrt{s_0} = 100$GeV.

We see that relative values of cross sections for different polar angles can be very large.

The above analysis was carried out in the lowest order approximation; however, there is no problem to perform the same analysis including all one-loop radiative corrections. In fact, the complete one-loop radiative corrections for the process $e^+e^- \to \bar{f}f$ were presented in [29] Sect. 6. Using these results one can calculate the differential cross section for arbitrary polarized initial particles for $e^+e^- \to \bar{p}p$

processes with the inclusion of all one-loop radiative corrections. Since however the final formula for the cross section with radiative corrections is rather complex we shall present the corresponding analysis in a separate publication. We carry out also the analysis of $d\sigma/d\Omega$ cross sections using the approximate analytic representations for form-factors.

The analysis carried out in [29] and [1] indicates that radiative corrections for many measurable quantities do not change significantly the lowest order predictions. Hence one may expect that the differential cross section (40) and Figs. 3–6 exhibit reasonably well the main features of annihilation of polarized positrons and electrons into $\bar{p}p$ pairs.

The predictions for $e^+e^- \to \bar{p}p$ processes presented here were obtained in the framework of BSM without any use of quark or gluon fields. The confirmation of these predictions by experimental data would strongly support the BSM framework. We stress that - as discussed in the introduction - there is no possibility to derive the corresponding formula in the framework of conventional SM because the hadronization amplitude $M_{qg,\bar{p}p}$ for transition from a quark-gluon system to a proton-antiproton pair cannot be calculated in a reliable manner.

7 Applications of Baryons Standard Model

We give now a brief description of electroweak, lepton-hadron and hadron-hadron processes which we actually analyze or plan to analyze soon in the framework of BSM

7.1 Electroweak Processes

The electroweak sector of BSM is the same as in SM and the kinematical properties of fundamental fermions in BSM are similar to those in QCD; consequently most of the predictions for pure electroweak processes in the BSM will have a similar analytical form as in SM and numerically will be as good as in SM. In fact the predictive power of BSM in electroweak sector will be even bigger than that of SM: for instance all SM predictions in the electroweak sector in the one-loop approximation depend on the value of quark masses; since quarks are unobserved the concept of quark masses is unclear and the concrete values of quark masses are to large extent arbitrary; as Prof. Lipkin often stresses any quark mass from zero to infinity is equally good; consequently radiative corrections in SM calculations contain essential uncertainties. Contrary, in BSM the masses of baryons which will appear in one-loop formulas for observed quantities are the definite masses of physical baryons like protons, neutrons, Λ-particles etc. This will give a definite value of radiative corrections coming from the strong sector and will allow to check better the BSM predictions in the electroweak sector.

We plan to investigate several typical electroweak processes like Bhabba scattering, e^+e^- annihilation into W^+W^- pair or the decay of the Higgs particle into the fermion-antifermion pair. These processes were carefully examined in SM

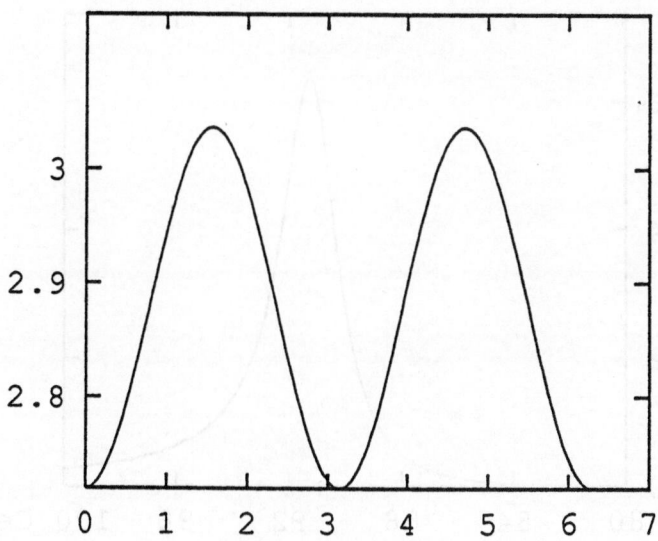

Fig. 3. $d\sigma/d\Omega(s_0, \theta_0, \varphi)$ cross section $\times 10^7$ for $\sqrt{s_0} = 90\,\text{GeV}$, $\theta_0 = \pi/10$, $P_L^+ = P_L^- = 0$, and $P_T^+ = P_T^- = 1$.

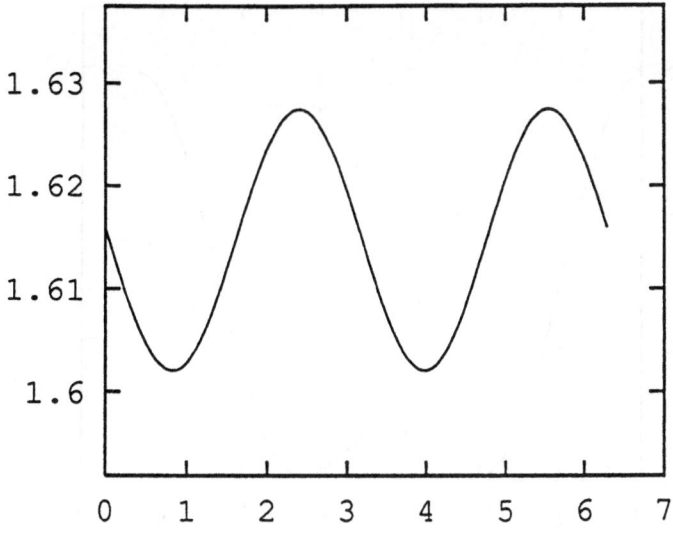

Fig. 4. $d\sigma/d\Omega(s_0, \theta_0, \varphi)$ cross section $\times 10^{10}$ for $\sqrt{s_0} = 78.17\,\text{GeV}$, $\theta_0 = \pi/10$, $P_L^+ = P_L^- = 0$ and $P_T^+ = P_T^- = 1$.

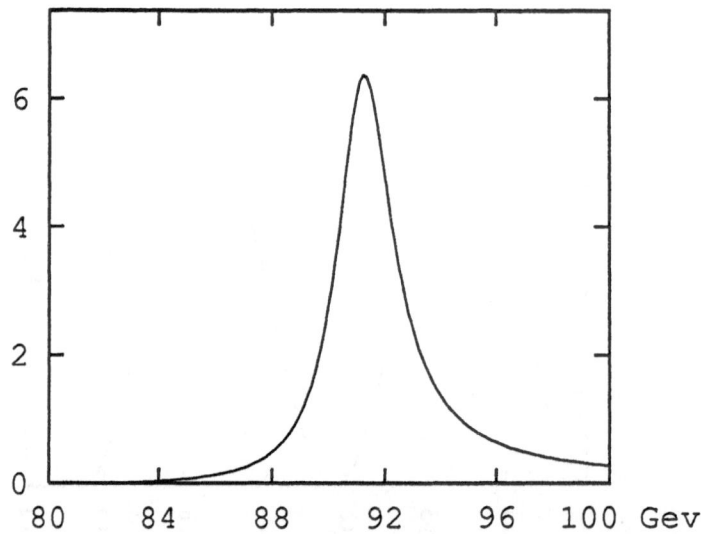

Fig. 5. $d\sigma/d\Omega(s,\theta_0,\varphi_0)$ cross section $\times 10^7$ as a function of energy for $\theta_0 = \pi/10$, $\varphi_0 = \pi/10$, $P_L^+ = P_L^- = 0$, and $P_T^+ = P_T^- = 1$.

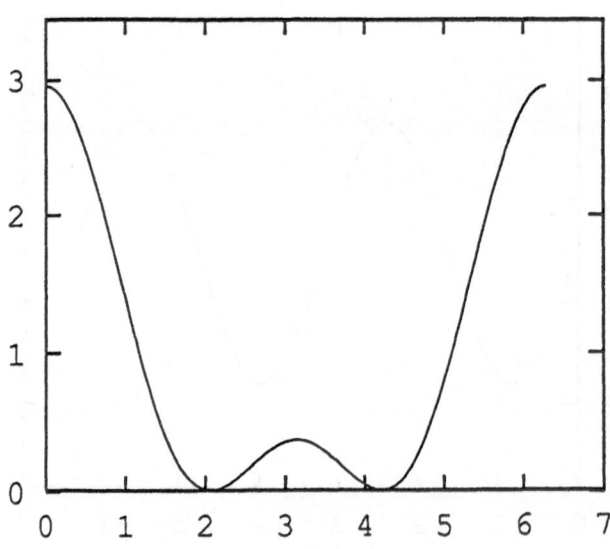

Fig. 6. $d\sigma/d\Omega(s_0,\theta,\varphi_0)$ cross section $\times 10^8$ as function of the polar angle θ for $\sqrt{s_0} = 100$GeV, $\varphi_0 = \pi/60$, $P_L^+ = P_L^- = 0$, and $P_T^+ = P_T^- = 1$.

including radiative corrections. As it was demonstrated the radiative corrections in some energy regions for these processes may constitute up to 50% of the lowest order result [47]. We shall calculate these processes with radiative corrections in BSM and compare predictions of BSM with experimental data and with SM predictions.

7.2 Asymptotic Freedom

It should be stressed that most of successful predictions of QCD and SM in hard collisions are derived from the asymptotic freedom property of QCD sector. It is noteworthy that in the strong sector determined by the SU(6) spontaneously broken gauge field theory model the asymptotic freedom can also be assured. In fact, as it we showed in [4], the simplest Higgs field representation which breaks SU(6) symmetry completely can be taken in the form

$$\Phi = \bigoplus_{k=1}^{6} H^k \,, \tag{45}$$

where H^k is the SU(6) vector representation. The most general Higgs potential for the representation (45) has the form:

$$V(\Phi) = -\left(\sum_{k=1}^{5} \mu_k^2 H^{k*} H^k\right) + \frac{\lambda}{2}\left(H^{k*} H^k\right)^2 + \frac{\rho}{2}\left(H^{k*} H^l\right)\left(H^{l*} H^k\right). \tag{46}$$

The one-loop renormalization group equations which determine the asymptotic freedom of the strong sector have the form (see e.g. [4, Sect. 5])

$$16\pi^2 \frac{dg^2}{dt} = -20.33g^4 \,,$$

$$16\pi^2 \frac{d\lambda}{dt} = 80\lambda^2 + 48\lambda\rho + 6\rho^2 - 35\lambda g^2 + \frac{19}{12}g^4 \,, \tag{47}$$

$$16\pi^2 \frac{d\rho}{dt} = 12\lambda\rho + 24\rho^2 - 35\lambda g^2 + 8g^4 \,.$$

It is evident that the gauge coupling constant is asymptotically free. To check whether the Higgs coupling constant can also be asymptotically free we set

$$\lambda = \Lambda g^2 \,, \quad \rho = R g^2 \,. \tag{48}$$

Using (47) we get for constant Λ and R the biquadratic system of equations

$$80\Lambda^2 + 48\Lambda R + 6R^2 - 14.67\Lambda + 6 + \frac{19}{12} = 0 \,, \tag{49}$$

$$12\Lambda R + 24R^2 - 35\Lambda + 20.33R + 8 = 0 \,. \tag{50}$$

The computer analysis shows that these equations have the two solutions

$$\Lambda^{(1)} = 1.12 \,, \quad R^{(1)} = -2.04 \tag{51a}$$

and

$$\Lambda^{(2)} = 0.17 , \quad R^{(2)} = -0.11 . \tag{51b}$$

Since gauge coupling is asymptotically free, both Higgs couplings by (48) are also asymptotically free.

It should be stressed that there are also other intervals of initial values of $\lambda(t_0)$ and $\rho(t_0)$ for which all couplings $\lambda(t), \rho(t)$ and $g(t)$ are asymptotically free; for instance if one takes $\rho(t_0)$ very small compared to $\lambda(t_0)$ and $g(t_0)$ then the terms ρ^2 and $\lambda\rho$ in (47) can be disregarded. It follows then from (47) that for

$$0.23g^2 < \lambda(t_0) \leq 0.39g^2$$

all one-loop β-functions for evolution of g, λ and ρ are negative. Hence we obtain another set of asymptotically free coupling constants in which both ρ and λ may be positive. It is interesting that small coupling constants ρ can generate large Higgs field expectation values. In fact, we have shown in [4, Sect. 3] that for the Higgs potential (46) the vacuum expectation values $\langle 0|H^k|0 \rangle = h_k \delta_{ki}$ of the Higgs fields are given by the formula

$$h_k = \left[\frac{1}{\rho} \left(\mu_k^2 - \frac{\lambda}{\rho + 6\lambda} \sum_{l=1}^{6} \mu_l^2 \right) \right]^{\frac{1}{2}} .$$

Thus small a coupling constant ρ can generate large vector meson masses as required by the considered model.

It follows from the above analysis that in the high energy limit the strong sector in BSM will behave as in the asymptotically free gauge field theory model with the SU(6) gauge group. Consequently all successful high energy limits predictions based on asymptotic freedom will persist also in BSM.

It is interesting however that the decrease of the running coupling constant (rcc) $g_{\mathrm{SU}(6)}^2$ with the increase of the scale μ will be much more quicker than in QCD.

In fact $\beta_0^{\mathrm{SU}(6)} = 20.33$ whereas for six flavors $\beta_0^{\mathrm{QCD}} = 7$. Hence in one-loop approximation the generator (i.e. the r.h.s.) of the renormalization group equation is in BSM approximately 3-times more negative than in case of QCD. Consequently by (47) the SU(6) rcc in BSM will decrease to zero much more quickly than QCD rcc. Since the decrease of the rcc determines the rate of decrease of physical quantities with energy $\sqrt{s}$ or momentum square transfer Q^2 the much faster decrease of rcc in BSM will imply the definite faster decrease of physical quantities. We propose in [48] some concrete experiments which could verify what is the decrease rate of rcc and considered physical quantities in BSM framework.

7.3 Lepton-Hadron Processes

The BSM allows for an effective analysis of lepton-hadron inclusive and exclusive processes like e.g. e^+e^- annihilation into $\bar{p}p$, $\overline{\Lambda}\Lambda$ or $\rho^+\rho^-$ pairs or ep elastic and inclusive scattering. As we discussed these processes could not be analyzed in SM in a satisfactory manner since there is no reliable description of hadrons in terms of quarks and gluons [8,36] or a reliable formalism for hadronization processes.

We presented in Sect. 6 for an illustration the analysis of the annihilation of polarized electrons and positrons into proton-antiproton pairs in the lowest order approximation. We are presently completing the analysis of this process including all one-loop radiative corrections for form factors and propagators as well as taking into account the analytic representation for form factors. The obtained formulas for cross sections show the rich dependence on the polarization of the incoming leptons and will provide then the acute test for BSM.

We also plan to analyze the annihilation of polarized electrons and positrons into the $\overline{\Lambda}\Lambda$ and $\rho^+\rho^-$ pairs.

We have also started the analysis of elastic scattering of polarized electrons on polarized protons in high energy limit, since we are convinced that richness of polarization effects in electron-proton scattering will provide the most severe test of BSM and other models. Since one expects that in the HERMES and the HERA experiment at DESY electrons might reach 92% polarization the predictions of our model may be soon tested.

8 Discussion

The BSM model represents a natural framework for a description of electroweak, lepton-hadron and hadron-hadron processes. The physical foundation of BSM are more natural than in SM. In fact in every model the strong sector must be included into the framework of pure electroweak theory in order to cancel out the lepton anomalies; the strong sector in QCD form cancels out these anomalies but introduces unobserved quarks and gluons; this leads to the serious conceptual and numerical difficulties of SM connected with the confinement of quarks and gluons, predictions of nonzero cross sections for unobserved quantities etc.. In case of BSM the baryon doublets from the strong sector also cancels out the electroweak anomalies; however all particles in BSM are observed. So the previous difficulties do not appear at all. I am personally convinced that whenever we formulate a physical model in terms of unobserved particles we shall run into difficulties since the concept of existing but unobserved particles and fields is unphysical.

It is remarkable that in the low energy region the BSM reduces to the VDM, which is very successful in the description of many low energy electroweak and strong processes [49]; on the other hand in the high energy limit the strong sector of BSM reduces to the asymptotically free SU(6) gauge field theory. Thus contrary to QCD the BSM provides the effective framework for analysis of low and high energy processes in electroweak and strong interactions.

The BSM allows to carry out the effective analysis of electroweak, lepton-hadron and hadron-hadron processes inclusive as well as exclusive. Thus all present experimental data from PETRA, TRISTAN, LEP, HERMES, HERA or Fermilab can be analyzed in BSM framework. As we discussed in Sect. 2 the lepton-hadron and hadron-hadron exclusive processes cannot be analyzed in a reliable manner in SM because this model does not provide a satisfactory dynamical description of hadrons in terms of quarks and gluons or a reliable description of the hadronization processes [3,8].

Acknowledgments

This work is dedicated to Prof. Dr. H.D. Doebner on the occasion of his sixtieth birthday. The author gratefully acknowledges the warm hospitality extended to him by Prof. Abdus Salam and Prof. Paolo Budinich at ICTP and SISSA. He is also grateful to Prof. A.Szymacha and Mr. M. Pawlowski for helpful discussions and to Dr. J. Kraskiewicz for computer drawings of curves in Sect. 6. The author thanks also to Miss. Claudia Parma for excellent typing.

References

1.   See e.g. i) C. Kiesling: "Tests of the Standard Theory of Electroweak Interactions", Springer Tracts in Modern Physics, vol. 112, (Springer-Verlag, 1988)
      ii) A. Ali: "Theory of Electrowek Interactions in e^+e^- Annihilation", lectures at Summer School on High Energy Physics and Cosmology, ICTP, Trieste, July 1990
2.   A.D. Krisch: preprint, University of Michigan UMHE 89–19, 1989
3.   S.M. Troshin, N.E. Tyurin: Int. J. Mod. Phys. 5 2689 (1990)
4.   R. Rączka: Nuovo Cim. A98 297 (1987)
5.   See e.g. the most recent excellent reviews:
      i) J.C. Collins, D.E. Soper, G. Sterman: "Factorization of Hard Processes in QCD", in *Perturbative Quantum Chromodynamics*, ed. by A.H. Mueller (World Scientific, Singapore 1989) p. 1
      ii) S.J. Brodsky, G.P. Lepage: "Exclusive Processes in Quantum Chromodynamics", ibid p. 93
      iii) S.J. Brodsky: preprint SLAC–PUB–5082, 1989
6.   See e.g. D.W. Duke, R.G. Roberts: Phys. Rep. 120 275 (1985)
7.   See e.g. A.L.E. Yaouans et al.: *Hadron Transitions in the Quark Model* (Gordon and Breach, NY, 1988)
8.   The EMC Collaboration: Nucl. Phys. B238 1 (1989)
9.   B. Anderson, G. Gustafson, G. Ingleman, T.Sjostrand: Phys. Rep. 97 31 (1983)
10.   A. Ali, E. Pietarinen, G. Kramer, J. Willrodt: Phys. Lett. B93 155 (1980); P. Hoyer et al.: Nucl. Phys. B161 349 (1979)
11.   A. Gaidot et al.: Phys. Lett. B57 389 (1975); B61 103 (1976)
12.   Proceeding of International Conference "Spin and Polarization Dynamics in Nuclear and Particle Physics", ICTP, Trieste, ed. by A.O. Barut et al. (World Scientific, Singapore, 1990)

13. K. Heller: Proceedings of 6th International Symposium on High Energy Spin Physics, Marseille, Sept. 1984, ed. by J. Soffer; Journal de Physique **46** C2-121 (1985)

14. F. Abe et al.: Phys. Rev. **D34** 1950 (1986)

15. P.E. Schlein: J. de Physique **46** C2-131 (1985)

16. D.G. Crabb et al.: Phys. Rev. Lett. **41** 1257 (1978);
    J.R. O'Fallon et al.: ibid. **39** 733 (1977)

17. G.L. Kane, J. Pumplin, W. Repko: Phys. Rev. Lett. **41** 1989 (1978)

18. P.H. Hansen et al.: Phys. Rev. Lett. **50** 802 (1983);
    D.C. Peaslee et al.: Phys. Rev. Lett. **51** 2359 (1983);
    P.R.Cameron et al.: Phys. Rev. **D32** 3070 (1985)

19. V.L. Solovianov: i) Proc. of the 7th International Symposium on High Energy Spin Physics, Protvino 1986, vol. I, p. 26;
    ii) Proc. of the Workshop on Experimental Program at UNK, Protvino 1987, p. 191

20. V. Yuan et al.: Phys. Rev. Lett. **57** 1680 (1986);
    S.Kistryn et al.: Phys. Rev. Lett. **58** 1616 (1987)

21. N. Lockyer et al.: Phys. Rev. **D30** 860 (1984)

22. SLAC-Yale Collaboration: Phys. Rev. Letts. **37** 1261 (1976); **41**, 70, (1978); **51**, 1135 (1983);
    see also the recent review by T. Sloan, G. Smadja, R. Voss: Phys. Rep. **162** 45 (1988)

23. G. Altarelli: Ann. Rev. of Nucl. and Particle Science **39** 357 (1989)

24. J. Ashman et al.: Phys. Lett. **B206** 364 (1988);
    see also Ref. [8]

25. J.E. Mandula: preprint Department of Energy, Division of High Energy Physics, Washington, 1990

26. A.V. Efremov, O.V. Teryaev: JINR report E2-88-278 (1988)

27. G. Altarelli, G.G. Ross: Phys. Lett. **B212** 391 (1988)

28. R.D. Carlitz, J.C. Collins, A.H. Mueller: Phys. Lett. **B214** 229 (1988)

29. W. Hollik: Fortschr. Physik **38** 165 (1990)

30. R. Rączka: "A proposal for an ultimate cross check of QCD and Standard Model", preprint SISSA Trieste, EP/1990 in preparation

31. G. t'Hooft: in *The whys of subnuclear physics, Eriche 1977*, ed. by A. Zichichi (Plenum Press, NY, 1979);
    see also N.N.Khuri, H.C.Ren: Ann. Phys. **189** 142 (1989)

32. See e.g. i) Nucl. Phys. B (Proc. Suppl.) **9** (1989);
    "Lattice 88", Proc. of the 1988 Symposium on Lattice Field Theory, FNAL Batavia 1988, ed. by A.S. Kronfeld and P.B.Mackenzie
    ii) C. Rebbi: "An Overview of Recent Developments in Lattice Gauge Theory", lectures on Summer School in High Energy Physics and Cosmology, ICTP, Trieste 1990.

33. S. Nadkarni: Phys. Rev. **D27** 917 (1983)

34. S. Nadkarni: Phys. Rev. **D33** 3738 (1986); ibid **D34** 3904 (1988)

35. J.E. Mandula, M. Ogilvie: Phys. Lett. **B185** 127 (1987)

36. J.E. Mandula, M. Ogilvie: preprint Brookhaven National Laboratory, BNL–40440, 1987

37. D. Weingarten: "Lattice Quantum Chromodynamics", Research report RC 14378, 1/30/89, IBM Research Division N.Y.

38.   N. Cabibbo: Phys. Rev. Lett. **10** 531 (1963)

39.   See e.g. M. Chaichian, N.F.Nelipa: *Introduction to Gauge Field Theories* (Springer-Verlag, Berlin, 1984)

40.   See e.g. A. Billoire, A. Morel: "Introduction to unified theories of weak, electromagnetic and strong interactions", preprint Saclay D Ph-T80-104(1980)

41.   V.S. Fadin, E.A. Kuraev, L.N.Lipatov: JETP **71** 840 (1976); ibid **79** 377 (1977)

42.   J. Bartels: Nucl. Phys. **B151** 293 (1979); ibid. **B175** 365 (1980)

43.   P.A. Rączka, R. Rączka: Nuovo Cim. **A97** 494 (1987)

44.   P.A. Rączka, M. Pindor: Nuovo Cim. **A103** 349 (1990)

45.   I.Ya. Pomeranchuk: JETP **34** 725 (1958)

46.   See e.g. J.G. Rushbrooke: preprint CERN EP/85-178 (1985)

47.   J. Fleischer, F. Jegerlehner, M. Zralek: "C-Particles and Fields", Z. Phys. **42** 409 (1989)

48.   R. Rączka: "Analysis of Asymptotic Freedom in Baryon Standard Model", in preparation

49.   See e.g. M. Gari, W. Krumpelmann: Z. Phys. **A322** 689 (1985); and references contained therein

Is the Physical Vacuum Really Lorentz-Invariant?

I.E. Segal

Massachusetts Institute of Technology

Abstract: It is argued that the *physical* vacuum in a nonlinear quantized field will in general not be *invariant* under the full symmetry group of the underlying equations or Lagrangian, but only *covariant.* Fixed point considerations generically break the symmetry down to an amenable subgroup.

1 Introduction

In nonlinear relativistic quantized field theory, it is often, indeed generally, supposed that the vacuum is surely invariant. In 'practical' quantum field theory, this is taken as physically evident, from the observer-independence of the results of the theory that is at the basis of relativity theory. In addition, it is 'demonstrated' mathematically, as a consequence of the commutativity of fields (or currents in the case of fermions) entering into the interaction hamiltonian density at relatively space-like points. This 'of course' implies the commutativity of the interaction hamiltonians themselves at relatively space-like separations. This in turn legitimizes the rearrangements of the terms defining the S-matrix, or the physical vacuum that it leaves invariant, in one Lorentz frame so as to attain the corresponding terms in any other Lorentz frame.

In axiomatic relativistic (or conformal) quantum field theory, corresponding properties are postulated. In its most straightforward form, there is supposed to be a vector v in the underlying Hilbert space, say $\mathcal{K}$, that is invariant under the unitary action Γ of the Poincaré (or conformal) group G on $\mathcal{K}$, and is the lowest eigenstate of the hamiltonian. In more sophisticated formulations, involving e.g. the avoidance of Hilbert space as an *a priori* aspect of the quantized field, using instead expectation value forms or states of C^*-algebras, it is still required that if E denotes the vacuum state, as an expectation value form, a corresponding property to the Lorentz-invariance of the vector v is valid regarding E. Thus, in the algebraic formulation, G typically acts as a group of automorphisms of the field observable algebra $\mathcal{A}$, say as $A \to A^g$ for $g \in G$ and the vacuum state E is supposed to satisfy the relation $E[A^g] = E[A]$.

For free fields, the mathematical objects v, Γ, E, etc. are explicitly describable, and Lorentz (or conformal) invariance is evident. This is consistent with the postulated more general invariance of v, etc., but lends only weak support to the idea, since the real issue is the case of interacting fields, which are not well approximated by or represented in terms of free fields. Stronger confirmation would come from the rigorous verification of this invariance in a nonlinear relativistic quantum field, but this is so far lacking. Heuristic or partial treatments of interacting fields have appeared consistent with Lorentz-invariance of the vacuum vector, but a rigorous and basically complete example of a nonlinear relativistic field remains to be established.

I argue here that this may represent an impossible goal, and is in any event an inappropriate one, from general mathematical and physical standpoints. Mathematically, there is no reason to expect the existence of a physical vacuum that is invariant under anything other than an amenable (or abelian, compact, solvable, etc.) subgroup of the fundamental symmetry group of the theory (i.e. the invariance group of the underlying wave equations, or of the Lagrangian). Careful analysis of the formal argument from which Lorentz-invariance is derived in practical quantum field theory shows that it is fallacious, by virtue of its cavalier treatment of the notion of nonlinear function of an interacting field. Physically, the notion that the vacuum vector v itself should be Lorentz-invariant is far more stringent than is required for the fundamental desideratum that the laws of physics, – or more concretely, the S-matrix elements between states labeled by corresponding quantum numbers, – should be independent of the observer, if admissible relative to the underlying symmetry group of the theory. These ideas are not entirely new [1], but they do not seem to have been considered, let alone absorbed, to judge by the general run of the literature, and it is important to get the basic concepts of quantum field straight if the subject is to attain rigorous foundation and more effective empirical use.

2 The Fallacy in the Formal Argument

This argument is so persuasive at first glance that I will start with it, although from the perspective of the vacuum as a fixed point of a group action, the general theory of such fixed points seems a more basic issue.

To fix the ideas and avoid irrelevant abstraction, consider the simplest type of nonlinear wave equation:

$$\Box\phi + m^2\phi + P'(\phi) = 0$$

in Minkowski space, where P is a given polynomial (e.g., $P(\phi) = g\phi^4$, to take a particularly familiar case). The quantized field Lagrangian involves various quadratic terms in ϕ, but the most singular term is the interaction Lagrangian, which it is no essential loss of generality to assume to be of higher order, and which is crucial for the dynamics of the interacting field. Thus the S-matrix may be described, in physical parlance, as the integral of the negative of its time-ordered exponential;

or, in the interaction representation, in which practical quantum field theory is most commonly developed, the temporal evolution of the interacting field from one space-like surface to a later one may be similarly described.

At first glance it seems obvious that if x and y are relatively space-like points of Minkowski space $\mathcal{M}_0$, so that $\phi(x)$ and $\phi(y)$ commute (according to rather universally accepted formulations), then so also do $P(\phi(x))$ and $P(\phi(y))$, whatever they may mean. In the case of the Klein-Gordon free field, with Wick's corresponding interpretation of the product of local field operators, they do in fact commute, – quite rigorously, with natural specification of their spatial averages and their formulation as self-adjoint operators in the quantized field Hilbert space $\mathcal{K}$. So why shouldn't it be assumed that they commute also in the case of interacting fields?

Well, in the interacting case there are no creation and annihilation operators (a priori, at least), in terms of which Wick products are defined. This doesn't mean that definitions of $P(\phi(x))$ in the interacting case are largely, or even primarily, discretionary. What validates Wick's definition is not merely that it provides a convenient standardization that eliminates the immediate divergences, but that it is *local*, dynamically *invariant*, and kills off expectation values that must vanish, if correspondence principle or scattering theory ideas are to apply. Indeed, these properties uniquely characterize the Wick products of free fields, and since they don't involve free field structure, they can be applied to give a physically natural definition for the local product of interacting quantum fields (e.g., [2]). The only difference is that the *physical vacuum* expectation values must vanish.

More specifically, what this boils down to is that however a physically renormalized version of the polynomial $P(\phi(x))$ in the interacting field $\phi(x)$ is reasonable defined, it needs at least to have the property that if $::P(\phi(x))::$ denotes this renormalized version of the otherwise divergent expression $P(\phi(x))$, it must satisfy the constrains that $E[::P(\phi(x))::] = 0$, where E denotes physical vacuum expectation values. This constraint is dynamically invariant for the nonlinear equation, as the condition $E_0[\cdot P(\phi(x))\cdot] = 0$, where E_0 denotes free vacuum expectation values, is not. (Other questions may be raised about the meaningfulness, or rather lack of it, for free vacuum conditions on an interacting field at a finite time, but it suffices here to consider just the vanishing expectation value condition.)

What this means is that $::P(\phi(x))::$ is not well-defined until the physical vacuum is known; but of course the physical vacuum depends on the contingent definition given for $::P(\phi(x))::$. This is a bit awkward but means only that one has a genuinely nonlinear situation, unlike the essentially linear theory, from an abstract differential equation standpoint, in which one plugs a free field into the otherwise bothersome interaction hamiltonian. The interdependence of the solution to the quantized wave equation and the (physical) vacuum of the corresponding quantized field is not different from what it would be if one were trying to solve a stochastic nonlinear wave equation without a given probability distribution on the Cauchy data space, but rather with a constraint of the type that the solution be temporally invariant (i.e. the distribution in the Cauchy data space is such).

This interdependence between the precise mathematical meaning of $::P(\phi(x))::$ and the physical vacuum means that the usual formal argument for the Lorentz invariance of the putative physical vacuum vector v of the interacting field is circular. The intuitive formula expressing Lorentz invariance of renormalized powers of an *interacting* quantum field:

$$::P(\phi(L^{-1}x)):: = \Gamma(L)::P(\phi(x))::\Gamma(L)^{-1},$$

where L is a Lorentz transformation makes sense only when a unitary operator $\Gamma(L)$ exists that implements the action of L; and this has no reason to be the case unless the vacuum expectation form is invariant under L, for which in turn there is no a priori reason. The idea behind the formula has however a natural interpretation if $\Gamma(L)$ is interpreted as a unitary transformation from $\mathcal{K}_E$ to $\mathcal{K}_{EL}$, and if the renormalizations on the right and left are with respect to the corresponding distinct vacua.

3 The Mathematical Meaning of Physical Lorentz-Invariance

By physical Lorentz-invariance I mean here the principle that observers that differ only by a Lorentz transformation will observe the same physical laws, e.g. the same S-matrix elements between corresponding states, i.e. states labeled by corresponding quantum numbers. These quantum numbers derive from the momenta relative to the frames of the respective observers, and so are distinct, but conjugate within the Lorentz group. Let me explain why the attainment of such Lorentz frame-independence does not in the least require Lorentz-invariance of the vacuum vector v, or of the corresponding expectation form E, where $E[A] = < v, Av >$ for any observable A.

Let us adopt the familiar phenomenological framework in which there is a given algebra $\mathcal{A}$ of field observables, on which the Poincaré group G acts as a group of automorphisms. This is a quite general and conservative framework, and in particular makes no assumption about an ad hoc field Hilbert space, or relation to a free field. How is the physical vacuum to be defined? A natural definition, validated by its applicability to free fields [3] and basic physical criteria of stability and invariance, is as a state E of the algebra $\mathcal{A}$ enjoying the following two properties. First, it is temporally invariant; i.e. if A^t denotes the effect on the observable A of the temporal displacement through times t, then $E[A^t] = E[A]$ for all A and t. This says only that E is a stationary state. That it is in fact a vacuum requires that for arbitrary A and B in $\mathcal{A}$, $E[A^t B]$ should be a positive-frequency function of t.

There is no requirement that E should be invariant under all of G, as far as the underlying physical ideas are concerned, nor is it a reasonable constraint to insist on from a mathematical standpoint, since it is to be expected that in general no such state exists, even when a vacuum in the more limited sense just indicated exists. The stationarity of E means that it is a fixed-point of the action of G on the state space S of $\mathcal{A}$, induced from its action as automorphisms of $\mathcal{A}$. The action of G

on S is an affine action on a convex set. General theory tells us that one can expect a fixed point (with standard assumptions as to regularity and compactness) when G is abelian, solvable, compact, or appropriately amenable. But when G contains a non-compact simple Lie group, there is no reason to expect a fixed point, as a consequence the of complicated structure of such groups 'near infinity'; and experience indicates that it is only in exceptionally simple cases that a fixed point exists. Thus it is distinctly over-optimistic to postulate the existence of a Lorentz- (or conformally-) invariant vacuum in the case of an interacting field, even when the group acts as a perfectly regular group of automorphisms of a corresponding algebra of observables. Even to assume the existence of states that are stationary relative to the entire group is unrealistic.

How then is physical Lorentz-invariance to be manifested? Since space translations commute with time translation in the Lorentz group, it is to be expected that a physical vacuum that is invariant under space translation exists (this is automatic if the vacuum is unique, by an easy argument). Moreover, by virtue of the compactness of the euclidean rotation group, the vacuum may also be supposed to be rotationally invariant (this will again be automatic if the original physical vacuum, defined only by considerations of temporal invariance, is unique). Thus the entire euclidean group E_3 in three dimensions will, together with time evolution, act by unitary transformations on the Hilbert space $\mathcal{K}_E$ attained by applying the canonical construction of an associated representation to E. In terms of this seven-dimensional group, $\mathbb{R}^1 \times E_3$, the usual relativistic quantum numbers are definable; energy, linear momenta, angular momenta.

A second observer who is related to the first observer by a Lorentz transformation L will have as his vacuum the state E^L given by the equation $E^L[A] = E[A^L]$, where $A \to A^L$ denotes the automorphism of $\mathcal{A}$ corresponding to L. The corresponding canonical Hilbert space $\mathcal{K}_E$ may be different from the Hilbert space $\mathcal{K}_{E_0}$ of the first observer, but there will nevertheless be a unique unitary transformation from the one Hilbert space to the other, that carries the vacuum vector v in $\mathcal{K}_E$ into the vacuum vector v ' in $\mathcal{K}_{E^L}$, and induces the transformation of self-adjoint operators in $\mathcal{K}_E$ that represent the usual relativistic quantum numbers in the first Hilbert space into the corresponding operators on the second Hilbert space. Put briefly, this means that there is Lorentz *covariance*, but not necessarily Lorentz *invariance*. There doesn't need to be any one Hilbert space for the quantized field associated with the respective observers.

If the S-matrix exists along normal lines, it too will be observer-dependent, but the S-matrix elements between states labelled by relativistic quantum numbers will be the same. Both observers will have the same underlying single-particle Hilbert spaces, the free field over the temporally asymptotic form of either field Hilbert space $\mathcal{K}_E$ or $\mathcal{K}_{E^L}$. The respective S-matrices will be unitarily equivalent by the simple change of Lorentz frame as implemented on the free field corresponding to the action in the single particle space. Thus the two observers will see the 'same physics'. But their respective Hilbert spaces, on which the interacting quantum field is represented, will only be unitarily equivalent via a transformation that connects the respective $\mathbb{R}^1 \times E_3$ groups, and not the ten-dimensional Poincaré group.

4 Discussion

The foregoing analysis applies equally to equilibrium states as well as vacua, and to other symmetry groups and space-times. For the fundamental physical purposes, the equilibrium state or vacuum need be invariant only under the 'Newtonian' group $\mathbb{R}^1 \times E_3$ that defines the local frame of the observer, and from which the quantum numbers used to describe states are derived. But it is reasonable to expect (e.g., when the state in question have appropriate unicity) that these states transform covariantly. Energy levels and S-matrix elements for states labelled by such quantum numbers should be the same for different observers, even though the quantum numbers themselves are only equivalent (via a unitary transformation from one Hilbert space onto a different Hilbert space), and not identical.

An illustrative example is the case of the conformally invariant wave equation $\Box \phi + g\phi^3 = 0$. The conformal invariance permits it to be transferred to the globally conformally invariant space-time $\mathcal{M}$ that is the universal cover of the conformal compactification of Minkowski space-time $\mathcal{M}_0$, which is Lorentz-covariantly imbedded in $\mathcal{M}$. In $\mathcal{M}$, the analog of the 'Newtonian' group N in terms of which direct observation is made in $\mathcal{M}_0$, is the isometry group K of the Einstein Universe $\mathbb{R}^1 \times S^3$, which is conformally equivalent to $\mathcal{M}$. For the free equation, i.e. with $g = 0$, there is a unique vacuum under each of the conformal group, the Poincaré group, N, and K, and they are all the same. For the nonlinear equation, there may be a K-invariant vacuum, definable as the ground state of the Einstein energy (which generates temporal evolution in $\mathbb{R}^1 \times S^3$). Because the Einstein energy is greater than the usual Minkowski (relativistic) energy in a positive-energy context (as putatively the case here), it may have a ground state without the Minkowski energy having one. It is unlikely that there exists a vacuum under the full conformal group, but a unique vacuum for K will be *covariant*. (K is here the maximal essentially compact subgroup of the conformal group, and all such subgroups are conjugate, like the different Newtonian groups in the Poincaré group P). It is even less likely that there exists a vacuum for P, but in the event that a unique vacuum exists for N, Poincaré covariance will apply, although not invariance. These aspects are relevant to constructive quantum field theory for conformally invariant equations, to which various structures based on a free-field analyzed do not in the least apply. A reasonable definition for generalized Wick renormalization, relative to an arbitrary vacuum, is developed in [4], extending the special case treated in [2].

References

1.   I.E. Segal: "Charactérisation mathématique des observables en théorie quantique des champs", in C.N.R.S. Coll. int. LXXV, Paris, 57–103 (1959)
2.   I.E. Segal: "Mathematical characterisation of the physical vacuum, III", J. Math. **6** 500–523 (1962)
3.   I.E. Segal: "Construction of nonlinear quantum processes, II", Inv. Math. **14** 211–242 (1971)
4.   J.C. Baez, I.E. Segal, Z. Zhou: "Introduction to algebraic and constructive quantum field theory", in press (Princeton University Press, 1990)

III

General Quantization Methods

Quantization, Coherent States and Diffeomorphism Groups

S. Twareque Ali [1] and Gerald A. Goldin [2]

[1]Department of Mathematics and Statistics, Concordia University,
Montréal, Québec, CANADA H4B 1R6
[2]Departments of Mathematics and Physics, Rutgers University,
New Brunswick, New Jersey 08903, U. S. A.

Abstract: We suggest extending the method of coherent-states quantization, which applies to homogeneous spaces for locally compact groups, to the case of infinite-dimensional groups such as diffeomorphism groups. Such a framework could unify a number of different approaches to quantum theory. We review some relevant results and examples, and take a first step in this program by demonstrating that the unitary representation of Diff($\mathbb{R}$) obtained from the "single-particle" coadjoint orbit is in fact square-integrable.

1 Introduction

The problem of quantization is a legacy from the early days of quantum mechanics. Although it is believed that quantum mechanics is the more fundamental theory, which classical mechanics only approximates, it is nonetheless customary (and useful in actual practice) to start with a classical model for a physical system and then to arrive at the quantum picture by appropriately *quantizing* the model. For this purpose we have the time-honored procedure of *canonical quantization,* which calls for replacing the classical observables of position and momentum respectively by the operators of multiplication by x and differentiation with respect to x (up to a constant). This procedure seemed natural to the pioneers of quantum mechanics as the correct way to pass from the classical approximation to the more accurate quantum description.

But with time it was recognized that more than ritual deference to historical convention underlies the quantization procedure. In a manner of speaking, the ghost of quantum mechanics hovers over the classical theory – indeed, to such an extent that the essence of the former can be captured by a deeper analysis of the latter. Of course there are limits to how much of the precise quantum content of a theory can be recovered from its classical approximation: for example, there are "purely quantum" effects which leave behind no classical traces. Nevertheless in most practical situations a classical model, or a related classical model, enables one to make the intuitive, often heuristic, leap from classical physics to the more exact quantum picture.

The effort to identify and describe precisely those fundamental characteristics or elements of a classical theory in which its quantum features are most fully (or most elegantly) encoded, has led to a number of different *quantization programs*. Apart from the original canonical quantization prescription, we have today several other broadly developed methods for making the transition from classical to quantum mechanics. Representative among these are the Mackey program, based on *systems of imprimitivity*, which requires the presence of an underlying symmetry group in the classical theory [1–3], and the related more general methods of *geometric quantization* developed by Auslander, Kirillov, Kostant, Souriau and others [4–6]; as well as the methods described by Berezin, in which classical observables are replaced by operator symbols, and the technique of *second quantization* in which an already-quantized system is treated as if certain objects it contains were classical observables, to be quantized again [7]. A method of second quantization was also developed by Segal [8], in which nonlinear field equations are quantized on infinite-dimensional manifolds.

Prof. H.-D. Doebner, with several of his students and collaborators, has developed over the years a highly interesting variant of the Mackey technique which also includes some aspects of geometric quantization [9–14]. The name *Borel quantization*, or more specifically *quantum Borel kinematics*, has been given to this approach, which not only exploits the Borel structure of the underlying particle configuration space (as does the Mackey theory), but also makes use of the geometry of this configuration space to (locally) mimic *group-like* actions: thus, in a sense, straddling the fence between the Mackey procedure and geometric quantization methods.

The present authors as well have sought to explore quantization and second-quantized field theories making use of group-theoretical methods and group actions on homogeneous spaces, in ways which (from different starting points) make contact with Prof. Doebner's work. We felt that an appropriate tribute to him on this occasion would be to present an account of this work highlighting those points of contact, in which we also take steps toward unifying our respective points of view.

One group-theoretical approach to the problem of quantization is the method of *coherent states*, also called *prime quantization* [15,16], which (it should be noted in addition) has some similarities with the Berezin technique. Traditionally, coherent states have been linked to certain types of group representations [17]. However, certain functional-analytic and measure-theoretic properties of these states can exist even in the absence of an underlying symmetry group action. It is, in fact, exactly these which allow one to specify a quantization procedure using coherent states. In Sects. 2 and 3, we outline this method of quantization and employ it in specific cases to illustrate its versatility.

Of special interest from the standpoint of coherent-states quantization is the situation where one assumes an underlying group which is infinite dimensional.

Even when the single-particle configuration space of a classical system possesses no global symmetry, for example when the system is neither translation- nor rotation-invariant, it is possible to describe corresponding quantum systems by introducing an infinite dimensional group – the group Diff(M) of diffeomorphisms (i.e., general coordinate transformations) of the underlying position space M – and studying its unitary representations. Such a perspective was taken some years ago by Goldin, Sharp and Menikoff, who studied local current algebras defined from second-quantized non-relativistic fields, and obtained various diffeomorphism group representations [18–22]. The results were more than a new formalism: they included independent, mathematically rigorous predictions of new possibilities, such as that of particles in two-dimensional space obeying *fractional statistics*, intermediate between bosons and fermions and transforming according to representations of braid groups. Such particles had already been proposed on general topological and geometric grounds by Leinaas and Myrheim [23,24], and were subsequently studied independently by Wilczek, who named them *anyons* [25,26].

Isham [27,28] also arrived at diffeomorphism group representations from considerations of the topology and geometry of quantum configuration spaces; while Prof. Doebner, in collaboration with Tolar and Angermann [10–12], did so independently within the framework of quantum Borel kinematics. Thus it is interesting to ask (as we do here) whether the quantum theories which such infinite-dimensional groups describe can be obtained through coherent-states quantization on appropriate homogeneous spaces of diffeomorphism groups.

It is likewise important in quantum field theory to consider groups of smooth mappings from the position-space manifold to a given compact Lie group, such as SU(2) or SU(3), under pointwise operations. The resulting groups, which are also infinite dimensional, are called *local current groups* or *gauge groups*, and their representations have been studied by a number of researchers [29–31]; their semi-direct products with diffeomorphism groups can describe quantum particles with internal degrees of freedom [19,21,32].

When the underlying position space is a one-dimensional manifold, there exists a unique, non-trivial central extension of each of these infinite-dimensional groups, and of the corresponding infinite-dimensional Lie algebras. In the case of Diff(S^1), the group of C^∞ (orientation preserving) diffeomorphisms of the circle, the central extension is the group for the *Virasoro algebra*; in the case of local current groups over S^1, the central extensions are *Kac-Moody groups*. These groups (and their semidirect products) are suitable for describing relativistic (conformal-invariant) field theories in two-dimensional space-time, and have many other applications to physics [33–39].

So we see that extending the method of coherent-states quantization to homogeneous spaces for infinite-dimensional groups would provide the opportunity to unify a number of approaches to quantum theory. We do not claim to offer such an extended theory here; rather, we give only an indication as to how it might be possible to proceed.

2 A Simple Coherent-States Example

Consider the well-known example of coherent states in the quantum mechanics of a single particle moving in one dimension. On the Hilbert space $\mathcal{H}$ of pure states of the system, we define the *creation* and *annihilation* operators $a^\dagger, a$, satisfying the commutation relation,

$$[a, a^\dagger] = I.$$

For each complex number $z \in \mathbb{C}$, we define a *coherent state*, a vector $|z\rangle \in \mathcal{H}$, by means of the eigenvalue equation,

$$a|z\rangle = z|z\rangle.$$

Using the orthonormal basis $\{|n\rangle\}_{n=0}^\infty$ in $\mathcal{H}$ defined by the normalized eigenstates of $a^\dagger a$,

$$a^\dagger a|n\rangle = n|n\rangle,$$

we can write the vectors $|z\rangle$ in the form [40]

$$
\begin{aligned}
|z\rangle &= \exp\{-|z|^2/2\}\exp\{za^\dagger\}|0\rangle \\
&= \exp\{-|z|^2/2\}\sum_{n=0}^\infty \frac{z^n}{\sqrt{n!}}|n\rangle,
\end{aligned}
\tag{1}
$$

where we have used the fact that $a^\dagger|n\rangle = \sqrt{n+1}\,|n+1\rangle$.

The term *coherent state* refers to the fact that in such states a quantized electromagnetic field has correlation functions which factor in the manner of classical fields. However, in this section we wish to bring out two other aspects of these states, using the above example: first, their relationship to the quantization map; and second, a group-theoretical property.

First, we relate these coherent states to the process of quantization. Write the complex number z in terms of its real and imaginary parts,

$$z = \frac{1}{\sqrt{2}}(q + ip),$$

where $(q, p) \in \mathbb{R}^2$. We shall consider (q, p) as representing a point in the classical phase space Γ of the single-particle system, q being its position and p its momentum coordinate. We shall also write

$$|z\rangle = |q, p\rangle.$$

Now we consider the measure-theoretic properties of the one-dimensional projection operators

$$F(q, p) = |q, p\rangle\langle q, p|.$$

For any Borel set B of Γ, define the bounded positive operator

$$a(B) = \int_B |q, p\rangle\langle q, p|\, \mathrm{d}\nu(q, p),$$
$$\mathrm{d}\nu(q, p) = \frac{\mathrm{d}q\, \mathrm{d}p}{2\pi}$$
,

where the integral is understood to converge in the weak operator topology. Then the operators $a(B)$ define a *normalized positive operator valued measure* (POV-measure), satisfying [15,40]:

$$a(\emptyset) = 0, \qquad a(\Gamma) = I, \tag{2}$$

and

$$a(\bigcup_{i \in J} B_i) = \sum_{i \in J} a(B_i), \tag{3}$$

where $\emptyset$ is the empty set in Γ, I the identity operator on $\mathcal{H}$, J a countable index set, B_i and B_j are disjoint sets for $i \neq j$ in J, and where the sum in (3) is assumed to converge weakly. (To demonstrate that $a(\Gamma) = I$ requires a straightforward calculation.) The (bounded) positive operator-valued function $(q, p) \longmapsto F(q, p)$ is called a *density function* for the POV-measure a, since for any Borel set $B \subseteq \Gamma$,

$$a(B) = \int_B |q, p\rangle\langle q, p|\, \mathrm{d}\nu(q, p).$$

We shall define integrals of ordinary (complex-valued) scalar functions f on the phase space Γ with respect to the POV-measure a, in the weak sense – i.e.,

$$\pi^*(f) = \int_\Gamma f(q, p)\, \mathrm{d}a(q, p) \tag{4}$$

exists (as a bounded operator on $\mathcal{H}$) iff $(\forall \Psi_1, \Psi_2 \in \mathcal{H})$ the complex integrals

$$\int_\Gamma f(q, p)\langle \Psi_1 | F(q, p)\Psi_2\rangle\, \mathrm{d}\nu(q, p)$$

exist and are finite. Of course, we can next extend the integral (4) to a class of functions on Γ wider than that which leads to bounded operators. But even before we do that, note that the map π^* (from functions to operators) given by (4) already defines a quantization.

To see this more clearly, let $\mathcal{C}_\infty(\Gamma)$ be the set of all complex-valued continuous functions on Γ which vanish at infinity. This set is a C^*-algebra [41] in a natural way (under pointwise addition and multiplication, with the uniform norm) and may be taken to be the underlying algebra of classical observables. Then π^* can be viewed as the map which takes classical observables into operators on the Hilbert space. Indeed,

$$\pi^* : \mathcal{C}_\infty(\Gamma) \longrightarrow \mathcal{L}(\mathcal{H})$$

is not only linear but positive; i.e.. if $f \in \mathcal{C}_\infty(\Gamma)$ is a positive function, then $\pi^*(f)$ is a positive operator. Also, if $\{f_n\}_{n=1}^\infty$ is a sequence of functions in $\mathcal{C}_\infty(\Gamma)$ such that $(\forall q, p \in \Gamma) f_n(q,p)$ is nondecreasing and converges to 1 (we denote this by $f_n \uparrow \mathbb{1}$ where $\mathbb{1}(q,p) \equiv 1$), then (weakly)

$$\pi^*(f_n) \uparrow I.$$

Next let f_{pos} and f_{mom} be the classical observables of position and momentum,

$$\left. \begin{aligned} f_{pos}(q,p) &= q, \\ f_{mom}(q,p) &= p. \end{aligned} \right\} \tag{5}$$

Then, extending π^* in (4) in a standard way [42], we can compute the two un-bounded, essentially self-adjoint operators,

$$Q = \pi^*(f_{pos}),$$
$$P = \pi^*(f_{mom}).$$

Now it can be checked [43,44] that Q and P satisfy the canonical Heisenberg commutation relations

$$[Q, P] = iI \qquad (\hbar = 1).$$

In other words, if we agree to call π^* a quantization map, then it does in fact achieve, as desired, a transformation of the set of classical observables into a set of operators on $\mathcal{H}$, obeying the correct commutation rule.

The question, of course, is how to arrive at this map, starting only with $\mathcal{C}_\infty(\Gamma)$ and the measure ν. The answer follows when we recognize that the POV-measure a sets up a unitary map between $\mathcal{H}$ and a certain Hilbert space of functions on Γ, which we now construct. Let

$$\tilde{\mathcal{H}} = L^2{}_\nu(\Gamma)$$

be the space of square-integrable, complex-valued functions on the phase space Γ with the usual inner product defined by ν; and define

$$\begin{aligned} W : \mathcal{H} &\longrightarrow L^2{}_\nu(\Gamma), \\ (W\Psi)(q,p) &= \langle q, p | \Psi \rangle. \end{aligned} \tag{6}$$

The fact that this map is unitary, from $\mathcal{H}$ onto a subspace $\mathcal{H}_K$ of $\tilde{\mathcal{H}}$, follows directly from (2). Let $\mathbb{P}_K$ be the orthogonal projection operator on $\tilde{\mathcal{H}}$ such that

$$\mathbb{P}_K \tilde{\mathcal{H}} = \mathcal{H}_K$$

Then $\mathbb{P}_K$, as an operator on $\tilde{\mathcal{H}}$, has an integral kernel K; i.e., we can write

$$(\mathbb{P}_K \tilde{\Phi})(q,p) = \int_\Gamma K(q,p; q',p') \tilde{\Phi}(q',p') \, d\nu(q',p') \tag{7}$$

$(\forall \tilde{\Phi} \in \tilde{\mathcal{H}})$, where

$$K(q,p; q',p') = \langle q,p | q',p' \rangle.$$

The kernel K has the following properties:

$$K(q,p; q',p') = \overline{K(q',p'; q,p)}, \tag{8}$$

$$K(q,p;q.p) > 0, \qquad (9)$$

$$K(q,p;q',p') = \int_\Gamma K(q,p;q'',p'')\,K(q'',p'';q',p')\,d\nu(q'',p''), \qquad (10)$$

reflecting in particular that the projection $\mathbb{P}_K$ satisfies

$$\mathbb{P}_K = (\mathbb{P}_K)^* = (\mathbb{P}_K)^2 .$$

Equation (10) is called the reproducing property, and because of it K is called a *reproducing kernel*. For $\Phi \in \mathcal{H}_K$, (7) implies the reproducing property

$$\int_\Gamma K(q,p;q',p')\Phi(q',p')\,d\nu(q',p') = \Phi(q,p), \qquad (11)$$

for (with respect to ν) almost all $(q,p) \in \Gamma$. However, since $K(q,p;q',p')$ is defined for *all* $(q,p),(q',p') \in \Gamma$, (11) can actually be used to *define* $\Phi(q,p)$ for all $(q,p) \in \Gamma$. In this way $\mathcal{H}_K$ becomes a space of actual functions, defined everywhere – not just equivalence classes of functions modulo measure zero sets. Moreover, once this is done, for each $(q,p) \in \Gamma$ the evaluation map $E_K(q,p) : \mathcal{H}_K \longrightarrow \mathbb{C}$ given by

$$E_K(q,p)\Phi = \Phi(q,p) \quad (\forall \Phi \in \mathcal{H}_K), \qquad (12)$$

is linear and continuous. Let $E_K(q,p)^* : \mathbb{C} \longrightarrow \mathcal{H}_K$ be the dual map. Then from (6) and (11), it can be seen that

$$WF(q,p)W^{-1} = E_K(q,p)^*E_K(q,p). \qquad (13)$$

Defining

$$\left.\begin{array}{rl} F_K(q,p) &= E_K(q,p)^*E_K(q,p), \\ a_K(B) &= Wa(B)W^{-1} = \int_B F_K(q,p)\,d\nu(q,p), \end{array}\right\} \qquad (14)$$

we see that $(\forall \Phi \in \mathcal{H}_K)$

$$\langle \Phi|a_K(B)\Phi\rangle = \int_B |\Phi(q,p)|^2\,d\nu(q,p). \qquad (15)$$

Thus, if we define

$$\pi_K^* : \mathcal{C}_\infty(\Gamma) \longrightarrow \mathcal{L}(\mathcal{H}_K),$$
$$\pi_K^* = W \circ \pi^*,$$

we have simply reexpressed the quantization map π_K^* in terms of operators on $\mathcal{H}_K$. One can then work directly with a_K to obtain, in parallel with (4),

$$\pi_K^*(f) = \int_\Gamma f(q,p)\,da_K(q,p). \qquad (16)$$

The point to note in this discussion is that the existence of the evaluation maps $E_K(q,p)$ in (12), and hence of a_K, is a consequence of the fact that $\mathcal{H}_K$ is a reproducing kernel subspace of $\mathcal{H}$. Moreover, the existence of the coherent states $|q,p\rangle$ themselves is a consequence of the reproducing property of K, a point which will be brought out more fully in the next section.

Writing

$$\xi_{q,p} = W|q,p\rangle,$$

it is not hard to see using (12) and (13) that ($\forall \lambda \in \mathbb{C}$)

$$\lambda \xi_{q,p} = E_K(q,p)^* \lambda,$$
$$K(q,p;q',p') = E_K(q,p)E_K(q',p')^*.$$

Looking back, the following scheme emerges for the quantization of the algebra $\mathcal{C}_\infty(\Gamma)$ of classical observables over the phase space Γ: Construct the Hilbert space $\tilde{\mathcal{H}} = L^2{}_\nu(\Gamma)$. Identify in $\tilde{\mathcal{H}}$ an appropriate reproducing kernel subspace $\mathcal{H}_K$, and determine its canonical evaluation maps $E_K(q,p)$. Use these to construct the normalized POV-measure a_K via (14). The quantization map π_K^* is then given by (16).

We conclude our discussion of the quantization map with a comment about the Hilbert space $\mathcal{H}_K$ of the quantized system. From (15) it would appear that the $a_K(B)$ are *localization operators* on phase space, provided we also interpret $|\Phi(q,p)|^2$ as a *localized density* on phase space, for the system in the state $\Phi \in \mathcal{H}_K$. Of course q and p cannot here be interpreted as the usual quantum mechanical position and momentum, since these are not simultaneously measurable. They can instead be interpreted as representing the *average* position and momentum (respectively), in an intrinsic sense [15,43,44]. In other words, it is the phase space itself of the physical system which becomes the localization manifold in this method of quantization. Such a phase space often arises as the cotangent bundle of a configuration-space manifold. It can also arise as a coadjoint orbit of an appropriate group.

Now we turn to the second important point of this example: a group-theoretical look at the coherent states $|q,p\rangle$. It is from this perspective that we shall later seek to make contact with diffeomorphism group representations in quantum mechanics.

First observe that the equation $a(\Gamma) = I$ in (2), when written out in full, is

$$\int_\Gamma |q,p\rangle\langle q,p| \frac{dq\,dp}{2\pi} = I,$$

which is just the statement that the set of vectors $\{|q,p\rangle\}$ forms an *overcomplete family of states* (OFS) in $\mathcal{H}$. Next, note that (1) can be written differently as

$$|z\rangle = \exp\{za^\dagger - \bar{z}a\}|0\rangle,$$

or if we introduce the position and momentum operators Q, P on $\mathcal{H}$,

$$Q = \frac{1}{\sqrt{2}}(a + a^\dagger),$$

$$P = \frac{1}{i\sqrt{2}}(a - a^\dagger),$$

in the form

$$|q,p\rangle = \exp\{i(pQ - qP)\}|0\rangle.$$

Thus we have the unitary operators on $\mathcal{H}$ given by

$$U(q, p) = \exp\{i(pQ - qP)\},\tag{17}$$

and these give us an irreducible, unitary representation of the Weyl-Heisenberg group, the group obtained by exponentiating the canonical commutation relation $[Q, P] = iI$. Actually, we have only represented a particular *section* in this group by $U(q, p)$, as the group itself depends on three real parameters – the coefficients of Q, P and I – and (17) represents only two. What is represented is a section of the group considered as a principal bundle over a certain homogeneous space; i.e., over a manifold on which the group elements act. The existence of the coherent states now embodies a mathematical property of the group representation – that it is *square integrable* in a sense to be defined in the next section.

To sum up, quantization using coherent states involves realizing the classical phase space as a homogeneous space of some appropriate symmetry group, and then looking for square-integrable representations of the group over this manifold.

3 Square-Integrable Group Representations and Coherent States

In this section we summarize the general features of square-integrable group representations, their relationship to POV-measures and reproducing-kernel Hilbert spaces, and the appearance of coherent states in this general context [45–50].

Let G be a Lie group (finite or infinite dimensional) and H a closed subgroup of G such that the coset space

$$X = G/H$$

is locally compact and carries a quasi-invariant measure ν (i.e., for every $g \in G$, the class of measure-zero sets in X is preserved under the transformation $g : X \longrightarrow X$). Note that for a finite-dimensional Lie group G, such a measure always exists on X; while for G infinite dimensional, we are interested here in closed subgroups having finite codimension in G. Let

$$\beta : X \longrightarrow G$$

be a *Borel section*, i.e., a Borel function such that $(\forall x \in X)\ p(\beta(x)) = x$, where $p(g) = gH \in X$. For any $g \in G$, denote by β_g the covariantly transformed section

$$\begin{aligned}\beta_g(x) &= g\beta(g^{-1}x)\\&= \beta(x)h(g, g^{-1}x)\end{aligned}\tag{18}$$

for all $x \in X$, where $x \longmapsto gx$ is the natural action of G on X and $h : G \times X \longrightarrow H$ is the cocycle

$$h(g, x) = \beta(gx)^{-1}g\beta(x);$$

thus h satisfies the *cocycle equation*

$$h(g'g, x) = h(g', gx)\, h(g, x).$$

Let ν_g be the transformed measure $d\nu_g(x) = d\nu(g^{-1}x)$; i.e.,

$$\int_X f(gx)\,d\nu(x) = \int_X f(x)\,d\nu_g(x)$$

where f is any measurable function on X. Then the Radon-Nikodym derivative

$$\lambda(g,x) = \frac{d\nu_g}{d\nu}(x)$$

defines $\lambda : G \times X \longrightarrow \mathbb{R}^+$ as a cocycle which can be chosen [51] so that it is defined for all $g \in G$ and $x \in X$, and so that the cocycle equation

$$\lambda(g_1 g_2, x) = \lambda(g_1, x)\lambda(g_2, g_1^{-1}x)$$

holds for all $g_1, g_2 \in G$ and $x \in X$ (not just almost everywhere). We shall assume that the cocycle λ has been so chosen.

Suppose next that we have a (weakly) continuous, unitary, irreducible representation $g \longmapsto U(g)$ of G on some (separable) Hilbert space $\mathcal{H}$. We say that the representation U is *square integrable mod* (H, β) if there exists a positive (self-adjoint) bounded operator F on $\mathcal{H}$ of rank $n < \infty$, and a positive (self-adjoint) bounded operator A on $\mathcal{H}$, with $\|A\| = 1$, having an inverse A^{-1} that is permitted to be unbounded, such that

$$\int_X F_\beta(x)\,d\nu(x) = A, \qquad (19)$$

where

$$F_\beta(x) = \lambda(\beta(x), x)U(\beta(x))FU(\beta(x))^*. \qquad (20)$$

The integral in (19) is assumed to converge weakly.

It is not hard to show from this definition that the property of square integrability depends only on the measure class of ν and not on the individual measure. Indeed, if ν' is another, equivalent measure on X having the Radon-Nikodym derivative cocycle

$$\lambda'(g,x) = \frac{d\nu'_g}{d\nu'}(x),$$

define F'_β by replacing λ with λ' in (20), and one easily sees that

$$F'_\beta(x)\,d\nu'(x) = F_\beta(x)\,d\nu(x).$$

Note too that if the representation U is square integrable mod (H, β), then it is also square integrable mod (H, β_g), for all $g \in G$, with β_g as in (18). Indeed, writing

$$A_g = U(g)AU(g)^*$$

and using the quasi-invariance of ν, it follows immediately that A_g has all the desired properties, and

$$\int_X F_{\beta_g}(x)\,d\nu(x) = A_g.$$

In other words, the definition of square integrability depends only on the *class* of sections β_g, $g \in G$, and not on any individual member of the class.

Assuming now that U is square integrable mod (H, β), coherent states are constructed as follows: Since F is a (finite) rank-n positive operator, there exists a set of n mutually orthogonal vectors η^i $(i = 1, 2, \ldots, n)$, such that

$$F = \sum_{i=1}^{n} |\eta^i\rangle\langle\eta^i|. \tag{21}$$

Set

$$\eta^i_{\beta(x)} = \lambda(\beta(x), x)^{\frac{1}{2}} U(\beta(x))\eta^i, \tag{22}$$

and denote by $\mathcal{S}_\beta$ the resulting family of states:

$$\mathcal{S}_\beta = \{\eta^i_{\beta(x)} \,|\, i = 1, 2, \ldots, n; \, x \in X\}. \tag{23}$$

Then, $\mathcal{S}_\beta$ is a family of *coherent states* for U. It constitutes an *overcomplete family* of vectors in $\mathcal{H}$, since in terms of $\eta^i_{\beta(x)}$, (19) assumes the form

$$\sum_{i=1}^{n} \int_X |\eta^i_{\beta(x)}\rangle\langle\eta^i_{\beta(x)}| \, \mathrm{d}\nu(x) = A. \tag{24}$$

Now consider the Hilbert space $\tilde{\mathcal{H}} = L^2_\nu(X, \mathbb{C}^n)$ of $\mathbb{C}^n$-valued functions on X, square integrable with respect to ν. It can be shown [50] in analogy with the example in Sect. 2 that the set of functions Ψ in $\tilde{\mathcal{H}}$ of the type

$$\Psi_i(x) = \langle\eta^i_{\beta(x)}|\phi\rangle \quad (\phi \in \mathcal{H}),$$

$i = 1, 2, \ldots, n$, becomes itself a closed Hilbert space $\mathcal{H}_K$ when equipped with the scalar product

$$\langle\Psi \,|\, \Phi\rangle_K = \sum_{i=1}^{n} \int_X \overline{\Psi_i(x)} (A_K^{-1}\Phi)_i(x) \, \mathrm{d}\nu(x) \tag{25}$$

for $\Phi, \Psi \in \mathcal{H}_K$, where

$$A_K = WAW^{-1}$$

and $W : \mathcal{H} \longrightarrow \mathcal{H}_K \subset \tilde{\mathcal{H}}$ is the *unitary map*

$$(W\phi)_i(x) = \langle\eta^i_{\beta(x)}|\phi\rangle$$

$(\forall \phi \in \mathcal{H})$; $x \in X$, and $i = 1, 2, \ldots, n$. Note that although $\mathcal{H}_K$ is a subset of $\tilde{\mathcal{H}}$, it is not necessarily a closed subspace of it; the norm defined by the inner product in (25) is not equivalent to the L^2-norm if A^{-1} is unbounded.

Finally, $\mathcal{H}_K$ carries a reproducing kernel K with components $K_{ij}(x, y)$, defined by

$$K_{ij}(x, y) = \langle\eta^i_{\beta(x)}|A^{-1}\eta^j_{\beta(y)}\rangle, \tag{26}$$

and having the three properties

$$K_{ij}(x, y) = \overline{K_{ji}(y, x)}, \tag{27}$$

$$K_{ii}(x,x) > 0, \tag{28}$$

$$\sum_{\ell=1}^{n} \int_X K_{i\ell}(x,y)K_{\ell j}(y,z)\,\mathrm{d}\nu(y) = K_{ij}(x,z), \tag{29}$$

generalizing (8)–(10). Equation (29) implies the reproducing property: for any $\Psi \in \mathcal{H}_K$ and $x \in X$,

$$\Psi_i(x) = \sum_{j=1}^{n} \int_X K_{ij}(x,y)\Psi_j(y)\,\mathrm{d}\nu(y).$$

If the group G is infinite dimensional, there do not in general exist invariant measures on either the group itself or on its quotient spaces. However there could still exist measures (on at least some quotient spaces) that are quasi-invariant. Then it might again be possible to construct coherent states, as indicated above, for particular unitary representations of G; and that is one reason for developing the framework for coherent-states quantization using only the quasi-invariance of the measure ν. An especially interesting example of such an infinite-dimensional group is the diffeomorphism group discussed in Sect. 1, which we now look at in more detail.

4 Diffeomorphism Group Representations and Quantum Mechanics

In this section we show how an extremely general framework for quantum theory can be established through the study of diffeomorphism group representations.

In 1968, Dashen and Sharp [52] considered the formal algebra obtained from the following set of (non-relativistic) densities and currents (at the fixed time t, for $x \in \mathbb{R}^3, m = \hbar = 1$):

$$\left. \begin{array}{c} \rho(x) = \psi^*(x)\psi(x), \\ J(x) = \frac{1}{2i}\left[\psi^*(x)\nabla\psi(x) - (\nabla\psi^*(x))\psi(x)\right]. \end{array} \right\} \tag{30}$$

When the field ψ is taken to satisfy equal-time canonical commutation ($-$) or anticommutation ($+$) relations,

$$[\psi(x), \psi^*(y)]_{\pm} = \delta(x - y), \tag{31}$$

the densities and currents obey a set of singular commutation relations,

$$\left. \begin{array}{c} [\rho(x), \rho(y)] = 0, \\ [\rho(x), J^k(y)] = -i\frac{\partial}{\partial x^k}[\delta(x - y)\rho(x)], \\ [J^j(x), J^k(y)] = -i\frac{\partial}{\partial x^k}[\delta(x - y)J^j(x)] + i\frac{\partial}{\partial y^j}[\delta(x - y)J^k(y)]. \end{array} \right\} \tag{32}$$

The density $\rho(x)$ and current $J(x)$ were subsequently interpreted as operator-valued distributions on test-function spaces of (respectively) $\mathcal{C}^\infty$ scalar functions and $\mathcal{C}^\infty$ vector fields on $\mathbb{R}^3$, vanishing at infinity. Then the resulting semidirect product Lie algebra of functions and vector fields can be exponentiated [19] to yield a semidirect product group of functions with diffeomorphisms of $\mathbb{R}^3$. The Lie algebra thus obtained is

$$\left.\begin{array}{l} [\rho(f_1),\,\rho(f_2)] = 0, \\ [\rho(f),\,J(g)] = i\rho(g\cdot\nabla f), \\ [J(g_1),\,J(g_2)] = -iJ([g_1,\,g_2]), \end{array}\right\} \tag{33}$$

where f, f_1, f_2 are C^∞ scalar functions on $\mathbb{R}^3$ of rapid decrease (together with all derivatives) at infinity; g, g_1, g_2 are C^∞ vector fields on $\mathbb{R}^3$, also of rapid decrease at infinity; and $[g_1, g_2] = g_1 \cdot \nabla g_2 - g_2 \cdot \nabla g_1$ denotes the *Lie bracket* of the vector fields.

In light of the work by Prof. Doebner and his collaborators on quantization on manifolds [9–14], we can regard this algebra more generally, as the semidirect product algebra of scalars and vector fields on a smooth, connected Riemannian manifold M. The operators $\rho(f)$ are then directly related to the quantum Borel kinematics, when the smooth functions f are taken to approximate indicator functions of Borel sets. If M is compact, we can of course remove the restrictions about behavior at infinity.

Equations (32) and (33) can be derived not only from formal calculations using (30) and (31), but also within an explicit representation of (31) by nonrelativistic Fock fields, where the commutators in (32) can be calculated. The same infinite-dimensional Lie algebra is obtained whether one starts with the Bose or the Fermi representation of the canonical fields, and no Schwinger terms occur in these representations. This is one way to see that (33) can describe nonrelativistic quantum systems. It is also interesting that if (31) is replaced by the more general q-commutator

$$[\psi(\boldsymbol{x}),\,\psi^*(\boldsymbol{y})]_q = \delta(\boldsymbol{x} - \boldsymbol{y}),$$

where $[A,\,B]_q = AB - qBA$, the formal calculations using (30) still result in (32) and consequently (33); though we now no longer have an immediate analogue of the Fock representation.

Here we shall focus only on the third equation of (33), which is to be regarded as a representation by self-adjoint operators of the Lie algebra Vect(M), whose elements are C^∞ vector fields on M of rapid decrease in all derivatives (no restriction at infinity is needed, of course, if M is compact). Given $g \in$ Vect(M), there exists a unique one-parameter group of C^∞ diffeomorphisms of M, $\phi_t : M \to M$, the composition of diffeomorphisms satisfying $\phi_{t_1} \circ \phi_{t_2} = \phi_{t_1+t_2}$, such that

$$\frac{\partial}{\partial t}\phi_t(\boldsymbol{x}) = g(\phi_t(\boldsymbol{x})),$$
$$\phi_{t=0}(\boldsymbol{x}) = \boldsymbol{x}.$$

Such a one-parameter subgroup of the diffeomorphism group is called a *flow*; it is smooth in t as well as in $\boldsymbol{x}$. Its existence globally depends on g being well-behaved at infinity, in which case $(\forall t)\,\phi_t(\boldsymbol{x}) \to \boldsymbol{x}$ rapidly as $|\boldsymbol{x}| \to \infty$. If g has compact support, then ϕ_t also has compact support; where the support of ϕ means the intersection of all closed sets $C \subseteq M$ such that $\phi(\boldsymbol{x}) = \boldsymbol{x}$ outside C. We write ϕ_t^g when we wish to stress the dependence of the flow on the vector field g.

Next consider the group Diff(M), consisting of all C^∞ diffeomorphisms from M to itself which become rapidly trivial at infinity, under the operation of composition of diffeomorphisms. We write $(\phi_1\phi_2)(\boldsymbol{x}) = \phi_2(\phi_1(\boldsymbol{x}))$ for $\phi_1, \phi_2 \in$ Diff(M). It is not difficult to make Diff(M) a topological group, by endowing it with a topology of uniform convergence in all derivatives. Then for any $g \in$ Vect(M), the flow ϕ_t^g is a continuous one-parameter subgroup of Diff(M). Thus, if V is a continuous unitary representation (CUR) of Diff(M) in a Hilbert space $\mathcal{H}$, the unitary subgroup $V(\phi_t^g)$ defines a CUR of $\mathbb{R}$ for each $g \in$ Vect(M), and (from Stone's theorem) we have the uniquely-defined self-adjoint generator

$$J(g) = \lim_{t \to 0} \frac{V(\phi_t^g) - I}{it}, \tag{34}$$

with $V(\phi_t^g) = e^{itJ(g)}$. Since $J(g)$ is in general an unbounded operator, it has a dense domain of definition, the vectors for which the limit in (34) exists, which is not all of $\mathcal{H}$; thus the third commutation relation of (33) makes sense only on some common dense invariant domain for the operators $J(g)$. However the unitary operators $V(\phi)$ are subject to no such restriction. The study of unitary representations of Diff(M) makes it easier to characterize the global properties of quantum systems; while the study of self-adjoint representations of Vect(M) highlights their local properties. Ultimately, it is the physical meaning of $J(g)$ as the momentum density averaged with the smooth vector field g, which together with (34) provides a quantum physical interpretation for a CUR of Diff(M). Note that the flows are (trivially) path-connected to the identity diffeomorphism; thus for our purposes it is sufficient to restrict attention to the *identity component* of the diffeomorphism group, which (taking M to be orientable) includes only orientation-preserving diffeomorphisms.

For any (bounded or unbounded) region B in M, we also have the closed subgroup of Diff(M) consisting of those diffeomorphisms having support in B, and the corresponding subalgebra of Vect(M) consisting of vector fields with support in B; providing a way to associate physical measurements with regions of M.

Many unitary equivalence classes of representations of Diff(M) have been obtained, and interpreted quantum-mechanically. Though Diff(M) itself is infinite dimensional, some of its representations describe quantum systems having finite-dimensional configuration-spaces; for example, systems of N particles ($N < \infty$) obeying Bose statistics, Fermi statistics, parastatistics, or (in $\mathbb{R}^2$) the statistics of anyons. Other representations are associated with infinitely many configuration-space degrees of freedom; for example, infinite Bose or Fermi gases, or representations describing vortex configurations. The techniques that have been brought to bear in constructing representations include semidirect products with nuclear spaces and induced representation theory, systems of imprimitivity, and the method of coadjoint orbits [18–22,53–57,66]; for a more detailed overview of some of these representations, see [22] or [58]. It is remarkable that such different quantum systems, many of which were originally obtained by representing distinct algebras of fields, arise through the classification of the CUR's of just the group Diff(M).

To consider a prototypical representation, let Δ be a measure space carrying an action of Diff(M); we write $(\phi, \gamma) \to \phi\gamma$ for $\phi \in$ Diff(M) and $\gamma \in \Delta$, with $\phi_1(\phi_2\gamma) = (\phi_1\phi_2)\gamma$. Here the space Δ is the quantum *configuration* space (not the phase space). Let μ be a measure on Δ quasi-invariant for Diff(M); i.e., such that every $\phi \in$ Diff(M) preserves the class of μ-measure zero sets in Δ. Let $\frac{d\mu_\phi}{d\mu}(\gamma)$ be the Radon-Nikodym derivative of the transformed measure μ_ϕ with respect to μ (as before, giving a cocycle with values in $\mathbb{R}^+$). Let $\mathcal{M}$ be a (finite- or infinite-dimensional) Hilbert space, and let $\mathcal{H} = L^2_\mu(\Delta, \mathcal{M})$ be the Hilbert space of square-integrable functions on Δ taking values in $\mathcal{M}$. For $\Psi \in \mathcal{H}$, we write

$$[V(\phi)\Psi](\gamma) = \alpha_\phi(\gamma)\Psi(\phi\gamma)\sqrt{\frac{d\mu_\phi}{d\mu}(\gamma)},$$

where $\alpha_\phi(\gamma)$ is a *unitary cocycle*; i.e., for each ϕ, $\alpha_\phi(\gamma) : \mathcal{M} \to \mathcal{M}$ is a unitary-operator-valued function defined almost everywhere on Δ, and $\forall \phi_1, \phi_2 \in$ Diff(M) the cocycle equation

$$\alpha_{\phi_1}(\gamma)\alpha_{\phi_2}(\phi_1\gamma) = \alpha_{\phi_1\phi_2}(\gamma)$$

holds almost everywhere. Under the right technical conditions, $V(\phi)$ defines a CUR of Diff(M), from which (34) yields a self-adjoint representation of Vect(M). When Δ is a homogeneous space for Diff(M) obtained as the quotient-space with respect to a closed subgroup H, various inequivalent cocycles can be obtained by *inducing* from CUR's of H, as in the examples which follow.

Some unitary representations of Diff(M)

We shall describe briefly several examples of CUR's of Diff(M).

Ex. 1. A Single Particle in M. Let $\Delta = M$, let $\mathcal{M} = \mathbb{C}$, let $\alpha_\phi \equiv 1$, let μ be a measure on M locally equivalent to Lebesgue measure, and let $\phi : M \to M$ be given by $(\phi, \boldsymbol{x}) \to \phi^{-1}(\boldsymbol{x})$. Then $\frac{d\mu_\phi}{d\mu}(\boldsymbol{x}) = \mathcal{J}_{\phi^{-1}}(\boldsymbol{x})$, the Jacobian of ϕ^{-1} at $\boldsymbol{x}$; and we have

$$V(\phi)\Psi(\boldsymbol{x}) = \Psi(\phi^{-1}(\boldsymbol{x}))\sqrt{\mathcal{J}_{\phi^{-1}}(\boldsymbol{x})}. \qquad (35)$$

This is the representation of Diff(M) describing the quantum mechanics of a single particle in M.

Ex. 2. Finitely Many Identical Bosons or Fermions. Write M^N for the Cartesian product $M \times M \times \ldots \times M$ of N copies of M, and denote by $\mathcal{D} \subset M^N$ the "diagonal" set of all N-tuples $(\boldsymbol{x}_1, \ldots, \boldsymbol{x}_N)$ such that $\boldsymbol{x}_i = \boldsymbol{x}_j$ for some $i \neq j$. Let Δ be the set of all subsets of M having exactly N distinct elements. Then Δ can be identified with $(M^N - \mathcal{D})/S_N$, where S_N is the symmetric group. The group Diff(M) acts (transitively) on Δ in the natural way, generalizing Example 1: for $\gamma \in \Delta$, $\gamma = \{\boldsymbol{x}_1, \ldots, \boldsymbol{x}_N\}$, we define $\phi\gamma = \{\phi^{-1}(\boldsymbol{x}_1), \ldots, \phi^{-1}(\boldsymbol{x}_N)\}$. Choosing μ on Δ locally equivalent to the product of Lebesgue-equivalent measures on M, it is quasi-invariant for the action of Diff(M) on Δ.

For a particular $\gamma_0 \in \Delta$, consider next the *stability subgroup* $K_{\gamma_0} \subset \mathrm{Diff}(M)$ consisting of all $\phi \in \mathrm{Diff}(M)$ such that $\phi\gamma_0 = \gamma_0$. Any such diffeomorphism must effect a permutation of the points $\boldsymbol{x}_j$ comprising γ_0, and we thus have a natural homomorphism h_{γ_0} from K_{γ_0} to S_N. In the case $M = \mathbb{R}^3$, S_N is in fact just the *fundamental group* $\pi_1(\Delta)$, consisting of homotopy classes of loops in Δ based at γ_0. It is of course well-known that the two one-dimensional unitary representations of S_N describe particle statistics: the trivial representation $T^B(\sigma) \equiv 1$ for $\sigma \in S_N$ describing bosons, and the alternating representation $T^F(\sigma) = \pm 1$ (in accordance with whether σ is the product of an even or an odd number of exchanges) describing fermions. Now if we take $\mathcal{M} = \mathbb{C}$, there exist two cohomology classes of cocycles, $\alpha_\phi^B(\gamma)$ and $\alpha_\phi^F(\gamma)$, satisfying appropriate technical assumptions (of nontrivial complexity), such that for $\phi \in K_{\gamma_0}$, $\alpha_\phi^B(\gamma_0) = T^B(h_{\gamma_0}(\phi))$ and $\alpha_\phi^F(\gamma_0) = T^F(h_{\gamma_0}(\phi))$; and where (loosely speaking, because of issues having to do with measure-zero sets) a corresponding rule holds for almost any choice of γ_0. We then obtain the induced CUR's V^B and V^F of $\mathrm{Diff}(M)$, equivalent respectively to the symmetric and antisymmetric tensor products of N copies of the CUR given by (35).

Thus representations of $\mathrm{Diff}(M)$ describe N bosons or N fermions in M [53].

Ex. 3. Anyons in $\mathbb{R}^2$. Suppose in Example 2 we consider specifically the two-dimensional case $M = \mathbb{R}^2$. The configuration space Δ now has a fundamental group $\pi_1(\Delta)$ larger than S_N; in fact, $\pi_1(\Delta) = B_N$, the *braid group* for N objects. If $\phi \in K_{\gamma_0}$, we can associate ϕ with a braid by considering a *path* ϕ_t in $\mathrm{Diff}(\mathbb{R}^2)$ that travels from the identity to ϕ; i.e., a continuous family of diffeomorphisms, not necessarily a flow, parameterized by $t \in [0,1]$), such that $\phi_{t=0}$ is the identity in $\mathrm{Diff}(\mathbb{R}^2)$, and $\phi_{t=1} = \phi$. The action of ϕ_t on the points comprising γ_0 gives an element of B_N. In this way we obtain a homomorphism h_{γ_0} from K_{γ_0} to B_N, rather than simply to S_N.

Now a one-dimensional unitary representation of B_N is given by $T^\theta(b) = e^{i\theta}$, where $b \in B_N$ is the braid consisting of the simple counterclockwise exchange of two points. Composing T^θ with h_{γ_0} gives us a CUR of K_{γ_0}. Again, there exist classes of cocycles corresponding to the representations T^θ, and corresponding CUR's of $\mathrm{Diff}(\mathbb{R}^2)$. When $\theta = 0$ we recover Bose statistics, and when $\theta = \pi$ we have Fermi statistics; but for intermediate values of θ, we have something new – the statistics of anyons in two-dimensional space, intermediate between bosons and fermions. Particles or excitations obeying such statistics, anticipated by Finkelstein and Rubinstein [59], were introduced explicitly by Leinaas and Myrheim [23–24] based on considerations of configuration-space topology; the diffeomorphism group representations outlined here led to many of their important physical properties [20–21,53,60], and removed the need for extra assumptions about the configuration space. Further development of the theory of anyons, including a conjectured relationship with fractional spin in two dimensions, dates from the work of Wilczek [25–26]; for recent reviews of their origins, properties and applications, see [61–62]. Replacing $\mathbb{R}^2$ by a different two-dimensional manifold (e.g. the two-sphere S^2) leads to variations of the braid statistics [63]. Mathematically, the key step is to be able to lift the action of $\mathrm{Diff}(M)$ (where dim $M = 2$) from Δ to the universal covering space of Δ.

Ex. 4. Paraparticles. Higher-dimensional unitary representations of S_N or B_N also induce representations of Diff(M) describing quantum systems. In Example 2, set $\mathcal{M}$ equal to $\mathbb{C}^n$ instead of $\mathbb{C}$, and replace T by an irreducible unitary representation of S_N of dimension $n > 1$. We can then take $\alpha_\phi(\gamma)$ to be a unitary cocycle acting in $\mathcal{M}$, with $\alpha_\phi(\gamma_0) = T(h_{\gamma_0}(\phi))$ for $\phi \in K_{\gamma_0}$. Thus we obtain CUR's of Diff(M) describing paraparticles [64]. In Example 3, we can similarly substitute for T^θ a higher-dimensional unitary representation of B_N, and obtain "para-braid statistics" [64–65].

Ex. 5. Towers of Spins. Finally, we indicate how particle spin can be described by means of diffeomorphism group representations [32]. This is most easily illustrated in the case of one particle, by setting $M = \mathbb{R}^3$ in *Ex. 1*. Let GL$^+(3, \mathbb{R})$ be the *general linear group* of 3×3 matrices with positive determinant. For $\boldsymbol{x}_0 \in \mathbb{R}^3$, let $K_{\boldsymbol{x}_0}$ denote the stability subgroup of Diff($\mathbb{R}^3$) at $\boldsymbol{x}_0$. Define $h_{\boldsymbol{x}_0}$: $K_{\boldsymbol{x}_0} \to$ GL$^+(3, \mathbb{R})$ by setting $[h_{\boldsymbol{x}_0}(\phi)]^k_j = \partial_j \phi^k(\boldsymbol{x}_0)$, the matrix of derivatives of ϕ at $\boldsymbol{x}_0$. We denote this matrix by $\mathcal{D}_\phi(\boldsymbol{x}_0)$. Then $h_{\boldsymbol{x}_0}$ defines a continuous group homomorphism.

Next we define a related homomorphism from $K_{\boldsymbol{x}_0}$ to the universal covering group $\overline{\text{GL}^+}(3, \mathbb{R})$ of GL$^+(3, \mathbb{R})$. Let Γ be a path in $\mathbb{R}^3$ from infinity to $\boldsymbol{x}_0$. As $\boldsymbol{x}$ moves along Γ, for $\phi \in K_{\boldsymbol{x}_0}$, consider $h_{\boldsymbol{x}}(\phi) = [\partial_j \phi^k(\boldsymbol{x})] = \mathcal{D}_\phi(\boldsymbol{x}) \in$ GL$^+(3, \mathbb{R})$. This defines a path in GL$^+(3, \mathbb{R})$ from the identity matrix to $\mathcal{D}_\phi(\boldsymbol{x}_0)$, and thus an element of $\overline{\text{GL}^+}(3, \mathbb{R})$ associated with ϕ. The result is a well-defined homomorphism $h'_{\boldsymbol{x}_0}$, independent of the particular choice of Γ.

Now let T be an irreducible CUR of $\overline{\text{GL}^+}(3, \mathbb{R})$ in $\mathcal{M}$, which we take to be infinite-dimensional. Again T composed with $h'_{\boldsymbol{x}_0}$ gives a CUR of the stability group $K_{\boldsymbol{x}_0}$, inducing a representation of Diff($\mathbb{R}^3$). However such a representation does not describe a single spin, but a *tower* of spins. The Iwasawa decomposition of $\overline{\text{GL}^+}(3, \mathbb{R})$ with respect to its maximal compact subgroup SU(2) permits a CUR of $\overline{\text{GL}^+}(3, \mathbb{R})$ to be described by an infinite collection of representations of SU(2), having integer or half-integer spin (but not both in the same irreducible representation of $\overline{\text{GL}^+}(3, \mathbb{R})$), and occurring with various multiplicities.

The general expression for the self-adjoint current density acting in $L^2_\mu(\mathbb{R}^3, \mathcal{M})$ becomes

$$J(g) = J_0(g)I + c(\nabla \cdot g)I + \frac{1}{2}(\nabla \times g) \cdot \Sigma + \frac{1}{2} \sum_{m=-2}^{2} G_{-m}(g)T_m, \qquad (36)$$

where each term has a physical meaning:

(a) $J_0(g)I$ is the orbital (kinetic) momentum, as in *Ex. 1*;

(b) $c \in \mathbb{R}$ is a probability diffusion coefficient, and the second term is a diffusion current of interest in the foundations of nonlinear quantum mechanics, obtained independently by Doebner and Tolar and Goldin, Menikoff, and Sharp [10,66–67]: setting $c = 0$, we still have a CUR induced by the covering group SL$^+(3, \mathbb{R})$ of the *special* linear group;

(c) Σ is a vector of the SU(2) generators in $\overline{GL^+}(3, \mathbb{R})$, and $\frac{1}{2}(\nabla \times g) \cdot \Sigma$ is the spin contribution to the current density;

(d) T_m ($m = -2, -1, 0, 1, 2$) are a basis of spin-changing generators of $\overline{GL^+}(3, \mathbb{R})$, and the $G_{-m}(g)$ are corresponding functional coefficients, so that the last term of (36) is the spin-changing current; this term cannot be set separately equal to zero without also setting the spin generators to zero.

It has been suggested [22,32] that such representations might have application to supermultiplets of hadrons lying on Regge trajectories, to excited states of nuclei [68], or to particles in a strong, nonuniform gravitational field.

In the above examples, a common feature is that the CUR of the diffeomorphism group is obtained by inducing from a representation of a discrete or locally compact group such as S_N, B_N, or $GL^+(3, \mathbb{R})$; these smaller groups play the role of *gauge groups* for the theory (they are structure groups for fiber bundles over Δ). Many other specific representations of diffeomorphism groups have been studied, including important cases where Δ is not a single orbit. However we omit their description here, since they are not as immediately relevant to the issue of coherent-states quantization.

In order to make contact with the theory of coherent states, we will need homogeneous spaces for Diff(M) with respect to which CUR's such as the above are square integrable. For this purpose we propose to make use of *coadjoint orbits* of Diff(M), which are also of interest in geometric quantization [56] and in the quantization of vorticity [69].

A coadjoint orbit

In the rest of this section, we briefly sketch how the single-particle representation written above is obtained through the coadjoint-orbit construction of representations of Diff(M). We take as our example the case $M = \mathbb{R}$. This will lay the groundwork for the final example in Sect. 5, in which we demonstrate this representation to be square integrable.

The adjoint representation of Vect(M) is defined for $g \in$ Vect(M) by $(\mathrm{Ad}\,g)h = [g, h]$, $h \in$ Vect(M). Exponentiating this representation from g to ϕ_t^g, we obtain the adjoint representation of Diff(M) in Vect(M),

$$(\mathrm{Ad}\,\phi)(h) = (\mathcal{D}_{\phi^{-1}}h) \circ \phi.$$

Now, the coadjoint representation acts on the dual space $\mathcal{V}'$ of Vect(M), the space of continuous linear functionals from Vect(M) to $\mathbb{R}$. Elements of $\mathcal{V}'$ are *generalized vector fields* on M; i.e., vector fields with components in the space $\mathcal{S}'$ of tempered distributions. For $A \in \mathcal{V}'$, we write $\langle A, g \rangle$ to denote the value of the functional A on the vector field g. The coadjoint representation of Diff(M) is then defined by

$$\langle A', h \rangle = \langle A, \mathrm{Ad}\,(\phi^{-1})h \rangle,$$

where $A' = (\mathrm{Coad}\,\phi)A,$.

For $M = \mathbb{R}$, the most elementary coadjoint orbit of $\mathrm{Diff}(\mathbb{R})$ can now be obtained by considering the generalized vector field $A_{(x,\lambda)} = \lambda\delta_x$, where δ_x is the functional that *evaluates* a function or vector field on $\mathbb{R}$ at the point $x \in \mathbb{R}$, and λ is a real coefficient; i.e., $\langle A_{(x,\lambda)}, g \rangle = \lambda g(x)$. Then we have

$$\left. \begin{aligned} \langle A', g \rangle &= \langle A_{(x,\lambda)}, (\mathcal{D}_\phi g) \circ \phi^{-1} \rangle \\ &= \lambda\phi'(\phi^{-1}(x))g(\phi^{-1}(x)), \end{aligned} \right\} \tag{37}$$

whence

$$A' = A_{(\phi^{-1}(x), [(\phi^{-1})'(x)]^{-1}\lambda)} \,. \tag{38}$$

Note that the action of $(\mathrm{Coad}\ \phi)$ given by (37) or (38) preserves the sign of λ, since $(\phi^{-1})'(x) > 0$. Thus (taking the case $\lambda > 0$), an element of the coadjoint orbit Γ is a pair (x, λ), where $x \in \mathbb{R}$ and $\lambda \in \mathbb{R}^+$, and a diffeomorphism ϕ acts on Γ in the coadjoint representation by

$$\begin{aligned} x' &= \phi^{-1}(x), \\ \lambda' &= \frac{1}{(\phi^{-1})'(x)}\lambda \,. \end{aligned}$$

The coadjoint orbit serves as a *reduced classical phase space*, on which quantization is now to take place. Note that Γ is here a submanifold of the cotangent bundle $T^*(\mathbb{R})$; for $(x, \lambda) \in \Gamma$, we think of $g(x)$ as an element of the tangent space $T_x(\mathbb{R})$, and λ as an element of the cotangent space $T_x^*(\mathbb{R})$.

The stability subgroup $K_{(x_0,\lambda_0)} \subset \mathrm{Diff}(\mathbb{R})$ associated with (x_0, λ_0) is now $\{\phi \mid \phi^{-1}(x_0) = x_0 \text{ and } (\phi^{-1})'(x_0) = 1\}$, which has codimension 2 in $\mathrm{Diff}(\mathbb{R})$. In the geometric quantization program, the next step is to choose a *polarization*; i.e., a subgroup H of $\mathrm{Diff}(\mathbb{R})$ intermediate between $K_{(x_0,\lambda_0)}$ and $\mathrm{Diff}(\mathbb{R})$, whose codimension is half that of $K_{(x_0,\lambda_0)}$. Here, we let $H = \{\phi \mid \phi^{-1}(x_0) = x_0\}$, relaxing the condition on $(\phi^{-1})'(x_0)$; H thus has codimension 1, and the quantum configuration space is just $\Delta = \mathrm{Diff}(\mathbb{R})/H \cong \mathbb{R}$. We think of the polarization as determining a *foliation* of Γ, and Δ as the set of *leaves* in the foliation.

The Lie algebra of K is the set of vector fields g on $\mathbb{R}$ for which both $g(x_0) = 0$ and $g'(x_0) = 0$, while the Lie algebra of H contains all vector fields vanishing at x_0, with no condition on the derivative. Now the generalized vector field $(x_0, \lambda_0) \in \mathcal{V}'$, restricted to the Lie algebra of H, is identically zero. Thus H satisfies trivially the important condition that (x_0, λ_0) vanish on the bracket of two elements in the Lie algebra of H, and (x_0, λ_0) exponentiates to yield a *character* of H – which, in this case, is identically one. This character in turn induces the CUR of $\mathrm{Diff}(\mathbb{R})$ given in *Ex. 1* above.

The question we shall want to ask, then, is whether coherent-states quantization can be applied to such a coadjoint orbit. For this we shall need the representation (35) to be square integrable with respect to Γ, as a homogeneous space for the diffeomorphism group, *modulo* an appropriate Borel section.

5 Quantization Using Coherent States

Drawing upon the simple example given in Sect. 2, we can now develop a general quantization procedure on phase space, using coherent states. Later in this section we shall give specific examples, including the case where the coadjoint orbit described in Sect. 4 is taken to be the phase space.

Assume, as before, that the classical algebra of observables $\mathcal{A}_{j\ell}$ is the C^*-algebra of all complex, continuous functions on the phase space Γ, which vanish at infinity. The phase space itself is to be identified with the cotangent bundle

$$\Gamma = T^*(M),$$

of a manifold M representing the position space of the system. Γ has a natural symplectic structure with a volume form which yields a measure. This is the measure which we take for ν. Next we construct the Hilbert space $\tilde{\mathcal{H}} = L_\nu^2(\Gamma, \mathbb{C}^n)$ and embed $\mathcal{A}_{j\ell}$ into the set of multiplication operators on $\tilde{\mathcal{H}}$. Thus, $f \in \mathcal{A}_{j\ell}$ is mapped to $\tilde{P}(f)$, where

$$(\tilde{P}(f)\tilde{\Psi})(\zeta) = f(\zeta)\tilde{\Psi}(\zeta)$$

($\forall \tilde{\Psi} \in \tilde{\mathcal{H}}$) and for almost all (with respect to ν) $\zeta \in \Gamma$. In a standard manner [41-42], the domain of the algebraic homomorphism $\tilde{P}$ may be extended to the set $L_\nu^\infty(\Gamma)$ of all bounded ν-measurable functions on Γ, and its range to a commutative von Neumann algebra. In that case, for any ν-measurable function $f \in L_\nu^\infty(\Gamma)$,

$$\tilde{P}(f) = \int_\Gamma f(\zeta)\, \mathrm{d}\tilde{P}(\zeta),$$

where the integration is with respect to the projection-valued measure $\tilde{P}$ defined for $B \in \mathcal{B}(\Gamma)$, the family of Borel sets of Γ, by

$$(\tilde{P}(B))\tilde{\Psi})(\zeta) = \chi_B(\zeta)\tilde{\Psi}(\zeta);$$
$$\chi_B(\zeta) = \begin{cases} 1 & \text{if } \zeta \in B, \\ 0 & \text{otherwise.} \end{cases}$$

A *(prime) quantization of the classical algebra* $\mathcal{A}_{j\ell}$ is then defined to be a positive linear map

$$\pi_K^* : L_\nu^\infty(\Gamma) \longrightarrow \mathcal{L}(\mathcal{H}),$$

where $\mathcal{L}(\mathcal{H})$ is the set of all bounded linear operators on the Hilbert space $\mathcal{H}$, satisfying: (i) the C^*-algebra generated by the set

$$\pi^*(\mathcal{A}_{j\ell}) = \{\pi^*(\{)|\{ \in \mathcal{A}_{j\ell}\}$$

is weakly dense in $\mathcal{L}(\mathcal{H})$; and (ii) $\mathcal{H} = \mathcal{H}_K$ is a reproducing-kernel Hilbert subspace of $\tilde{\mathcal{H}}$. It follows from the general theory of reproducing-kernel Hilbert spaces [15–16,50] that $\mathcal{H} = \mathcal{H}_K \subset L_\nu^2(\Gamma, \mathbb{C}^n)$ has a canonically associated POV-measure a_K, defined on the Borel sets of Γ, such that ($\forall B \in \mathcal{B}(\Gamma)$)

$$a_K(B) = \int_B F(\zeta)\, \mathrm{d}\nu(\zeta),$$

where
$$a_K(\Gamma) = I,$$

and $(\forall \zeta \in \Gamma)$ $F(\zeta)$ is a positive, self-adjoint, rank-n operator. Moreover, in terms of a_K, the mapping,

$$\pi_K^*(f) = \mathbb{P}_K \tilde{P}(f) \mathbb{P}_K$$
$$= \int_\Gamma f(\zeta) \, da_K(\zeta), \qquad (39)$$

where $\mathbb{P}_K$ is the projection operator

$$\mathbb{P}_K \tilde{\mathcal{H}} = \mathcal{H}_K,$$

gives a prime quantization if Γ does not contain a part consisting of discrete points.

If the phase space Γ arises as a homogeneous space of a symmetry group G, for example if Γ is a coadjoint orbit [70] of G, then a quantization of $\mathcal{C}_\infty(\Gamma)$ would, in view of the discussion in Sect. 3, amount to looking for representations of G which are square integrable with respect to Γ. We illustrate this point with some examples.

Free non-relativistic particle of mass $m > 0$

The kinematical symmetry group is the extended Galilei group $\tilde{G}$ [71]. This is an 11-parameter Lie group, with general element

$$g = (\theta, b, \boldsymbol{a}, \boldsymbol{v}, R),$$

consisting of a phase term $\theta \in \mathbb{R}$, a time translation $b \in \mathbb{R}$, a space translation $\boldsymbol{a} \in \mathbb{R}^3$, a velocity boost $\boldsymbol{v} \in \mathbb{R}^3$ and a spatial rotation R having three real parameters. The product rule is

$$g_1 g_2 = (\theta_1, b_1, \boldsymbol{a}_1, \boldsymbol{v}_1, R_1)(\theta_2, b_2, \boldsymbol{a}_2, \boldsymbol{v}_2, R_2)$$
$$= (\theta_1 + \theta_2 + \omega(g_1, g_2), b_1 + b_2, \boldsymbol{a}_1 + R_1 \boldsymbol{a}_2 + \boldsymbol{v}_1 b_2, \boldsymbol{v}_1 + R_1 \boldsymbol{v}_2, R_1 R_2),$$

where
$$\omega(g_1, g_2) = m[\tfrac{1}{2} \boldsymbol{v}_1^2 b_2 + \boldsymbol{v}_1 \cdot R_1 \boldsymbol{a}_2].$$

Consider the subgroup $\Theta \otimes T \otimes SO(3)$ of $\tilde{G}$, consisting of all phase and time translations and spatial rotations. An element $g \in \tilde{G}$ has a coset decomposition of the form:

$$g = (\theta, b, \boldsymbol{a}, \boldsymbol{v}, R)$$
$$= (0, 0, \boldsymbol{a} - \boldsymbol{v}b, \boldsymbol{v}, I)(\theta, b, \boldsymbol{0}, \boldsymbol{0}, R),$$

corresponding to the quotient $\tilde{G}/\Theta \otimes T \otimes SO(3)$. Thus the homogeneous space

$$\Gamma = \tilde{G}/\Theta \otimes T \otimes SO(3)$$

has the global coordinatization, $(q, p) \in \mathbb{R}^6$, in terms of which the *invariant* measure is $dq\,dp$. We identify $\Gamma \cong T^*(\mathbb{R}^3)$ with the phase space of the system (it is a coadjoint orbit of $\tilde{G}$ [49]) and consider the section

$$\left. \begin{array}{l} \beta : \Gamma \longrightarrow \tilde{G}, \\ \beta(q, p) = (0, 0, q, \frac{p}{m}, I). \end{array} \right\} \tag{40}$$

The classical algebra of observables $\mathcal{A}_{|\ell}$, is the C^*-algebra of phase-space functions $f : \Gamma \longrightarrow \mathbb{C}$.

The Hilbert space $\tilde{\mathcal{H}} = L^2_{dq\,dp}(\Gamma)$ has reproducing kernel subspaces constructed as follows [16,72]: Let ϵ be a square-integrable, rotationally-invariant function on $\mathbb{R}^3$, normalized to unity; i.e.

$$\int_{\mathbb{R}^3} |\epsilon(\boldsymbol{k})|^2 \, d\boldsymbol{k} = 1, \tag{41}$$

$$\epsilon(R\boldsymbol{k}) = \epsilon(\boldsymbol{k}), \quad R \in \mathrm{SO}(3),$$

the latter equation holding for almost all $\boldsymbol{k} \in \mathbb{R}^3$. For $\ell = 0, 1, 2, 3, \ldots$, define the kernel $K_{e,\ell} : \Gamma \times \Gamma \longrightarrow \mathbb{C}$ by

$$K_{e,\ell}(q, p; q', p') = \frac{2\ell + 1}{(2\pi)^3} \int_{\mathbb{R}^3} \exp\left[i\boldsymbol{k} \cdot (q - q')\right] \mathcal{P}_\ell \left[\frac{(\boldsymbol{k} - \boldsymbol{p}) \cdot (\boldsymbol{k} - \boldsymbol{p}')}{\|\boldsymbol{k} - \boldsymbol{p}\| \|\boldsymbol{k} - \boldsymbol{p}'\|}\right]$$
$$\times \overline{\epsilon(\boldsymbol{k} - \boldsymbol{p})} \epsilon(\boldsymbol{k} - \boldsymbol{p}') \, d\boldsymbol{k}, \tag{42}$$

where $\mathcal{P}_\ell$ is the Legendre polynomial of order ℓ. It is straightforward to check that $K_{e,\ell}$ satisfies all the properties (27)–(29) of a reproducing kernel. Let $\mathbb{P}_{e,\ell}$ be the associated projection operator

$$(\mathbb{P}_{e,\ell}\tilde{\Psi})(q, p) = \int_{\Gamma} K_{e,\ell}(q, p; q', p') \tilde{\Psi}(q', p') \, dq' dp', \tag{43}$$

where $\tilde{\Psi} \in \tilde{\mathcal{H}}$, and let $\mathcal{H}_{e,\ell}$ be the subspace of $\tilde{\mathcal{H}}$,

$$\mathcal{H}_{e,\ell} = \mathbb{P}_{e,\ell}\tilde{\mathcal{H}}.$$

It is clear from (41) and (42) that the projected functions in $\mathcal{H}_{e,\ell}$,

$$\Psi_{e,\ell} = \mathbb{P}_{e,\ell}\tilde{\Psi} \quad (\tilde{\Psi} \in \tilde{\mathcal{H}})$$

are continuous in the variables (q, p).

A prime quantization of the classical algebra can now be carried out as outlined earlier in this section. For $f \in \mathcal{A}_{|\ell}$, we define the operator $\tilde{P}(f)$ on $\tilde{\mathcal{H}}$ as,

$$(\tilde{P}(f)\tilde{\Psi})(q, p) = f(q, p)\tilde{\Psi}(q, p).$$

The quantized version $\pi^*_{e,\ell}(f)$ of f is then the operator

$$\pi^*_{e,\ell}(f) = \mathbb{P}_{e,\ell}\tilde{P}(f)\mathbb{P}_{e,\ell}, \tag{44}$$

on $\mathcal{H}_{e,\ell}$. In particular, if f^i_{pos} and f^i_{mom}, $i = 1, 2, 3$, are the classical position and momentum observables (see (5)),

$$
\left.\begin{array}{l}
f^i_{\text{pos}}(\boldsymbol{q},\boldsymbol{p}) = q^i\,, \\
f^i_{\text{mom}}(\boldsymbol{q},\boldsymbol{p}) = p^i\,,
\end{array}\right\} \tag{45}
$$

then an application of (44), using (43), yields

$$
Q^i \equiv \pi^*_{e,\ell}(f^i_{\text{pos}}) = q^i + i\frac{\partial}{\partial p^i}\,,
$$

$$
P^i \equiv \pi^*_{e,\ell}(f^i_{\text{mom}}) = -i\frac{\partial}{\partial q^i}\,,
$$

so that indeed,

$$
[Q^k, P^j] = i\delta_{kj}\,, \quad k,j = 1,2,3\,.
$$

Let us point out at this stage the important relationship between this quantization map

$$
\pi^*_{e,\ell} : \mathcal{A}_{|\ell} \longrightarrow \mathcal{L}(\mathcal{H}_{\mathbb{1},\ell})\,,
$$

and the square integrability of a certain representation of $\tilde{G}$. As shown in [72], $\mathcal{H}_{e,\ell}$ carries a unitary irreducible representation of $\tilde{G}$ (corresponding to a particle of mass m and spin ℓ) given by the unitary operators $U_{e,\ell}(g)$ on $\mathcal{H}_{e,\ell}$, for $g = (\theta, b, \boldsymbol{a}, \boldsymbol{v}, R) \in \tilde{G}$:

$$
\left.\begin{array}{l}
(U_{e,\ell}(g)\Psi)(\boldsymbol{q},\boldsymbol{p}) = \exp\left[i\{\theta + (P^2/2m)b + m\boldsymbol{v}\cdot(\boldsymbol{q} - \boldsymbol{a})\}\right] \\
\qquad\qquad \times \Psi(R^{-1}(\boldsymbol{q} - \boldsymbol{a}), R^{-1}(\boldsymbol{p} - m\boldsymbol{v}))\,,
\end{array}\right\} \tag{46}
$$

where $P^2 = -\nabla^2_q$, for $\Psi \in \mathcal{H}_{e,\ell}$. Using the Borel section (40) and the vector $\eta^{e,\ell} \in \mathcal{H}_{e,\ell}$ given by

$$
\eta^{e,\ell}(\boldsymbol{q},\boldsymbol{p}) = K_{e,\ell}(\boldsymbol{q},\boldsymbol{p}; \boldsymbol{0},\boldsymbol{0})\,,
$$

we can generate the family of coherent states (see (22)–(23)),

$$
\mathcal{S}_{e,\ell} = \{\eta^{e,\ell}_{\beta(\boldsymbol{q},\boldsymbol{p})} = U_{e,\ell}(\beta(\boldsymbol{q},\boldsymbol{p}))\eta^{e,\ell} \,|\, (\boldsymbol{q},\boldsymbol{p}) \in \Gamma\}\,.
$$

It is then straightforward to verify that

$$
\int_\Gamma |\eta^{e,\ell}_{\beta(\boldsymbol{q},\boldsymbol{p})}\rangle\langle\eta^{e,\ell}_{\beta(\boldsymbol{q},\boldsymbol{p})}|\, \mathrm{d}\boldsymbol{q}\, \mathrm{d}\boldsymbol{p} = \mathbb{P}_{e,\ell}\,,
$$

which should be compared with (24).

Thus, the representation $U_{e,\ell}$ in (46) of the extended Galilei group $\tilde{G}$ is square integrable $\mathrm{mod}\,(H, \beta)$ (see Sect. 3), where H is the subgroup $\Theta \otimes T \otimes \mathrm{SO}(3)$ of $\tilde{G}$, and β is the section defined in (40).

Free spin-0 relativistic particle of mass $m > 0$, in 2-dimensional space-time

This example is based on Refs. [48], [50] and [73]. The restriction to spin 0 and 2-dimensional space-time is for neatness and notational convenience only.

The kinematical symmetry group is $\mathcal{P}_+^\uparrow(1,1)$, the Poincaré group in 1 space and 1 time dimension. Its elements g consist of a space-time translation a and a Lorentz boost Λ,

$$g = (a, \Lambda),$$

with

$$a = (a^0, \boldsymbol{a}) \in \mathbb{R}^2,$$

$$\Lambda \equiv \Lambda_p = \frac{1}{m} \begin{pmatrix} p^0 & \boldsymbol{p} \\ \boldsymbol{p} & p^0 \end{pmatrix}, \qquad m > 0, \quad p \in \mathcal{V}_m^+.$$

We can also write a^1 for $\boldsymbol{a}$, p^1 for $\boldsymbol{p}$, etc. Here

$$\mathcal{V}_m^+ = \{(p^0, \boldsymbol{p}) \in \mathbb{R}^2 \,|\, p^0 > 0, \quad (p^0)^2 - \boldsymbol{p}^2 = m^2\}$$

is the *forward mass hyperbola*. The group multiplication law is

$$\begin{aligned} g_1 g_2 &= (a_1, \Lambda_1)(a_2, \Lambda_2) \\ &= (a_1 + \Lambda_1 a_2, \Lambda_1 \Lambda_2). \end{aligned}$$

Furthermore, the elements Λ_p of the Lorentz group act on $\mathcal{V}_m^+$ in the natural manner,

$$k \longrightarrow k' = \Lambda_p k, \qquad k \in \mathcal{V}_m^+.$$

This action is transitive, and the measure $d\mu = d\boldsymbol{k}/k^0$ is invariant under it.

Consider next the subgroup T of time translations,

$$T = \{g \in \mathcal{P}_+^\uparrow(1,1) \,|\, g = ((a^0, \boldsymbol{0}), I_2), a^0 \in \mathbb{R}\},$$

where I_2 is the 2×2 identity matrix. The quotient space

$$\Gamma = \mathcal{P}_+^\uparrow(1,1)/T$$

can be identified with a coadjoint orbit [49,74] having an invariant measure; the orbit has a global parameterization $(\boldsymbol{q}, \boldsymbol{p}) \in \mathbb{R}^2$, in terms of which the invariant measure is $d\boldsymbol{q}\, d\boldsymbol{p}$. Let us also define a Borel section

$$\left. \begin{aligned} \beta_0 &: \Gamma \longrightarrow \mathcal{P}_+^\uparrow(1,1), \\ \beta_0(\boldsymbol{q}, \boldsymbol{p}) &= ((0, \boldsymbol{q}), \Lambda_p), \quad p = (\sqrt{\boldsymbol{p}^2 + m^2}, \boldsymbol{p}). \end{aligned} \right\} \tag{47}$$

Any other section $\beta : \Gamma \longrightarrow \mathcal{P}_+^\uparrow(1,1)$ is related to β_0 in the manner

$$\beta(\boldsymbol{q}, \boldsymbol{p}) = \beta_0(\boldsymbol{q}, \boldsymbol{p})((f(\boldsymbol{q}, \boldsymbol{p}), 0), I_2), \tag{48}$$

where $f : \Gamma \longrightarrow \mathbb{R}$ is a Borel function. Actually, it will be useful for our purposes to restrict β to the class of *affine sections*, for which $f(\boldsymbol{q}, \boldsymbol{p})$ assumes the form

$$f(\boldsymbol{q}, \boldsymbol{p}) = \phi(\boldsymbol{p}) + \boldsymbol{q} \cdot \theta(\boldsymbol{p}), \tag{49}$$

ϕ being an arbitrary real-valued Borel function of p and θ a Borel function satisfying

$$\left| \theta(p) - \frac{p}{m} \right| < \frac{p^0}{m} . \tag{50}$$

The condition (50) ensures that the space-time translation part of $\beta(q, p)$ is a spacelike vector [50].

Suppose now that we take Γ to be the phase space for our classical, relativistic particle of mass m and spin 0. As before, the classical algebra of observables consists of functions on Γ. To quantize it we have to look for representations of $\mathcal{P}_+^\uparrow(1,1)$ which are square integrable $\mathrm{mod}\,(T, \beta)$, where β needs to be specified. Let $\mathcal{H}_W$ be the Hilbert space

$$\mathcal{H}_W = L_\mu^2(V_m^+)$$

with $\mathrm{d}\mu = \mathrm{d}k/k^0$, and on it consider the unitary irreducible representation of $\mathcal{P}_+^\uparrow(1,1)$ given by the operators $U(a, \Lambda)$ [48]:

$$(U(a, \Lambda_p)\Psi)(k) = e^{ik \cdot a} \Psi(\Lambda_p^{-1}k), \qquad \Psi \in \mathcal{H}_W,$$
$$k \cdot a = k^0 a^0 - \mathbf{k} \cdot \mathbf{a}.$$

This representation is square integrable mod (T, β), for any β in the affine class [50]; i.e., satisfying (48)- -(50). Indeed, if $\eta \in \mathcal{H}_W$ is any vector which is in the domain of the operator $(P^0)^{\frac{1}{2}}$, where

$$(P^0\Psi)(k) = k^0 \Psi(k)$$

(note that this domain is dense in $\mathcal{H}_W$), then η can be used to construct a family of coherent states

$$\mathcal{S}_\beta = \{ \eta_{\beta(q,p)} = U(\beta(q,p))\eta \,|\, (q,p) \in \Gamma \},$$

for any affine section β. Furthermore,

$$\int_\Gamma |\eta_{\beta(q,p)}\rangle\langle\eta_{\beta(q,p)}| \, \mathrm{d}q \, \mathrm{d}p = A_\beta^\eta,$$

where A_β^η is the positive, bounded, invertible operator

$$(A_\beta^\eta\Psi)(k) = A_\beta^\eta(k)\Psi(k),$$
$$A_\beta^\eta(k) = \int_{V_m^+} \mathcal{A}_\beta(k, p) |\eta(p)|^2 \, \frac{\mathrm{d}p}{p^0},$$
$$\mathcal{A}_\beta(k, p) = 2\pi \frac{(\Lambda_k^{-1}p)^0}{k^0 - \theta[(\Lambda_p^{-1}k)^1] \cdot p} .$$

In the particular case where $\beta = \beta_0$ given by (47), and η satisfies the normalization condition

$$\langle \eta | P^0 \eta \rangle = \frac{m}{2\pi} ,$$

one gets

$$\int_\Gamma |\eta_{\beta_0(q,p)}\rangle\langle\eta_{\beta_0(q,p)}| \, \mathrm{d}q \, \mathrm{d}p = I.$$

Hence the Hilbert space $\tilde{\mathcal{H}} = L^2_{\mathrm{d}\boldsymbol{q}\,\mathrm{d}\boldsymbol{p}}(\Gamma)$ has reproducing kernel subspaces

$$\mathcal{H}_\eta = \mathbb{P}_\eta \tilde{\mathcal{H}},$$

where the projection operator $\mathbb{P}_\eta$ has the reproducing kernel (see (26) and (43))

$$K_\eta(\boldsymbol{q},\boldsymbol{p};\boldsymbol{q}',\boldsymbol{p}') = \langle \eta_{\beta_0(\boldsymbol{q},\boldsymbol{p})} | \eta_{\beta_0(\boldsymbol{q}',\boldsymbol{p}')} \rangle \,.$$

The mapping

$$W_\eta : \mathcal{H}_W \longrightarrow \mathcal{H}_\eta,$$
$$(W_\eta \Psi)(\boldsymbol{q},\boldsymbol{p}) = \langle \eta_{\beta_0(\boldsymbol{q},\boldsymbol{p})} | \Psi \rangle \,,$$

is a Hilbert space isometry. Thus, defining

$$F_\eta(\boldsymbol{q},\boldsymbol{p}) = W_\eta | \eta_{\beta_0(\boldsymbol{q},\boldsymbol{p})} \rangle \langle \eta_{\beta_0(\boldsymbol{q},\boldsymbol{p})} | W_\eta^* \,,$$

any classical observable $f \in \mathcal{A}_{|\ell}$ can be quantized by the prescription (see (39))

$$\pi_\eta^* : \mathcal{A}_{|\ell} \longrightarrow \mathcal{L}(\mathcal{H}_\eta),$$
$$\pi_\eta^*(f) = \int_\Gamma f(\boldsymbol{q},\boldsymbol{p})\,\mathrm{d}\boldsymbol{q}\,\mathrm{d}\boldsymbol{p}\,.$$

It can then be checked that the classical observables f_{pos} and f_{mom} of position and momentum respectively (see (5) and (45)) are mapped by this quantization to the *Newton-Wigner position and momentum operators* [15].

Square-integrability of the one-particle representation of Diff($\mathbb{R}$)

In Sect. 4 we sketched several unitary representations of Diff(M). The simplest of these, with $M = \mathbb{R}$, acts in $\mathcal{H} = L^2_{\mathrm{d}x}(\mathbb{R})$ and describes a single quantum particle (see (35)):

$$V(\phi)\Psi(x) = \Psi(\phi^{-1}(x))\sqrt{\mathcal{J}_{\phi^{-1}}(x)}\,,$$
$$\mathcal{J}_\phi(x) = \frac{\mathrm{d}\phi}{\mathrm{d}x}\,.$$

We also saw how this representation could be obtained directly from a coadjoint orbit Γ of Diff($\mathbb{R}$). Here we ask about the square integrability of V; more specifically, the existence of a Borel section $\beta : \Gamma \to$ Diff($\mathbb{R}$), *modulo* which V is square integrable.

For $y \in \mathbb{R}$ and $\lambda \in \mathbb{R}^+$, we shall write $\beta(y,\lambda) \in$ Diff($\mathbb{R}$), so that the diffeomorphism $\beta(y,\lambda) : \mathbb{R} \to \mathbb{R}$ takes on values $\beta(y,\lambda)(x)$ for $x \in \mathbb{R}$. In the notation of Sect. 4, we choose $x_0 = 0$ and $\lambda_0 = 1$ (for specificity); then the stability subgroup K is just $\{\phi \mid \phi(0) = 0, \phi'(0) = 1\}$. Now for $\psi \in$ Diff($\mathbb{R}$), the coset $p(\psi) = \psi K$ is the set of all diffeomorphisms ϕ that map the particular point $y = \psi^{-1}(0)$ to 0, and for which $\phi'(y)$ has the fixed value $\lambda = \psi'(\psi^{-1}(0)) = [(\psi^{-1})'(0)]^{-1}$. Thus, the condition that β be a section requires

$$\left.\begin{aligned}\beta(y,\lambda)(y) &= 0, \\ \frac{\mathrm{d}\beta(y,\lambda)}{\mathrm{d}x}(x)\Big|_{x=y} &= \lambda,\end{aligned}\right\} \tag{51}$$

or equivalently.

$$\beta(y,\lambda)^{-1}(0) = y,$$
$$\frac{\mathrm{d}[\beta(y,\lambda)]^{-1}}{\mathrm{d}x}(0) = \frac{1}{\lambda}.$$

To satisfy (51), and to provide the properties we later need, let $\alpha(\lambda, x)$ be a real-valued function for $\lambda > 0$ and $-1 \le x \le 1$, measurable in λ and C^∞ in x, with $\alpha > 0$ and $\alpha(\lambda, 0) = \lambda$, and satisfying also

$$\int_0^{\pm 1} \alpha(\lambda, x)\,\mathrm{d}x = \pm 1 \quad (\forall \lambda). \tag{52}$$

Now, for $(y, \lambda) \in \Gamma$, define:

$$\beta(y,\lambda)(x) = \int_0^{x-y} \alpha(\lambda, x')\mathrm{d}x', \quad x \in [y-1, y+1]$$

$$\beta(y,\lambda)(x) \to x \,(\text{smoothly, rapidly}) \text{ as } |x| \to \infty, \quad x \notin [y-1, y+1]$$

so that $(\forall y, \lambda)$ $\beta(y, \lambda)$ is a C^∞ diffeomorphism of $\mathbb{R}$, with the correct behavior as a function of x at infinity, satisfying when $x \in [y-1, y+1]$,

$$\beta(y,\lambda)'(x) = \alpha(\lambda, x-y).$$

Then $\beta(y,\lambda)(x)|_{x=y} = 0$ automatically, while $\beta(y,\lambda)'(x)|_{x=y} = \alpha(\lambda, 0) = \lambda$ as desired. Furthermore, it is easy to show from (52) that

$$\beta(y,\lambda)^{-1}(1) = y + 1,$$
$$\beta(y,\lambda)^{-1}(-1) = y - 1,;$$

i.e., since $\beta(y, \lambda)$ is a strictly increasing function of x,

$$\beta(y,\lambda)^{-1} : [-1, 1] \longrightarrow [y-1, y+1]$$

is surjective.

Now we are ready to construct our family of coherent states, and explore the further properties of β which are needed for the representation V to be square integrable. For convenience, we take $\mathrm{d}\nu = \mathrm{d}y\mathrm{d}\lambda$ in Γ. Let the vector $\eta(x)$ be the indicator function $\chi_{[-1,1]}(x)$ in $\mathcal{H}$, which is 1 for $x \in [-1, 1]$ and 0 otherwise. Then for $(y, \lambda) \in \Gamma$,

$$\eta_{\beta(y,\lambda)}(x) = [V(\beta)\eta](x)$$
$$= \eta(\beta(x))\sqrt{\beta'(x)}$$
$$= \chi_{[y-1,y+1]}(x)\sqrt{\alpha(\lambda, x-y)}.$$

For $\Phi \in \mathcal{H}$, consider (see (21)–(24))

$$\int_\Gamma \langle \Phi \,|\, \eta_{\beta(y,\lambda)} \rangle \langle \eta_{\beta(y,\lambda)} \,|\, \Phi \rangle \mathrm{d}\nu(y, \lambda). \tag{53}$$

The essential condition for square integrability of V mod (K, β) is now the condition that the operator A in (19) and (24) be bounded; i.e., that the integral (53) be finite. But (53) becomes

$$\int_\Gamma \overline{\Phi(x)}\Phi(x')\alpha(\lambda,x-y)^{\frac{1}{2}}\alpha(\lambda,x'-y)^{\frac{1}{2}}\chi_{[y-1,y+1]}(x)\chi_{[y-1,y+1]}(x')\,\mathrm{d}x\mathrm{d}x'\mathrm{d}y\mathrm{d}\lambda\,,$$

or with the changes of variable $z=x-y$, $z'=x'-y$,

$$\int_\Gamma \overline{\Phi(z+y)}\Phi(z'+y)\alpha(\lambda,z)^{\frac{1}{2}}\alpha(\lambda,z')^{\frac{1}{2}}\chi_{[-1,1]}(z)\chi_{[-1,1]}(z')\,\mathrm{d}z\mathrm{d}z'\mathrm{d}y\mathrm{d}\lambda\,. \qquad (54)$$

Integrating first with respect to y, we have ($\forall z,z'$) the inequality

$$\left|\int_\Gamma \overline{\Phi(z+y)}\Phi(z'+y)\mathrm{d}y\right| \le \|\Phi\|^2\,;$$

thus (54) is bounded by

$$\|\Phi\|^2 \int_0^\infty \left[\int_{-1}^1 \alpha(\lambda,z)^{\frac{1}{2}}\mathrm{d}z\right]^2 \mathrm{d}\lambda\,.$$

Setting

$$\kappa(\lambda) = \left[\int_{-1}^1 \alpha(\lambda,z)^{\frac{1}{2}}\mathrm{d}z\right]^2, \qquad (55)$$

the needed condition is that

$$\int_0^\infty \kappa(\lambda)\mathrm{d}\lambda < \infty\,,$$

which is satisfied by many functions of λ; for example,

$$\kappa(\lambda) = \frac{\lambda}{1+\lambda^4}.$$

The key fact, then, is that (for fixed λ) it is possible to choose a C^∞ function $\alpha(\lambda,z)$ on $[-1,1]$, with $\alpha(\lambda,z) > 0$ and $\alpha(\lambda,0) = \lambda$, in such a manner as to satisfy (52) while the value of (55) is arbitrarily small – equaling the desired value $\kappa(\lambda)$. This is accomplished by selecting a small subinterval $I_w \subset [-1,1]$, of width $w \approx \kappa(\lambda)$, permitting α to become large on I_w so that $\alpha \approx w^{-1}$, and letting α be negligibly small outside I_w so that (52) still holds. Then (55) has the approximate value w. On an even smaller subinterval J that includes the point $z = 0$, where J is of negligible width in comparison with both w and λ (so as not to destroy (52) and (55)), $\alpha(\lambda,z)$ is adjusted so that $\alpha(\lambda,0) = \lambda$.

Clearly our ability to manipulate $\alpha(\lambda,z)$ in a very unconstrained way, deriving from the infinitely many degrees of freedom present in the group Diff($\mathbb{R}$), is what enables us to ensure the convergence of (53), and hence the boundedness of

$$\int_\Gamma |\eta_{\beta(y,\lambda)}\rangle\langle\eta_{\beta(y,\lambda)}|\,\mathrm{d}\nu(y,\lambda) = A\,.$$

Finally, it is important that A be invertible; i.e., if $A\Phi = 0$ for $\Phi \in \mathcal{H}$, then $\Phi = 0$. We omit the details of this argument, which is based on the observation that application of the operator A to Φ leads to convolution integrals which vanish only when Φ is indeed the zero vector.

Noting that Γ is here the group manifold of the affine group G, it is an interesting if elementary observation that a similar argument *fails* for the analogous (reducible) representation of G on $\mathbb{R}$. Indeed, let $\phi_{(y,\lambda)} : \mathbb{R} \to \mathbb{R}$ act by $\phi_{(y,\lambda)}(x) = \lambda x - y$, and write the representation V of G in $\mathcal{H}$,

$$V(\phi_{(y,\lambda)})\Psi(x) = \Psi(\lambda^{-1}(x+y))\sqrt{\lambda^{-1}}.$$

Defining $\beta : \Gamma \to G$ trivially, by $\beta(y,\lambda) = \phi_{(y,\lambda)}$, we have

$$\beta(y,\lambda)^{-1} : [-1,1] \to [\lambda^{-1}(y-1), \lambda^{-1}(y+1)].$$

Again letting $\eta(x) = \chi_{[-1,1]}$, we have

$$\eta_{\beta(y,\lambda)}(x) = \chi_{[\lambda^{-1}(y-1),\lambda^{-1}(y+1)]}(x)\sqrt{\lambda^{-1}}.$$

In this situation (53) becomes

$$\int_\Gamma \overline{\Phi(x)}\,\Phi(x')\lambda^{-1}\,\chi_{[\lambda^{-1}(y-1),\lambda^{-1}(y+1)]}(x)\cdot$$

$$\chi_{[\lambda^{-1}(y-1),\lambda^{-1}(y+1)]}(x')\,\mathrm{d}x\mathrm{d}x'\mathrm{d}y\mathrm{d}\lambda,$$

or with the changes of variable $z = \lambda x - y, z' = \lambda x' - y$,

$$\int_\Gamma \overline{\Phi(\frac{z+y}{\lambda})}\,\Phi(\frac{z'+y}{\lambda})\lambda^{-1}\chi_{[-1,1]}(z)\chi_{[-1,1]}(z')\lambda^{-2}\mathrm{d}z\mathrm{d}z'\mathrm{d}y\mathrm{d}\lambda.$$

But now, integrating first over y and then over z and z' leaves us with a factor proportional to

$$\int_0^\infty \lambda^{-1}\mathrm{d}\lambda$$

which diverges at both limits of integration. Thus we see that the extra degrees of freedom in $\text{Diff}(\mathbb{R})$ are essential to obtaining a section for which the integral (53) converges.

To sum up, we have reviewed how the method of coherent-states quantization has direct application to quantum mechanics for the Galilei and the Poincaré groups. In exploring the example of a coadjoint orbit for the diffeomorphism group, we have given some indication of how the infinite-dimensionality of the group actually contributes to our ability to obtain a family of coherent states. Though our choice of measurable section does not lend itself to immediate physical interpretation, the idea of using indicator functions to generate the coherent states is to us physically suggestive. There would appear to be considerable promise in developing further the theory of coherent-states quantization for infinite-dimensional groups such as diffeomorphism groups.

References

1.   G.W. Mackey: Ann. Math. **55** 101 (1952)
2.   G.W. Mackey: *Induced Representations of Groups and Quantum Mechanics* (Benjamin, New York, 1968)
3.   G.W. Mackey: *The Theory of Unitary Group Representations* (University of Chicago Press, Chicago, 1976)
4.   L. Auslander, B. Kostant: Inventiones Math. **14** 255 (1971)
5.   D.J. Simms, N.M.J. Woodhouse *Lectures on Geometric Quantization* (Springer, Berlin, 1976)
6.   J. Sniatycki: *Geometric Quantization and Quantum Mechanics* (Springer, Berlin, 1976)
7.   F.A. Berezin: *The Method of Second Quantization* (Academic Press, New York, 1966)
8.   I.E. Segal: J. Math. Phys. **1** 468 (1960)
9.   H.-D. Doebner, J. Tolar: J. Math. Phys. **16** 975 (1975)
10.   H.-D. Doebner, J. Tolar: "On Global Properties of Quantum Systems", in *Symposium on Symmetries in Science, Carbondale, Illinois 1979*, ed. by B. Gruber and R. S. Millman (Plenum, New York, 1979) pp. 475–486
11.   B. Angermann, H.-D. Doebner: Physica A**114** 433 (1982)
12.   B. Angermann, H.-D. Doebner, J. Tolar: "Quantum Kinematics on Smooth Manifolds", in *Nonlinear Partial Differential Operators and Quantization Procedures*, ed. by S. I. Andersson and H.-D. Doebner, Lecture Notes in Mathematics **1037** (Springer, Berlin, 1983), pp. 171–208
13.   S.T. Ali, H.-D. Doebner: "Quantization, Topology and Ordering", in *The Physics of Phase Space*, ed. by Y. S. Kim and W. W. Zachary, Lecture Notes in Physics **278** (Springer, Berlin, 1987), pp. 330–346
14.   H.-D. Doebner, H.J. Elmers, W.F. Heidenreich: J. Math. Phys. **30** 1053 (1989)
15.   S.T. Ali: Riv. Nuovo Cimento **8** 1 (1985)
16.   S.T. Ali, H.-D. Doebner: Phys. Rev. A**41** 1199 (1990)
17.   A.M. Perelomov, *Generalized Coherent States and their Applications* (Springer, Berlin, 1986)
18.   G.A. Goldin, D.H. Sharp: "Lie Algebras of Local Currents and their Representations", in *Group Representations in Mathematics and Physics: Battelle Seattle 1969 Rencontres*, ed. by V. Bargmann, Lecture Notes in Physics **6** (Springer, Berlin, 1970) pp. 300–311
19.   G.A. Goldin: J. Math. Phys. **12** 462 (1971)
20.   G.A. Goldin, R. Menikoff, D.H. Sharp: J. Math. Phys. **22** 1664 (1981)
21.   G.A. Goldin, R. Menikoff, D.H. Sharp: Phys. Rev. Lett. **51** 2246 (1983)
22.   G.A. Goldin "Diffeomorphism Groups, Semidirect Products and Quantum Theory", in *Fluids and Plasmas: Geometry and Dynamics*, ed. by J.E. Marsden, Contemp. Math. **28** (Am. Math. Soc., Providence, 1984) pp. 189–207
23.   J.M. Leinaas, J. Myrheim: Nuovo Cimento B**37** 1 (1977)
24.   J.M. Leinaas: Fortschritte der Physik **28** 579 (1980)
25.   F. Wilczek: Phys. Rev. Lett. **48** 1144 (1982)
26.   F. Wilczek: Phys. Rev. Lett. **49** 957 (1982)
27.   C.J. Isham: "Quantum Field Theory and Spatial Topology", in *Conference on Differential Geometric Methods in Theoretical Physics: Trieste 1981*, ed. by G. Denardo and H.-D. Doebner (World Scientific, Singapore, 1983), pp. 171–185

28. C.J. Isham: "Topological and Global Aspects of Quantum Theory", in *Relativity, Groups and Topology II, Procs. 1983 Les Houches Summer School*, ed. by B. DeWitt and R. Stora (North Holland, Amsterdam, 1984) pp. 1059–1290
29. R.S. Ismagilov: Mat. Sbornik **100** 117 (1976)
30. A.M. Vershik, I.M. Gel'fand, M.I. Graev: Dokl. Akad. Nauk. SSSR. **232** 745 (1977)
31. J. Mickelsson: *Current Algebras and Groups* (Plenum, New York, 1989)
32. G.A. Goldin, D.H. Sharp: Commun. Math. Phys. **92** 217 (1983)
33. V.G. Kac: Funct. Anal. Appl. **1** 328 (1967)
34. R.V. Moody: Bull. Am. Math. Soc. **73** 217 (1967)
35. R.V. Moody: J. Algebra **10** 211 (1968)
36. M.A. Virasoro: Phys. Rev. D1 2933 (1970)
37. P. Goddard, D. Olive: Int. J. Mod. Phys. A1 303 (1986)
38. A. Pressley, G. Segal: *Loop Groups* (2nd ed., Clarendon Press, Oxford, 1988)
39. C. Itzykson, H. Saleur, J.-B. Zuber: *Conformal Invariance and Applications to Statistical Mechanics* (World Scientific, Singapore, 1988) and papers therein
40. R.J. Glauber: Phys. Rev. **131** 2766 (1963)
41. M. Takesaki: *Theory of Operator Algebras I* (Springer, New York, 1979)
42. S.K. Berberian: *Notes on Spectral Theory* (Van Nostrand, Princeton, N.J., 1966)
43. S.T. Ali, E. Prugovečki: J. Math. Phys. **18** 219 (1977)
44. S.T. Ali, G.G. Emch: J. Math. Phys. **27** 2936 (1986)
45. M. Duflo, C.C. Moore: J. Funct. Anal. **21** 209 (1976)
46. A. Carey: Rep. Math. Phys. **14** 247 (1978)
47. A. Grossmann, J. Morlet, T. Paul, J. Math. Phys. **26** 2473 (1985)
48. S.T. Ali, J.P. Antoine: Ann. Inst. Henri Poincaré **51** 23 (1989)
49. S. De Bièvre: J. Math. Phys. **30** 1401 (1989)
50. S.T. Ali, J.-P. Antoine, J.-P. Gazeau: "Square Integrability for Group Representations on Homogeneous Spaces, I and II", preprints (UCL-IPT-89-18 and UCL-IPT-89-19, Louvain-la-Neuve, 1989)
51. V.S. Varadarajan: *Geometry of Quantum Theory, Vol. II* (Van Nostrand, Princeton, N.J., 1968)
52. R.F. Dashen, D.H. Sharp: Phys. Rev. **165** 1857 (1968)
53. G.A. Goldin, R. Menikoff, D. H. Sharp: J. Math. Phys. **21** 650 (1980)
54. I.M. Gel'fand, N.Ya. Vilenkin: *Generalized Functions, Vol. IV* (Academic, New York, 1964)
55. A.B. Borisov: J. Phys. A **12** 1625 (1979)
56. A.A. Kirillov: Ser. Math. Sov. **1** 351 (1981)
57. P. Stovicek: "Quantization and Systems of Imprimitivity for the Group of Diffeomorphisms", in *Procs. of the XV Int'l. Conf. on Differential Geometric Methods in Theoretical Physics (Clausthal, 1986)*, ed. by H.-D. Doebner and J. D. Hennig (World Scientific, Singapore, 1987) pp. 208–218
58. G.A. Goldin, D.H. Sharp: "Diffeomorphism Groups and Local Symmetries: Some Applications in Quantum Physics", in *Symmetries in Science III*, ed. by B. Gruber and F. Iachello, (Plenum, New York, 1989) pp. 181–205
59. D. Finkelstein, J. Rubinstein: J. Math. Phys. **9** 1762 (1968)
60. G.A. Goldin, D.H. Sharp: Phys. Rev. D28 830 (1983)
61. G. S. Canright, S.M. Girvin: Science **247** 1197 (1990)
62. G.A. Goldin: "Predicting Anyons: The Origins of Fractional Statistics in Two-Dimensional Space", in *Symmetries in Science V*, ed. by B. Gruber. L. C. Biedenharn, and H.-D. Doebner (Plenum, New York, 1991, in press)

63.   G.A. Ringwood, L. M. Woodward: Phys. Rev. Lett. **53** 1980 (1983)
64.   G.A. Goldin, "Parastatistics, Θ-Statistics, and Topological Quantum Mechanics from Unitary Representations of Diffeomorphism Groups", in *Procs. of the XV Int'l. Conf. on Differential Geometric Methods in Theoretical Physics (Clausthal, 1986)*, ed. by H.-D. Doebner and J. D. Hennig (World Scientific, Singapore, 1987) pp. 197–207
65.   G.A. Goldin, R. Menikoff, D.H. Sharp: Phys. Rev. Lett. **54** 603 (1985)
66.   G.A. Goldin, R. Menikoff, D. H. Sharp "Induced Representations of Diffeomorphism Groups Described by Cylindrical Measures", in *Measure Theory and its Applications: Procs. of the 1980 Conference*, ed. by G. A. Goldin and R. F. Wheeler (Northern Illinois Univ. Dept. of Math. Sci., DeKalb, Illinois, 1980) pp. 207–218
67.   G.A. Goldin: "On the Distinguishability of Particles Described by Unitary Representations of Diff(M)", in *Infinite Dimensional Lie Algebras and Quantum Field Theory*, ed. by H.-D. Doebner, J. D. Hennig, and T. D. Palev (World Scientific, Singapore, 1988) pp. 28–41
68.   L. Weaver, L.C. Biedenharn, R.Y. Cusson: Ann. Phys. **77** 250 (1973)
69.   G.A. Goldin, R. Menikoff, D.H. Sharp: Phys. Rev. Lett. **58** 2162 (1987)
70.   A.A. Kirillov: *Elements of the Theory of Representations* (Springer, Berlin, 1976)
71.   J.-M. Lévy-Leblond: "Galilei Group and Galilean Invariance", in *Group Theory and its Applications, Vol.II*, ed. by E.M. Loebl (Academic Press, New York, 1971) pp. 221–299
72.   S.T. Ali, E. Prugovečki: Acta Appl. Mat. **6** 19 (1986)
73.   S.T. Ali, E. Prugovečki: Acta Appl. Mat. **6** 47 (1986)
74.   S.T. Ali, J.-P. Antoine, J.-P. Gazeau Ann. Inst. Henri Poincaré **52** 83 (1990)

Borel Quantization and the Origin of Topological Effects in Quantum Mechanics

J. Tolar

Department of Physics
Faculty of Nuclear Science and Physical Engineering
Czech Technical University
Břehová 7, CS-115 19 Prague, Czechoslovakia

Abstract: The review is devoted to topological global aspects of quantal description for systems whose configuration spaces are connected finite-dimensional differentiable manifolds without boundary. The treatment concentrates on quantizations of kinematical observables (positions and momenta) and attempts to motivate the global approach based on a generalization of imprimitivity systems called *quantum Borel kinematics*. These are classified by means of global invariants (quantum numbers of a topological origin).

1 Introduction

The successful development of quantum theory in this century shows convincingly that it provides perhaps the most universal language for the description of physical phenomena. In quantum theory, as in any other physical theory, two fundamental aspects can be distinguished: the mathematical formalism and the physical interpretation.

At the basis of the most common *mathematical formalism* of quantum mechanics lies the notion of a complex separable Hilbert space $\mathcal{H}$ of, in general, infinite dimension. Normed vectors in $\mathcal{H}$ correspond to pure states of a quantum system, whereas quantal observables are represented by self-adjoint operators in $\mathcal{H}$.

However, only the rules of a *physical interpretation* enable one to use quantum theory for the description of physical systems. The principal general rule is Born's statistical interpretation of the wave function. For each physical system, or at least for a certain class of them, it is further necessary to specify which operators in $\mathcal{H}$ are associated with physical observables measured by certain measuring devices. This means in particular that at least the operators of kinematical observables (position and momentum), and the dynamical evolution law of the system are to be specified.

The primary aim of *quantization* [1] of a given classical system is to associate self-adjoint operators with classical observables. As a rule, two main methods are used. The first one is based on Bohr's *correspondence principle*: the physical meaning of quantum operators is found by looking at their classical counterparts. In this way non-relativistic quantum mechanics was formulated by quantization of classical Hamiltonian mechanics, quantum theory of electromagnetic field by quantization of the Maxwell theory, etc. [3].

The correspondence principle can, of course, be the leading rule for quantization, if the observables already existed in a classical form. What can be done in the case of quantum observables without a classical analogue like the spin? Here the second method is often applicable, which uses *invariance principles* connected with the symmetries of the system. By Noether's theorem the operators corresponding to conserved quantities can be found as generators of some projective representation of the symmetry group in $\mathcal{H}$. As a far reaching application of this approach let us mention the relativistic quantum theory of elementary particles based on the irreducible unitary representations of the Poincaré group [4].

Both methods were used from the very first days of quantum theory, always taking into account specific physical properties of the systems considered. The first method usually appears in non-relativistic quantum mechanics as *canonical quantization* [3], for systems with the Euclidean configuration space $\mathbb{R}^n$. The position coordinates q_j and the canonically conjugate momenta p_k are quantized into self-adjoint position and momentum operators Q_j, P_k (in a separable Hilbert space $\mathcal{H}$), satisfying canonical commutation relations. This was originally discovered and mathematically formulated independently by W. Heisenberg and E. Schrödinger in 1925-26. The uniqueness of the mathematical formulation up to unitary equivalence was then guaranteed by the Stone-von Neumann Theorem.

Quantum mechanics on $\mathbb{R}^n$ became very soon a successful theory which has been able to correctly describe experimental findings in vast areas of quantum physics. However, there appeared problems where it was necessary to look for a formulation of quantum mechanics when the configuration space of a system was non-Euclidean. For instance, in connection with the studies of rotational spectra of molecules and of deformed nuclei, *quantum rotators* were introduced as fundamental quantum models with configuration spaces S^1 (the circle), S^2 (the 2-sphere) and SO(3) (the rotation group). The textbook treatment of these models (quantum mechanics of angular momentum) presents a successful application of the approach via invariance principles.

There were also attempts to enforce canonical quantization in cases where, like for rotators, global Cartesian coordinates do not exist on M. A formal quantization of generalized coordinates q_j and conjugate momenta p_k was suggested [5] on a manifold M with the Riemann structure g (metric tensor g_{jk} with determinant $g > 0$)

$$Q_j = q_j, \quad P_k = -i\hbar \frac{\partial}{\partial q_k} - \frac{i\hbar}{4} \frac{\partial}{\partial q_k}(\ln g). \tag{1}$$

---

[1] For the general concept of quantization and the relations between the quantum and classical descriptions we refer to [1,2].

Note that the additional term in P_k's makes them symmetric operators in $\mathcal{H} = L^2(M, d\mu)$ with respect to the Riemann measure $d\mu = \sqrt{g}\, d^n q$ on M. The main difficulty encountered here is that operators (1) are not globally defined since, in general, q_j are only *local coordinates*. It is therefore desirable to invent quantization methods which employ *global geometric objects*.

A deep and in its time not completely understood and recognized accomplishment in this direction was Dirac's investigation [6] of a quantum charged particle (charge ϵ) in the external magnetic field of a point-like magnetic monopole (magnetic charge g). If the singular *Dirac monopole* is placed at the origin of a Cartesian coordinate system in $\mathbb{R}^3$, one deals in fact with quantum mechanics on a topologically non-trivial effective configuration manifold $\mathbb{R}^3 \backslash \{(0,0,0)\}$ (the three-dimensional Euclidean space with the origin excluded). Here the formalism of quantum mechanics in connection with non-trivial topology of the configuration manifold leads to an unexpected topological quantum effect originating from a peculiar behaviour of the phase of a wave function: Dirac discovered that a quantal description exists only under the condition that the dimensionless quantity $eg/2\pi\hbar$ is an integer.

Another phenomenon of this kind is the Aharonov-Bohm effect [7]. Its origin can be traced to a shift of the phase of a wave function due to an external magnetic flux imposed on a charged particle. Here the effective Aharonov-Bohm configuration space is $\mathbb{R}^3 \backslash \mathbb{R}$, the three-dimensional Euclidean space with a straight line excluded. In both mentioned cases the topologies of the configuration spaces differ from the trivial topology of the Euclidean space.

These remarks about the early history of quantum mechanics clearly point to the need for a systematic development of *global quantization methods*. For systems with sufficiently *symmetric* configuration or phase spaces, two modern approaches in the theory of group representations can be applied:

1. method of *systems of imprimitivity* on homogeneous configuration manifolds [8,9];
2. method of *coadjoint orbits* which play the role of homogeneous phase spaces [10,11].

In the case of configuration or phase manifolds *without* geometric symmetries, two programs of global quantization were suggested:

3. *quantum Borel kinematics* on configuration manifolds [12,13] which extends the notion of Schrdinger systems [14,15];
4. *geometric quantization* on symplectic phase manifolds [10,16].

These methods have been elaborated to differing degrees of sophistication and have, in general, different classes of classical systems as their domains of applicability. For physical applications it is important that they yield both important classification theorems and explicit relations for quantization of kinematical observables.

Concerning other quantization methods respecting global properties of configuration or phase spaces we should especially mention:

5. the *Feynman path integral* method (it was used, e.g., in [17] for $M = SO(3)$ and in [18] for configuration spaces of identical particles);
6. *quantization as deformation* of classical mechanics [19];
7. *Dirac quantization* of systems with constraints in phase space [20].

This article is devoted to a survey of the approach 3. – **quantum Borel kinematics** – which yields a large class of quantizations for systems admitting localization on connected smooth finite-dimensional configuration manifolds without boundary.[2]

2 Quantization on Configuration Manifolds
2.1 Vector Fields on Manifolds, Localization

Generally, for a given smooth manifold M there is, a priori, no geometric symmetry group. As indicated in [12,13], the investigation of vector fields on M is a meaningful starting point. We denote by $\mathcal{X}(M)$ the Lie algebra of smooth vector fields on M, by $\mathcal{X}_0(M)$ its subalgebra of compactly supported vector fields, by $\mathcal{X}_c(M)$ the family of all complete vector fields, $\mathcal{X}_0(M) \subset \mathcal{X}_c(M)$. The flow ϕ^X of a complete vector field X represents a one-parameter group of diffeomorphisms $\{\phi_t^X\}_{t \in \mathbb{R}}$ of M, also called a *dynamical system* on M. And, vice versa, every dynamical system is a flow of some (uniquely determined) complete vector field

$$[X\psi](u) = [\frac{\mathrm{d}}{\mathrm{d}t}(\psi \circ \phi_t^X)(u)]_{t=0}.$$

The family of dynamical systems on M will be denoted by $D(M)$. The following theorem summarizes some well-known facts from differential geometry [21,22].

Theorem 1 *Let $f : M \to M'$ be a diffeomorphism. Then $f' : \mathcal{X}(M) \to \mathcal{X}(M')$, where $(f'.X)_{f(u)} = \mathrm{d}f_u(X_u)$, is an isomorphism of Lie algebras; the restriction $f' : \mathcal{X}_0(M) \to \mathcal{X}_0(M')$ is also a Lie algebra isomorphism; $f' : \mathcal{X}_c(M) \to \mathcal{X}_c(M')$ is a bijection. The mapping*

$$f^D : D(M) \to D(M') : \{\phi_t\}_{t \in \mathbb{R}} \mapsto \{f \circ \phi_t \circ f^{-1}\}_{t \in \mathbb{R}}$$

is bijective and $f^D(\phi^X) = \phi^{f'.X}$.

Localization of a quantum system on M is treated in terms of a *projection-valued measure* $E : B \mapsto E(B)$ mapping Borel subsets B of M ($B \in \mathcal{B}(M)$) into projection operators $E(B)$ in a separable Hilbert space $\mathcal{H}$ subject to the usual axioms of localization. [3] For a given subset $B \in \mathcal{B}(M)$, the projection $E(B)$ corresponds to a measurement which determines whether the system is localized in B; its eigenvalues 1 (0) correspond to situations when the system is found completely inside (outside) B, respectively.

---

[2] We consider only *paracompact* manifolds which, by Whitney's embedding theorem, can be regarded as submanifolds of $\mathbb{R}^n$.
[3] Axioms of localization are:

2.2 Classical and Quantum Borel Kinematics

Definition 1 The pair $(\mathcal{B}(M), \mathcal{X}_c(M))$ with a flow model

$$B \mapsto \phi_t^X(B) = \{ m \in M;\ \phi_{-t}^X(m) \in B \} \qquad (2)$$

of shifts along $X \in \mathcal{X}_c(M)$ is called *classical Borel kinematics*.

For every $\phi^X \in D(M)$ the manifold M becomes an $\mathbb{R}$-space (i.e. a G-space for the group $G = \mathbb{R}$). Attempting to generalize Mackey's quantization on homogeneous spaces [8,9], we require that there exist: a Hilbert space $\mathcal{H}$, a projection-valued measure E on M, and unitary representations V^X in $\mathcal{H}$ of the flow models (2) such that

$$V^X(t)\, E(B)\, V^X(-t) = E(\phi_t^X . B), \qquad (3)$$

where the objects $\mathcal{H}$, E do not depend on the choice of $\phi^X \in D(M)$. Equation (3) is just a quantal expression of the classical flow model (2). Generalized momentum operators can then be introduced via Stone's Theorem as (essentially self-adjoint) infinitesimal generators $P(X)$ of the one-parameter groups of unitary operators in $\mathcal{H}$,

$$V^X(t) = \exp(-iP(X)t/\hbar), \qquad t \in \mathbb{R}.$$

After these preparations we can state

Definition 2 **Quantum Borel kinematics** is a pair (V, E), where E is a projection-valued measure on M in a separable Hilbert space $\mathcal{H}$, and V associates with each $\phi^X \in D(M)$ a homomorphism V^X from $\mathbb{R}$ into the group of all unitary operators in $\mathcal{H}$ such that the following conditions are satisfied:

(i) Equation (2) holds for all $t \in \mathbb{R}$, $X \in \mathcal{X}_c(M)$;

(ii) The mapping $P \colon \mapsto P(X)$ from the Lie algebra $\mathcal{X}_0(M)$ into the space of essentially self-adjoint operators with common invariant dense domain in $\mathcal{H}$ is a Lie algebra homomorphism:

$$P(X + aY) = P(X) + aP(Y), \qquad (4)$$

$$[P(X), P(Y)] = -i\hbar\, P([X, Y]); \qquad (5)$$

(iii) Locality condition: if two flows $\phi^{X_i} \in D(M)$, $i = 1, 2$, after restriction on the set $(-a, a) \times B$, $a > 0$, $B \in \mathcal{B}(M)$, coincide, then the mappings $\mathbb{R} \times \mathcal{H} \colon (t, \psi) \mapsto$

$$E(B_1 \cap B_2) = E(B_1) \cdot E(B_2),$$
$$E(B_1 \cup B_2) = E(B_1) + E(B_2) - E(B_1 \cap B_2),$$
$$E(\bigcup_{i=1}^{\infty} B_i) = \sum_{i=1}^{\infty} E(B_i) \text{ for mutually disjoint } B_i \in \mathcal{B}(M),$$
$$E(M) = \mathbb{1}.$$

$V^{X_i}(t)\,\psi$ coincide on the domain $(-a, a) \times \mathcal{H}_B$, where $\mathcal{H}_B$ is the subspace of $\mathcal{H}$ projected out by $E(B)$.

If in (ii) only linearity (4) is required, we shall call (V, E) a *generalized system of imprimitivity*.

Remark. It follows from condition (iii) that if V^X is known for all $X \in \mathcal{X}_0(M)$, then it is determined for all $X \in \mathcal{X}_c(M)$; (iii) further implies that $P(X)$ are differential operators. Condition (ii) may sometimes be too restrictive since (5) excludes a non-vanishing external gauge field on M; in this connection we refer to [15,27].

The projection-valued measure E induces in a natural way a quantization Q of classical (smooth) real functions $f \colon M \to \mathbb{R}$ on configuration space (e.g. coordinate functions, potentials, etc.). Not necessarily bounded, self-adjoint quantum position operators $Q(f)$ are uniquely determined by their spectral decompositions

$$Q(f) = \int_{-\infty}^{\infty} \lambda \, dE_{\lambda}^f \,,$$

where the spectral function E_{λ}^f is given by spectral measure $E^f(\Delta) = E(f^{-1}(\Delta))$ on subsets $\Delta = (-\infty, \lambda)$ of $\mathbb{R}$. Equation (3) is then replaced by

$$V^X(t)\,Q(f)\,V^X(-t) = Q(f \circ \phi_t^X), \tag{6}$$

where $f \in C^\infty(M, \mathbb{R})$, and implies a generalization of the Heisenberg commutation relations in terms of coordinate-independent objects

$$[Q(f), P(X)] = i\hbar\, Q(X.f) \quad \text{on} \quad \mathcal{D} \subset \mathcal{H} \tag{7}$$

(it is assumed that operators $P(X), Q(f)$ have a common invariant dense domain $\mathcal{D}$ in $\mathcal{H}$). If an obvious relation

$$[Q(f), Q(g)] = 0 \quad \text{on} \quad \mathcal{D}, \tag{8}$$

is still added, then (8), (7) and (5) define a *Schrödinger system* in the sense of [14].

3 Construction of Quantum Borel Kinematics

3.1 Quasi-Invariant Measures

The question of existence and uniqueness of a measure which is *quasi-invariant* with respect to all those diffeomorphisms $\phi_1 \colon u \mapsto \phi(1, u)$ for which $\phi \in D(M)$, is answered by

Theorem 2 *The family of quasi-invariant measures on $\mathcal{B}(M)$ is non-empty and, moreover, all measures in this family are mutually equivalent and form a unique invariant measure class. After completion, those subsets in M which have measure zero are exactly measure zero sets in the sense of Lebesgue.*

The invariant measure class is called the *Lebesgue measure class*.

Proof. The fact that the family of sets of zero measure in the sense of Lebesgue is invariant under diffeomorphisms is well known. The existence part of the theorem may be seen as follows. Having embedded M in $\mathbb{R}^m$ (Whitney's Theorem), we can consider a tubular neighborhood M^ϵ of M in the normal bundle. Denoting by $\pi: M^\epsilon \to M$ the associated submersion, we can define $\mu(B) = \lambda^m(\pi^{-1}(B))$ for $B \in \mathcal{B}(M)$, with λ^m being the Lebesgue measure in $\mathbb{R}^m$; then μ is quasi-invariant. The assertion about uniqueness of the invariant measure class for $M = R^n$ follows from the fact that the family of diffeomorphisms ϕ_1 includes all translations and the assertion for the group of translations is known ([23], Chap.2). In general, M can be covered by a countable family of open sets, each of which is diffeomorphic to R^n and so the assertion is true again. □

3.2 Projection-Valued Measures

Let E be a projection-valued measure on M in a separable Hilbert space $\mathcal{H}$. It is known ([23], Chap.9) that there exist two sequences $\{\mathcal{K}_r\}$, $\{\nu_r\}$, $r = \infty, 1, 2, \ldots$, the first one consisting of Hilbert spaces, the other of measures on M such that $\dim \mathcal{K}_r = r$ and ν_r, ν_s are disjoint for $r \neq s$. [4] The projection-valued measure E is equivalent to the measure E^0 which acts via multiplication by indicator functions of subsets B on the elements of the Hilbert space $\mathcal{H}^0 = \bigoplus_r \mathcal{H}_r$, where $\mathcal{H}_r$ are the Hilbert spaces of vector-valued functions from M to $\mathcal{K}_r$, i.e. $\mathcal{H}_r = L^2(M, \mathcal{K}_r, \nu_r)$. The measures ν_r are determined uniquely up to equivalence. If only one ν_r is non-zero, the projection-valued measure E is called **homogeneous**. Due to the transitivity of actions of the family $D(M)$ the following theorem holds (for details see [23], Chap.9):

Theorem 3 *If (V, E) is a generalized system of imprimitivity, then E is homogeneous. The unique non-zero measure ν_r belongs to the Lebesgue measure class on M.*

3.3 Quantum Borel r-Kinematics

As a corollary to Theorem 3 also quantum Borel kinematics possess homogeneous projection-valued measures. Hence **quantum Borel r-kinematics** (QBK^r) on M are defined as quantum Borel kinematics with a *homogeneous* projection-valued measure E of degree $r = \infty, 1, 2, \ldots$. In [13] the classes of unitarily equivalent differentiable QBK^1's were classified. The classes of unitarily equivalent differentiable QBK^r's for finite $r = 2, 3, \ldots$ can be constructed in an analogous way [15,25], but only partial classification has been achieved up to now. It was shown, however, that every (differentiable) QBK^r is equivalent to a canonical one.

The construction of the canonical QBK^r contains the following ingredients:

---

[4] That is, there exist $B_r, B_s \in \mathcal{B}(M)$, $B_r \cup B_s = M$, $B_r \cap B_s = \emptyset$, $\nu_r(B_s) = \nu_s(B_r) = 0$.

1. Lebesgue measure ν on M;
2. Hermitean vector bundle $\eta = (\mathcal{E}, \pi, M; \mathbb{C}^r)$ over M with fibres diffeomorphic to $\mathbb{R}^r$ equipped with Hermitean inner product $< \cdot, \cdot >$;
3. Hermitean flat connection ∇ on η, i.e. a connection compatible with the inner product [5] and with *vanishing curvature*. [6]

The Hilbert space $\mathcal{H}$ is realized as the Hilbert space $L^2(\eta, < \cdot, \cdot >, \nu)$ of sections of η, i.e. (measurable) mappings $\sigma: M \to \mathcal{E}$ such that $\pi \circ \sigma = \mathrm{id}_M$ and with finite norm induced by the inner product

$$(\sigma, \tau) = \int_M < \sigma(u), \tau(u) > \, \mathrm{d}\nu(u).$$

Position operators $E(B)$ in $\mathcal{H}$ have the unique form

$$E(B)\sigma = \chi_B \cdot \sigma,$$

where χ_B is the indicator function for a Borel subset B. Then also smooth position operators $Q(f)$ have the usual form of the Schrödinger representation

$$Q(f)\sigma = f \cdot \sigma, \qquad f \in C^\infty(M, \mathbb{R}). \tag{9}$$

The most general form of *generalized momentum operators* $P(X)$ corresponding to complete vector fields X is [25]

$$P(X)\sigma = -i\hbar \nabla_X \sigma - \frac{i\hbar}{2} Q(\mathrm{div}_\nu X)\sigma + \omega(X) \cdot \sigma, \quad \sigma \in \mathcal{D}, \tag{10}$$

where $\mathrm{div}_\nu X = [\mathrm{d}\rho_t^X / \mathrm{d}t]_{t=0}$, $\rho_t^X \cdot \nu = \nu \circ \phi_t^X$; furthermore, $\omega(X)$ is a Hermitean $k \times k$-matrix-valued linear differential operator $\omega: \mathcal{X}(M) \to \mathrm{Sec}^\infty(\mathrm{Hom}(\eta, \eta))$ which, by (ii) and (iii), has to be of the form

$$\omega(X) = (\mathrm{div}_\nu X) \cdot \mathbf{\Phi},$$

where $\mathbf{\Phi}$ is a $\hat{\nabla}$-parallel section of $\mathrm{Hom}(\eta, \eta)$, $\hat{\nabla}$ being the connection on $\mathrm{Hom}(\eta, \eta)$ induced by ∇.

The whole variety of quantizations could be read off the formula (10) for possible $P(X)$. In order to get a more transparent result we define a QBK^r to be *of type 0* if [25]

$$\mathbf{\Phi} = c.\mathrm{id}_{\mathrm{Sec}\eta}, \qquad c \in \mathbb{R}.$$

Then we obtain an identical formula for $P(X)$ as in QBK^1 [13]:

$$P(X)\sigma = -i\hbar \nabla_X \sigma + (-\frac{i\hbar}{2} + c)(\mathrm{div}_\nu X) \cdot \sigma, \quad \sigma \in \mathcal{D}. \tag{11}$$

As proved in [25], on every smooth manifold M there exists a (differentiable) QBK^r of type 0. Let us note that for $r = 1$, the type 0 QBK^1's *classify all possible Borel quantizations* [13]; this is not the case, however, for $r > 1$.

---

[5] $X < \sigma, \tau > = < \nabla_X \sigma, \tau > + < \sigma, \nabla_X \tau >$.
[6] The property of flatness of ∇ derives from (5).

Quantum Borel kinematics are rather diverse – even for the trivial configuration space $\mathbb{R}^n$ – as the following example shows [15]. The origin of the divergence term in (11) can be traced there in a particularly clear way.

Example 1   Let $M = \mathbb{R}^n$ with a fixed basis. To every vector field $X = \sum_k X_k \, \partial/\partial x_k$ we relate a matrix-valued function

$$A(X)\colon \mathbb{R}^n \to \mathbb{R}^{n,n}, \quad [A(X)_u]_{i,j} = \frac{\partial X_i(u)}{\partial x_j}.$$

It is straightforward to verify

$$[A(X), A(Y)] - X.A(Y) + Y.A(X) = -A([X,Y]). \tag{8}$$

Let L be a skew-Hermitean representation of the Lie algebra $\mathrm{gl}^+(n, \mathbb{R})$ in a Hilbert space $\mathcal{H}^L$. We define operators $Q(f)$, $P(X)$ in $\mathcal{H} = L^2(\mathbb{R}^n, \mathcal{H}^L, dx_1 \ldots dx_n)$ by

$$Q(f)\psi = f.\psi, \quad P(X)\psi = -i\hbar(X + \frac{1}{2}\mathrm{div}X - L[A(X)].$$

Then, using (9) and the identity $X.L(A(Y)) = L(X.A(Y))$, the pair (P, Q) can be shown to be a Schrödinger system. Choosing the representation L in $\mathcal{H}^L = \mathbb{C}^1$ to be given by $L(A) = -ic\,\mathrm{tr}A$, where c is a real constant, we obtain $L(A(X)) = -ic\,\mathrm{div}X$. This is just the divergence term in (11) for $r = 1$, extensively discussed in [1,13]. [7]

Remark.   Consider the surjective mapping

$$p\colon \mathrm{gl}(n, \mathbb{R}) \to \mathrm{sl}(n, \mathbb{R})\colon A \mapsto A - \frac{1}{n}(\mathrm{tr}A)\mathbb{1}_n,$$

where $\mathbb{1}_n$ is the unit $n \times n$-matrix. This mapping permits to associate with every representation L' of $\mathrm{sl}(n, \mathbb{R})$ a representation $L = L' \circ p$ of $\mathrm{gl}(n, \mathbb{R})$. Then our mapping P is the infinitesimal form of a representation of the group of diffeomorphisms of $\mathbb{R}^n$ induced from $\mathrm{SL}(n, \mathbb{R})$ (see [1,24] for $n = 3$).

3.4 Classification of Quantum Borel Kinematics of Type 0

Finally, a complete classification of QBK^r's of type 0 is essentially the question of the topology of M. The corresponding investigations of [15,25] can be summarized in

Theorem 4   *The set of classes of unitarily equivalent QBK^r's of type 0 on M can be bijectively mapped onto the set of pairs (L, c), where $c \in \mathbb{R}$ and L denote the isomorphism classes of flat $\mathbb{C}^r$-bundles over M.*

---

[7] For general results concerning the "divergence-like" terms for arbitrary r (including $r = \infty$) see [26].

Since there is a one-to-one correspondence between the isomorphism classes of flat $\mathbb{C}^r$-bundles over M and flat $U(r)$-principal bundles over M, we can use Milnor's lemma [28]:

Lemma 1 *A $U(r)$-principal bundle over M admits a flat connection if and only if it is induced from the universal covering bundle of M by a homomorphism of the fundamental group $\pi_1(M)$ into $U(r)$.*

Thus, disregarding the real constant c, the set of inequivalent quantizations of type 0 with $\mathbb{C}^r$-valued wave functions is isomorphic to the set

$$\text{Hom}(\pi_1(M), U(r))$$

of r-dimensional unitary representations of the fundamental group of M.

In the case $r = 1$, i.e. of quantizations with complex-valued wave functions, the topological part of the classification reduces to $\text{Hom}(\pi_1(M), U(1))$, i.e. to the set of one-dimensional unitary representations of $\pi_1(M)$ [11,12,13,25,27,29].[8] Since the commutator subgroup $\Gamma(\pi_1(M))$ (generated by elements $aba^{-1}b^{-1}$) belongs to the kernels of all such one-dimensional representations and since the singular homology group $H_1(M, \mathbb{Z})$ is isomorphic to $\pi_1(M)/\Gamma(\pi_1(M))$ (the Hurewicz isomorphism), inequivalent QBK^1's are labeled by elements of the character group of $H_1(M, \mathbb{Z})$. Finally, let us describe the general structure of the Abelian group $H_1(M, \mathbb{Z})$ for compact M. It has a decomposition $H_1(M, \mathbb{Z}) = F \oplus T$, where the free Abelian group F is $F = \mathbb{Z} \oplus \cdots \oplus \mathbb{Z}$ (b_1 terms), with b_1 being the *first Betti number* of M, and the torsion Abelian group is $T = Z_{\tau_1} \oplus \cdots \oplus Z_{\tau_k}$ with Z_{τ_i} being cyclic groups of orders τ_i (*torsion coefficients*) such that $\tau_{i+1}/\tau_i = $ positive integer. Thus the characters of $H_1(M, \mathbb{Z})$ can be parametrized by $(b_1 + k)$-tuples

$$[e^{2\pi i\theta_1}, \cdots, e^{2\pi i\theta_{b_1}} \; ; \; e^{2\pi i m_1/\tau_1}, \cdots, e^{2\pi i m_k/\tau_k}]$$

with the numbers $\theta_l \in [0,1)$, $l = 1, \ldots, b_1$, and $m_i = 0, 1, \ldots, \tau_i - 1$, $i = 1, \ldots, k$, classifying inequivalent quantum Borel 1-kinematics on M.

Acknowledgements

The results summarized in this article grew out of a fruitful collaboration with Prof. Dr. H.D. Doebner at his Institute in Clausthal where, during pleasant stays in a number of years, I could enjoy warm hospitality and a stimulating research environment. I would find it a proper tribute to him if the review succeeded to put this part of his work in a proper perspective. Prof. Doebner has always been deeply interested in the general problem whether there exist relations between the global properties of configuration spaces and some physical properties of systems localized on them. As the article attempts to show, there is a fine interplay between quantum

---

[8] This result was obtained independently also in the Feynman path integral approach [18].

principles on the one hand, and the geometry and topology of the underlying manifold on the other.

References

1. S.T. Ali, G.A. Goldin: "Quantization, coherent states and diffeomorphism groups" (this volume)
2. C. Fronsdal: "Some ideas about quantization", Repts. Math. Phys. **15** 111–145 (1978)
3. P.A.M. Dirac: *The Principles of Quantum Mechanics*, 4th ed. (Oxford University Press, Oxford, 1958)
4. E.P. Wigner: "On unitary representations of the inhomogeneous Lorentz group", Ann. Math. **40** 149-204 (1939)
5. W. Pauli: *Wellenmechanik*, Handbuch der Physik **24** I (1933) p.120
6. P.A.M. Dirac: "Quantized singularities in the electromagnetic field", Proc. Roy. Soc. (London) A **133** 60 (1931)
7. Y. Aharonov, D. Bohm: "Significance of electromagnetic potentials in the quantum theory", Phys. Rev. **115** 485 (1959)
8. G.W. Mackey: *Induced Representations of Groups and Quantum Mechanics* (Benjamin, New York, 1968)
9. H.D. Doebner, J. Tolar: "Quantum mechanics on homogeneous spaces", J. Math. Phys. **16** 975-985 (1975)
10. A.A. Kirillov: *Elements of Representation Theory* (Nauka, Moscow, 1972; in Russian).
11. B. Kostant: "Quantization and unitary representations: Part I. Prequantization", Lecture Notes in Mathematics, Vol. 170 (Springer-Verlag, Berlin, 1970, pp.87-208)
12. H.D. Doebner, J. Tolar: "On global properties of quantum systems", in *Symmetries in Science*, ed. by B. Gruber, R. S. Millman (Plenum, New York, 1980, p. 475-486)
13. B. Angermann, H.D. Doebner, J. Tolar: "Quantum kinematics on smooth manifolds", Lecture Notes in Mathematics, Vol. 1037, (Springer-Verlag, Berlin, 1983, pp. 171-208)
14. I.E. Segal: "Quantization of nonlinear systems", J. Math. Phys. **1** 468-488 (1960)
15. P. Šťovček, Diploma Thesis (Technical University, Prague, 1981, in Czech)
16. J.-M. Souriau: *Structure des Systèmes Dynamiques* (Dunod, Paris, 1970)
17. L.S. Schulman, Phys. Rev. **176** 1558 (1968)
18. M.G.G. Laidlaw, C. Morette-DeWitt: "Feynman functional integrals for systems of indistinguishable particles", Phys. Rev. D **3**, 1375 (1970)
19. D. Sternheimer: "Deformation Theory and Quantization", Ann. Phys. (New York) **111** 61-110, 111-152 (1978)
20. P.A.M. Dirac: *Lectures on Quantum Mechanics* (Belfer Graduate School of Science, Yeshiva University, New York, 1964)
21. K. Nomizu: *Lie Groups and Differential Geometry* (The Mathematical Society of Japan, 1956)
22. S. Kobayashi, K. Nomizu: *Foundations of Differential Geometry*, I (Interscience-Wiley, New York, 1963)
23. V.S. Varadarajan: *Geometry of Quantum Theory II.Quantum Theory of Covariant Systems* (Van Nostrand Reinhold Co., New York, 1970)

24   G.A. Goldin, R. Menikoff, D.H. Sharp: "Induced representations of the group of diffeomorphisms of $\mathrm{I\!R}^3$", J. Phys. A: Math. Gen. **16** 1827 (1983)

25.   U.A. Mller: "Zur Quantisierung physikalischer Systeme mit inneren Freiheitsgraden", Ph.D. Thesis (Technical University, Clausthal, 1988)

26.   P. Šťovček: "Systems of imprimitivity for the group of diffeomorphisms I, II", Ann. Global Anal. Geom. **5** 89-95 (1987), **6** 31-37 (1988)

27.   H.D. Doebner, J. Tolar: "Symmetry and topology of the configuration space and quantization", in *Symmetries in Science II* edited by B. Gruber and R. Lenczewski (Plenum, New York, 1986, pp.115-126)

28.   J. Milnor: "On the existence of a connection with curvature zero", Comment. Math. Helv. **32** 215-223 (1958)

29.   B. Angermann, H.D. Doebner: "Homotopy groups and quantization of localizable systems", in *Proc. Xth Coll. Group Theoretical Methods in Physics*, Physica A **114** 433 (1982)

IV

Representations of Groups and Algebras

Symmetries of Quantum Group Coupling Coefficients*

L.C. Biedenharn, M.A. Lohe**

Department of Physics, Duke University
Durham, NC 27706, U.S.A.

Abstract: Quantum groups are new algebraic symmetry structures which have recently found application in many diverse fields of physics. We discuss the simplest of such structures: $SU_q(2)$ – the q-analog of the familiar quantal angular momentum group $SU(2)$ – and determine in complete detail the symmetries, under exchange of q-angular momenta, of the (q-3j) and (q-6j) coefficients.

1 Introduction

Quantum groups are remarkable mathematical structures that emerged as algebraic abstractions from quantum inverse scattering theory [1] and, almost simultaneously, solvable statistical mechanical models [2]. The term "quantum group" was coined by Drinfel'd [3], but more properly a quantum group is an algebra, a deformation of the universal enveloping algebra of an underlying classical Lie group – hence the alternative designation as a quantized universal enveloping algebra (QUE-algebra) or, more simply, as a q-analog or q-group. (The continuous deformation parameter is suggestively written as $q = e^{\hbar}$ so that for $\hbar \to 0$ we obtain the 'classical', undeformed Lie group.) In the last few years the number of applications of quantum groups has grown substantially [4].

The present note is concerned with the simplest quantum group, $SU_q(2)$, a q-analog to the familiar quantal angular momentum group $SU(2)$. The basic structural properties of $SU_q(2)$ are remarkably close to those of $SU(2)$: all unitary irreducible representations (irreps) are finite dimensional and indeed the dimensions are precisely the same, despite the deformation in the defining commutation relations (Sect. 2). There are, however, important qualitative differences stemming

---

* Supported, in part, by the National Science Foundation and the Department of Energy.

** Permanent address: Northern Territory University, P.O. Box 40146, Casuarina, NT, Australia, 0811.

from the fact that the addition of q-angular momenta is no longer commutative. (Technically this appears as a non-commutative co-product, see below). This introduces subtle changes in obtaining an analog to tensor operator structures (generalized Wigner-Eckart theorem) but the (q-3j) and (q-6j) coefficients can indeed be defined. As a consequence of the non-commutative co-product, however, Weyl's fundamental result on the reciprocity between the symmetric group (S_n) and the unitary group ($U(n)$) – in which irreps of both are classified by the same Young frames – is modified [5]. The reciprocity is now between $U_q(n)$ and the braid group (B_n), the latter group being of considerable current interest as the symmetry underlying "anyons" [6]. The purpose of the present paper is to examine the simplest occurrence of this relationship: the symmetry properties of q-Wigner-Clebsch-Gordan (q-WCG or (q-3j)) coefficients under the exchange of q-angular momenta.

It is yet another instance of the surprising relationship between mathematics and physics to note that the (q-6j) coefficients had been anticipated by Askey and Wilson [7], quite independent of any symmetry (q-group) structure that now can be seen to explain their properties. These authors defined the (q-6j) coefficients in terms of q-analogs of hypergeometric functions and this route provides the most expeditious way to the results we seek.

2 The Quantum Group $SU_q(2)$

The Lie algebra of $SU_q(2)$ is generated by the three operators $J_\pm^q, J_z^q$ satisfying the commutation relations

$$[J_z^q, J_\pm^q] = \pm J_\pm^q, \tag{1}$$

and

$$[J_+^q, J_-^q] = \frac{q^{J_z^q} - q^{-J_z^q}}{q^{\frac{1}{2}} - q^{-\frac{1}{2}}}, \quad q \in \mathbb{R}^+. \tag{2}$$

These defining relations for $SU_q(2)$ differ from those of ordinary angular momentum ($SU(2)$) in two ways:

(a) The commutator in (2) is not $2J_z$ as usual, but an infinite series (for generic q) involving all odd powers $(J_z^q)^1, (J_z^q)^3, \ldots$. Each such power is a linearly independent operator in the enveloping algebra; accordingly, the Lie algebra of $SU_q(2)$ is not of finite dimension.

(b) For $q \to 1$, the right hand side of (2) $\to 2J_z$. Thus we recover in the limit the usual Lie algebra of $SU(2)$.

The differences noted in (a) and (b) are expressed by saying that the quantum group $SU_q(2)$ is a deformation of the enveloping algebra of $SU(2)$. Let us introduce the notation

$$[n]_q \equiv \frac{q^{\frac{n}{2}} - q^{-\frac{n}{2}}}{q^{\frac{1}{2}} - q^{-\frac{1}{2}}},$$

$$= q^{\frac{(n-1)}{2}} + q^{\frac{(n-3)}{2}} + \ldots q^{-\frac{(n-1)}{2}}, \quad n \in \mathbb{Z}. \tag{3}$$

Anticipating that $2J_z^q$ has integer eigenvalues, we can then write (2) as: $[J_+^q, J_-^q] = [2J_z^q]_q$. These q-numbers, $[n]_q$, obey the rule: $[-n]_q = (-1)[n]_q$, with $[0]_q = 0$ and $[1]_q = 1$. Note that $[n]_q = [n]_{q^{-1}}$, so that the defining relations (1) and (2) are invariant to $q \leftrightarrow q^{-1}$. (This accounts for the convention using $q^{\frac{1}{2}}$ in (3).)

We have noted already that the addition of q-angular momenta is not commutative. This is expressed algebraically by defining a (non-commutative) *co-product*, Δ:

$$\Delta(J_\pm^q) \equiv q^{-J_z^q/2} \otimes J_\pm^q + J_\pm^q \otimes q^{+J_z^q/2}, \tag{4a}$$

$$\Delta(J_z^q) \equiv 1 \otimes J_z^q + J_z^q \otimes 1. \tag{4b}$$

(an alternative, distinct, co-product can be defined with q^{-1} replacing q.) For completeness we note that the following formulae for the *antipode*, S, and *co-unit*, ε, endow $SU_q(2)$ with the structure of a Hopf algebra [8,9]:

$$S(q^{\pm J_z^q}) = q^{\mp J_z^q}, \qquad S(J_\pm^q) = -q^{\pm 1} J_\pm^q \tag{5a}$$

$$\varepsilon(q^{\pm J_z^q}) = 1, \qquad \varepsilon(J_\pm^q) = 0. \tag{5b}$$

(This additional structure will, however, not be used below.)

A physicist confronted with this new symmetry algebra, (1) and (2), would seek to understand it by constructing representations of the J_i^q as finite dimensional matrices. The most expeditious way to do this is to use a variant of the familiar Jordan-Schwinger map [10] and realize J_i as bilinear boson operators, employing now *q-bosons*, defined to obey a deformation of the Heisenberg commutation relations [11]–[13]. In this way, one can easily construct explicit finite dimensional bases for all unitary $SU_q(2)$ irreps. The next step is to construct tensor operators and a suitable variant of the fundamental Wigner-Eckart theorem. This task is not completely straightforward, as one can see by noting that a basic tool for ordinary Lie algebraic analysis *fails: the generators under commutation do not realize the adjoint representation*. (This is clear from the remark that the adjoint representation has dimension 3, whereas under commutation the generators yield unlimitedly many independent operators.)

The required tensor operator structure has been proved by Biedenharn and Tarlini [14]. It will suffice for the purpose of the present paper to simply cite the relevant results. First we formalize the Hilbert space on which the tensor operators are to act: this is the *model space* [15] **M**, defined to be the direct sum of vectors carrying unitary irreps of $SU_q(2)$, each equivalence class of irreps *occurring once and only once*.

Definition Let **T** denote the vector space of operators mapping the *model space* **M** of $SU_q(2)$ into itself: **T: M** → **M**. An irreducible q-tensor operator is a set of operators, $\{t_{\Xi,\xi}\} \in$ **T** which carries a finite dimensional irrep Ξ, with vectors ξ, of $SU_q(2)$. That is:

$$J_\alpha^q(t_{\Xi,\xi}) = \sum_{\xi'} \langle \Xi, \xi' | J_\alpha^q | \Xi, \xi \rangle t_{\Xi,\xi'}, \tag{6}$$

where J_α^q is a generator of $SU_q(2)$, $J_\alpha^q(t_{\Xi,\xi})$ denotes an action of J_α^q on $\mathbf{T}$, and $\langle\ldots\rangle$ denotes the matrices of the generators for the irrep Ξ.

Theorem [14] *If $\{t_{\Xi,\xi}\}$ is a q-tensor operator of the quantum group G_q such that the co-product of G_q is compatible with the action $J_\alpha^q(t_{\Xi,\xi})$ then the matrix elements of $\{t_{\Xi,\xi})\}$ in $\mathbf{M}$ are proportional to the q-WCG coefficients of G_q with the constant of proportionality an invariant, and conversely.*

Corollary

(a) *There exists an algebra of q-tensor operators:*

$$\mathbf{T}\otimes\mathbf{T}\xrightarrow{W}\mathbf{T},$$

carrying products of irreducible q-tensor operators into irreducible q-tensor operators. (The map W is defined by the q-WCG coefficients).

(b) *In particular, Cor.(a) implies that there exists a product carrying an irreducible q-tensor operator into an invariant. Thus a norm exists and irreducible unit q-tensor operators $\hat{\mathbf{T}}$ are well-defined whose matrix elements, by the fundamental tensor operator theorem, are q-WCG coefficients.*

(c) *Denote the mapping in Cor.(a) by $\mathbf{T}\textcircled{w}\,\mathbf{T}\in\mathbf{T}$, the invariant product in Cor.(b) by $\mathbf{T}\cdot\mathbf{T}$. Then the (6-j) operators are defined by: $\hat{\mathbf{T}}\cdot(\hat{\mathbf{T}}\textcircled{w}\hat{\mathbf{T}})$. Similarly (3n-j) operators can be defined.*

We have already remarked that the generators are not an irreducible tensor operator. Adjoint tensor operators $t_{\ell,m}$ can, however, be constructed, and by using the q-WCG coefficients one can define the quadratic invariant (Casimir) operator:

$$I_2 = \mathbf{t}_1\cdot\mathbf{t}_1, \tag{7}$$

such that, acting on an irrep space,

$$I_2 \longrightarrow [2j]_q[2j+2]_q.$$

It is interesting to note that the Casimir operator (7) is positive semidefinite and that the eigenvalues involve q-integers, for both j integral and half-integral, in sharp contrast to Casimir operators obtained by less systematic methods. (Note that the q-integer property is preserved under operations of the algebra.)

Lusztig [16] and Rosso [17] have proved that for generic q, all unitary irreps of $SU_q(2)$ are finite dimensional, and have the same dimensionality as for $q = 1$. (Using the Casimir operator, the standard angular momentum proof [10] of finite dimensionality actually carries over directly to $SU_q(2)$.)

The (q-3j) and (q-6j) coefficients can be easily determined by q-boson techniques, but there is no need to do so, since these results have been anticipated by Askey and Wilson [7] and have been proved recently by many authors by a variety of techniques (as cited below).

Remark. We have explicitly assumed that q is generic and real, but in some physical applications (conformal field theory, for example) it is important to take q to be a root of unity, $(q^N = 1)$. For generic q, taking $[n] = 0$ implies $n = 0$, but this is no longer true for q a root of unity, and this fact leads to striking changes in the theory. We refer to Pasquier and Saleur [18] for further details.

3 q-Wigner-Clebsch-Gordan Coefficients

The decomposition of the tensor product of two irreducible finite dimensional unitary representations of $SU_q(2)$ has been carried out by Kirillov and Reshetikhin [19] and also Vaksman [20], and the explicit coupling coefficients $_qC^{j_1 j_2 j}_{m_1 m_2 m}$ calculated in several forms, including q-analogs of the Racah and van der Waerden forms. Let V^j denote the space carrying an irreducible representation of $SU_q(2)$, parametrized by half-integers j. As shown by Jimbo [21], the tensor product of two representations is completely reducible and can therefore be decomposed into a sum of irreducible components:

$$V^{j_1} \otimes V^{j_2} = \sum_j \oplus V^j \qquad (8)$$

where $|j_1 - j_2| \leq j \leq j_1 + j_2$. Let $|jm\rangle$ be a weight basis in the irreducible component V^j and denote by $|(j_1 j_2)jm\rangle$ a basis in $V^{j_1} \otimes V^{j_2}$. The q-WCG coefficients $_qC^{j_1 j_2 j}_{m_1 m_2 m}$ are defined by

$$|(j_1 j_2)jm\rangle = \sum_{m_1, m_2} {}_qC^{j_1 j_2 j}_{m_1 m_2 m} |j_1 m_1\rangle |j_2 m_2\rangle . \qquad (9)$$

The explicit form for these coefficients is [19]:

$$
\begin{aligned}
&{}_qC^{j_1 j_2 j}_{m_1 m_2 m} \\
&= \delta_{m,m_1+m_2} q^{\frac{1}{4}(j_1+j_2-j)(j_1+j_2+j+1)+\frac{1}{2}(j_1 m_2 - j_2 m_1)} \\
&\times \Delta(j_1 j_2 j)([j_1+m_1]![j_1-m_1]![j_2+m_2]![j_2-m_2]![j+m]![j-m]![2j+1])^{\frac{1}{2}} \\
&\times \sum_n \frac{(-)^n q^{-\frac{n}{2}(j_1+j_2+j+1)}}{[n]![j_1+j_2-j-n]![j_1-m_1-n]![j_2+m_2-n]![j-j_2+m_1+n]![j-j_1-m_2+n]!}
\end{aligned}
$$

$$(10)$$

where the triangle function Δ is given by

$$\Delta(abc) = \left(\frac{[a+b-c]![a-b+c]![-a+b+c]!}{[a+b+c+1]!} \right)^{\frac{1}{2}} \qquad (11)$$

and the symbol $[n]$ is defined in (3), and $[n]! \equiv [n][n-1]\ldots[1]$. (For convenience we drop the subscript q on $[n]_q$ in (3) from now on.)

For $q \to 1$ we find $[n] \to n$ and so the coefficients in (10) reduce to the usual Clebsch-Gordan coefficients $C^{j_1 j_2 j}_{m_1 m_2 m}$ in the van der Waerden form. Kirillov and

Reshetikhin have also given q-analog expressions of the Racah and Majumdar formulas, and a different derivation of the van der Waerden form has been given by Ruegg [22].

In order to derive symmetry properties of $_qC^{j_1 j_2 j}_{m_1 m_2 m}$ we express these coefficients in terms of basic hypergeometric functions $_3\phi_2$. These functions have been extensively studied in the mathematical literature and possess properties which generalize those of the classical hypergeometric functions. These properties have been described and developed by Bailey [23] and Slater [24], and more recently by Askey [25] and Milne [26]. A detailed account which summarizes modern developments has been provided by Gasper and Rahman [27] . Let

$$
(a;q)_n \; = \; \begin{cases} 1 & n = 0 \\ (1-a)(1-aq)\ldots(1-aq^{n-1}) & n > 0. \end{cases}
\tag{12}
$$

then the basic hypergeometric function $_{p+1}\phi_p$ is defined by

$$
_{p+1}\phi_p\left(\begin{matrix} a_1\ a_2\ \ldots\ a_{p+1} \\ b_1\ b_2\ \ldots\ b_p \end{matrix} ; q, z \right) = \sum_{n=0}^{\infty} \frac{(a_1;q)_n(a_2;q)_n\ \ldots\ (a_{p+1};q)_n}{(q;q)_n(b_1;q)_n\ \ldots\ (b_p;q)_n} z^n.
\tag{13}
$$

We can express the symbol $(a;q)_n$ in terms of the notation (3) by means of the formula

$$
(a;q)_n \; = \; (q^\alpha;q)_n \; = \; (1-q)^n q^{\frac{n}{4}(n+2\alpha-3)}([\alpha])_n
\tag{14}
$$

where we have put $a = q^\alpha$ and

$$
([\alpha])_n \; = \; [\alpha][\alpha+1]\ldots[\alpha+n-1].
\tag{15}
$$

We can write the function $_{p+1}\phi_p$ therefore in the form

$$
_{p+1}\phi_p\left(\begin{matrix} q^{\alpha_1} q^{\alpha_2} \ldots q^{\alpha_{p+1}} \\ q^{\beta_1} q^{\beta_2} \ldots q^{\beta_p} \end{matrix} ; q, z \right) = \sum_{n=0}^{\infty} \frac{([\alpha_1])_n([\alpha_2])_n\ldots([\alpha_{p+1}])_n q^{\frac{n\sigma}{2}}}{[n]!([\beta_1])_n([\beta_2])_n\ldots([\beta_p])_n} z^n
$$

$$
= {}_{p+1}\phi_p\left(\begin{matrix} q^{\alpha_1} q^{\alpha_2} \ldots q^{\alpha_{p+1}} \\ q^{\beta_1} q^{\beta_2} \ldots q^{\beta_p} \end{matrix} ; q^{-1}, zq^\sigma \right)
\tag{16}
$$

where

$$
\sigma \; = \; \sum_{i=1}^{p+1} \alpha_i - \sum_{i=1}^{p} \beta_i - 1.
\tag{17}
$$

For $q = 1$ this function reduces to

$$
_{p+1}F_p\left(\begin{matrix} \alpha_1 \ldots \alpha_{p+1} \\ \beta_1 \ldots \beta_p \end{matrix} ; z \right).
\tag{18}
$$

In the case that $\sigma = 0$ in (17) and $z = q$ the function $_{p+1}\phi_p$ is called *balanced* or *Saalschützian*. If one of the numerator parameters $\alpha_1, \ldots \alpha_{p+1}$ is a negative integer $-m$, the series in (16) terminates, and in this case any one of the denominator parameters $\beta_1 \ldots \beta_p$ can also be negative provided it is less than $-m$.

Now we can express $_qC^{j_1 j_2 j}_{m_1 m_2 m}$ in terms of a $_3\phi_2$ function by using the identities

$$[a-n]! = \frac{(-1)^n[a]!}{([-a])_n}, \quad [a+n]! = [a]!([a+1])_n \tag{19}$$

and we find

$$_qC^{j_1 j_2 j}_{m_1 m_2 m}$$

$$= \delta_{m,m_1+m_2} q^{\frac{1}{4}(j_1+j_2-j)(j_1+j_2+j+1)+\frac{1}{2}(j_1 m_2 - j_2 m_1)}$$

$$\times \left(\frac{[2j+1][j_1+m_1]![j_2-m_2]![j+m]![j-m]!}{[j_1+j_2+j+1]![j_1+j_2-j]![j-j_1+j_2]![j+j_1-j_2]![j_2+m_2]![j_1-m_1]!} \right)^{\frac{1}{2}}$$

$$\times ([j-j_2+m_1+1])_{j_1-m_1}([j-j_1-m_2+1])_{j_2+m_2}$$

$$_3\phi_2 \left(\begin{matrix} q^{-j_1-j_2+j} \ q^{-j_1+m_1} \ q^{-j_2-m_2} \\ q^{j-j_2+m_1+1} \ q^{j-j_1-m_2+1} \end{matrix} ; q,q \right).$$

$$\tag{20}$$

Similarly, the Racah form can be written

$$_qC^{j_1 j_2 j}_{m_1 m_2 m}$$

$$= \delta_{m,m_1+m_2}(-)^{j_1-m_1} q^{\frac{1}{4}[j_2(j_2+1)-j_1(j_1+1)-j(j+1)]+\frac{1}{2}m_1(m_1+1)}$$

$$\times \left(\frac{[j+m]![j_2-m_2]![j_1+m_1]![2j+1]}{[j_2+m_2]![j_1-j_2+j]![j_2-j_1+j]![j_1+j_2-j]![j_1+j_2+j+1]![j_1-m_1]![j-m]!} \right)^{\frac{1}{2}}$$

$$\times [j_2+j-m_1]!([j_2-j+m_1+1])_{j_1-m_1} \ _3\phi_2 \left(\begin{matrix} q^{j_1+m_1+1} \ q^{-j+m} \ q^{-j_1+m_1} \\ q^{-j_2-j+m_1} \ q^{j_2-j+m_1+1} \end{matrix} ; q,q \right).$$

$$\tag{21}$$

Because the numerator parameters in (20) are negative the series terminates and although the denominator parameters $j - j_2 + m_1 + 1$, $j - j_1 - m_2 + 1$ could become negative, they are always larger than the numerator parameters, and so no denominator zeroes occur and the $_3\phi_2$ function is well defined. A similar remark holds also for (21).

The utility of (20) or (21) lies in the fact that properties of the q-WCG coefficients can be derived directly from the well-known properties of basic hypergeometric functions. For $q = 1$, symmetries of the Clebsch-Gordan coefficients have been derived using transformation laws of the $_3F_2$ series found originally by Thomae [28] (see for example Biedenharn and Louck [29] and Huszár [30]). The symmetries form a group of order 72 and in fact we require only one transformation of the $_3F_2$ functions together with permutation symmetries in order to derive all 72 symmetries. This transformation can be obtained merely by reversing the order of summation in the terminating $_3F_2$ function. The q-analog of this transformation is obtained similarly, by reversing the order of summation, and is sufficient to enable us to derive the q-analog symmetries. The transformation is:

$$_3\phi_2 \left(\begin{matrix} q^{-\alpha} \ q^{-\beta} \ q^{-n} \\ q^{\delta+1} \ q^{\varepsilon+1} \end{matrix} ; q,z \right) =$$

$$= \frac{([\alpha - n + 1])_n ([\beta - n + 1])_n (-z)^n}{([\delta + 1])_n ([\varepsilon + 1])_n}$$

$$\times q^{-\frac{n}{2}(\alpha + \beta + \delta + \varepsilon + n + 3)} \,_3\phi_2 \left(\begin{matrix} q^{-\delta - n} & q^{-\varepsilon - n} & q^{-n} \\ q^{1 + \alpha - n} & q^{1 + \beta - n} \end{matrix} ; q^{-1}, z^{-1} \right)$$

$$(22)$$

where n is a positive integer.

We can now derive the following relations for the q-Wigner-Clebsch-Gordan coefficients:

$$_q C^{j_1 j_2 j}_{m_1 m_2 m} = (-)^{j_1 + j_2 - j} {}_{q^{-1}} C^{j_1 j_2 j}_{-m_1, -m_2, -m} \qquad (23a)$$

$$_q C^{j_1 j_2 j}_{m_1 m_2 m} = (-)^{j_1 + j_2 - j} {}_{q^{-1}} C^{j_2 j_1 j}_{m_2 m_1 m} \qquad (23b)$$

$$_q C^{j_1 j_2 j}_{m_1, m_2, m_1 + m_2} = {}_q C^{\frac{1}{2}(j_1 + j_2 + m_1 + m_2), \frac{1}{2}(j_1 + j_2 - m_1 - m_2), j}_{\frac{1}{2}(j_1 - j_2 + m_1 - m_2), \frac{1}{2}(j_1 - j_2 - m_1 + m_2), j_1 - j_2} \qquad (23c)$$

$$_q C^{j_1 j_2 j}_{m_1 m_2 m} = (-)^{j_2 + m_2} q^{-\frac{m_2}{2}} \sqrt{\frac{[2j + 1]}{[2j_1 + 1]}} \, {}_{q^{-1}} C^{j j_2 j_1}_{-m, m_2, -m_1} \qquad (23d)$$

$$_q C^{j_1 j_2 j}_{m_1 m_2 m} = (-)^{j_1 - m_1} q^{\frac{m_1}{2}} \sqrt{\frac{[2j + 1]}{[2j_2 + 1]}} \, {}_{q^{-1}} C^{j_1 j j_2}_{m_1, -m, -m_2} \qquad (23e)$$

$$_q C^{j_1 j_2 j}_{m_1 m_2 m} = (-)^{j_2 + m_2} q^{-\frac{m_2}{2}} \sqrt{\frac{[2j + 1]}{[2j_1 + 1]}} \, {}_{q^{-1}} C^{j_2 j j_1}_{-m_2 m m_1} . \qquad (23f)$$

The last two relations can be derived from $(23a, b, d)$ and are included for convenience only. $(23a, b, d)$ are obtained from the transformation (22) with a suitable identification of parameters, namely $n = j_1 + j_2 - j$ for $(23a, b)$ and $n = j_2 + m_2$ for $(23d)$. The symmetry $(23c)$ does not require use of (22), but follows simply from the invariance of the basic hypergeometric series under permutations of its numerator or denominator parameters. This also explains why we have $q \to q^{-1}$ in $(23a, b, d)$ but not $(23c)$. By combining $(23e)$ with $(23a)$ we obtain

$$_q C^{j_1 j_2 j}_{m_1 m_2 m} = (-)^{j - j_2 - m_1} q^{\frac{m_1}{2}} \sqrt{\frac{[2j + 1]}{[2j_2 + 1]}} \, {}_q C^{j_1 j j_2}_{-m_1 m m_2} , \qquad (24)$$

a relation which was found, together with $(23b)$, by Kirillov and Reshetikhin (correcting, however, a minus sign). These symmetry relations have also been obtained by Nomura [31] directly from the formula (10).

The formulation of the coefficients in terms of $_3\phi_2$ functions is also useful in order to relate the different forms in (20) and (21). We can obtain (21) directly from (20) with the help of the transformation

$$_3\phi_2 \begin{pmatrix} q^{\alpha_1} \ q^{\alpha_2} \ q^{-n} \\ q^{\beta_1} \ q^{\beta_2} \end{pmatrix} ; q, q $$

$$= \frac{([\beta_1 - \alpha_1])_n q^{\frac{n\alpha_1}{2}}}{([\beta_1])_n} {}_3\phi_2 \begin{pmatrix} q^{\alpha_1} \ q^{\beta_2-\alpha_2} \ q^{-n} \\ q^{\beta_2} \ q^{\alpha_1-\beta_1+1-n} \end{pmatrix} ; q, q^{\alpha_2-\beta_1+1} \qquad (25)$$

which can be found in Gasper and Rahman [27] (Eq. 3.2.2), and is a special case of an identity due to Sears [32] relating two $_4\phi_3$ functions. In (25), choose $\alpha_1 = -j_1-j_2+j$, $\alpha_2 = -j_2 = m$, $n = j_1-m_1$, $\beta_1 = j-j_2+m_1+1$, $\beta_2 = j-j_1-m_2+1$ and then reverse the order of summation using (22), and we obtain in this way the q-analog of the Racah form (21).

A third form has been given also by Kirillov and Reshetikhin for the q-WCG coefficients, which is the q-analog of a form found by Majumdar [33]. By using the transformation (25) on the expression (20) we obtain this form directly. Specifically, put $n = j_1 + j_2 - j$, $\alpha_1 = -j_2 - m_2$, $\alpha_2 = -j_1 + m_1$, $\beta_2 = j - j_2 + m_1 + 1$, $\beta_1 = j - j_1 - m_2 + 1$ in (25), then this transformation takes (20) into the Majumdar form given in Ref. 19.

4 (q-6j) or q-Racah Coefficients

As noted by Kirillov and Reshetikhin [19] the q-analog of the Racah coefficients can also be expressed in terms of basic hypergeometric functions. For $q = 1$ the expression of Racah coefficients in terms of $_4F_3$ functions with unit argument has been given by Rose [34], Erdélyi [35], and also Jahn and Howell [36], although the identification was incomplete in that the formulas were not valid over the full domain of definition of the quantum numbers occurring in the Racah coefficient. It was shown by Biedenharn and Louck [29] and also Rao and Venkatesh [37]–[39] that three different $_4F_3$ functions are necessary in order to represent the Racah coefficients over the full range of parameters. The generalization of these formulas to the q-analog case is immediate, and so we can represent the q-Racah coefficients as one of three $_4\phi_3$ functions.

The q-analogs of the Racah coefficients are obtained by decomposing the tensor product $V^{j_1} \otimes V^{j_2} \otimes V^{j_3}$ of three irreducible representations of $SU_q(2)$. This can be done in two possible ways (by decomposing first either $V^{j_1} \otimes V^{j_2}$ or $V^{j_2} \otimes V^{j_3}$) and the two sets of orthogonal bases obtained thereby are linearly dependent. This relation yields the q-Racah coefficients which, using the orthogonality properties of the q-WCG coefficients, can be obtained in the form:

$$([2e + 1][2f + 1])^{\frac{1}{2}} \begin{Bmatrix} a & b & e \\ d & c & f \end{Bmatrix}_q (-)^{a+b+c+d} {}_qC^{afc}_{\alpha,\gamma,\alpha+\gamma}$$

$$= \sum_{\beta\delta} {}_qC^{bdf}_{\beta\delta\gamma} \ {}_qC^{edc}_{\alpha+\beta,\delta,\alpha+\gamma} \ {}_qC^{abe}_{\alpha,\beta,\alpha+\beta} . \qquad (26)$$

An explicit expression for the (q-6j) symbols is the following [19]:

$$\begin{Bmatrix} a & b & e \\ d & c & f \end{Bmatrix}_q$$

$$= \Delta(abe)\,\Delta(cde)\,\Delta(acf)\,\Delta(bdf)$$

$$\times \sum_n \frac{(-)^n [n+1]!}{[n-a-b-e]!\,,[n-a-c-f]!\,,[n-b-d-f]!\,,[n-d-c-e]!}$$

$$\times \frac{1}{[a+b+c+d-n]!\,,[a+d+e+f-n]!\,,[b+c+e+f-n]!} \tag{27}$$

where the summation is over integers n satisfying

$$\max(a+b+e,\, a+c+f,\, b+d+f,\, d+c+e)$$
$$\le n \le \min(a+b+c+d,\, a+d+e+f,\, b+c+e+f) \tag{28}$$

as is determined from the denominator factors in (27). *We note that, unlike the q-Wigner-Clebsch-Gordan coefficient, (27) is invariant under $q \leftrightarrow q^{-1}$.*

In order to identify (27) with a $_4\phi_3$ function it is necessary to change the summation index n so as to accord with the definition (16) for $p = 3$, that is, we put $m = n - \min(a+b+c+d,\, c+d+e+f,\, b+c+e+f)$, and the three possibilities for m give the three expressions of the coefficient in terms of $_4\phi_3$. The identity which enables us to make this identification is

$$\sum_n \frac{(-)^n [n+1]!}{[n-\alpha_1]!\,,[n-\alpha_2]!\,,[n-\alpha_3]!\,,[n-\alpha_4]!\,,[\beta_1-n]!\,,[\beta_2-n]!\,,[\beta_3-n]!}$$

$$= \frac{(-)^{\beta_1}[\beta_1+1]!}{[\beta_2-\beta_1]!\,,[\beta_3-\beta_1]!\,,[\beta_1-\alpha_1]!\,,[\beta_1-\alpha_2]!\,,[\beta_1-\alpha_3]!\,,[\beta_1-\alpha_4]!}$$

$$\times {}_4\phi_3 \left(\begin{matrix} q^{\alpha_1-\beta_1}\ q^{\alpha_2-\beta_1}\ q^{\alpha_3-\beta_1}\ q^{\alpha_4-\beta_1} \\ q^{-\beta_1-1}\ q^{\beta_2-\beta_1+1}\ q^{\beta_3-\beta_1+1} \end{matrix} ; q, q \right) \tag{29}$$

where

$$\beta_1 \ge \max(\alpha_1, \alpha_2, \alpha_3, \alpha_4)$$
$$\beta_1 \ge 0,\ \beta_2 \ge \beta_1,\ \beta_3 \ge \beta_1$$
$$\alpha_1 + \alpha_2 + \alpha_3 + \alpha_4 = \beta_1 + \beta_2 + \beta_3 \tag{30}$$

which can be obtained directly from the definition (16). Now we choose

$$\beta_1 = \min(a+b+c+d,\, a+d+e+f,\, b+c+e+f) \tag{31}$$

with β_2, β_3 being the remaining two parameters and $(\alpha_1, \alpha_2, \alpha_3, \alpha_4)$ equal to any permutation of $(a+b+c,\, c+d+e,\, a+c+f,\, b+d+f)$. These parameters satisfy the constraints (30), in particular the constraint which ensures that the $_4\phi_3$ in (29) is balanced.

The (q-6j) symbols satisfy all the same 144 symmetry relations as for $q = 1$. These symmetries all follow by termwise rearrangement of factors in the series (27) and are also directly visible from the expression (29). All 144 symmetries are equivalent to permutations of the numerator or denominator parameters amongst themselves, exactly as for the $q = 1$ case. The (q-6j) symbols also satisfy the same orthogonality relations as for $q = 1$, and in fact their definition has already been anticipated by Askey and Wilson [7], who obtained q-analogs of the orthogonal Racah polynomials, as mentioned in the Introduction. Evidently, the quantum group $SU_q(2)$ underlies the properties of these polynomials, which were found from other considerations.

5 Asymptotic Limit of the q-6j Symbol

It was known very early [40] that the Wigner-Clebsch-Gordan coefficients can be regarded as a certain limit of Racah coefficients [10]. This fact has numerous consequences and in particular shows how properties of Wigner-Clebsch-Gordan coefficients can be derived from those of Racah coefficients. In order to obtain the q-analog result of this limit relation let us first introduce the following notation for the q-Racah coefficient (as in Ref. 10):

(a) $_qW^{abc}_{\rho\sigma\tau}(j) \equiv 0$ unless $\rho + \sigma = \tau$;

(b) $_qW^{abc}_{\rho\sigma\tau}(j) \equiv 0$ unless each of the triples $(j - \tau, a, j - \sigma)$, $(j - \sigma, b, j)$,
$\qquad\qquad\qquad\qquad$ $(j - \tau, c, j)$ consists of nonnegative integers and
$\qquad\qquad\qquad\qquad$ half-integers that satisfy the triangle conditions;

(c) $_qW^{abc}_{\rho\sigma\tau}(j) \equiv ([2c+1][2j - 2\sigma + 1])^{\frac{1}{2}} (-)^{a+b+\tau} \begin{Bmatrix} j - \tau & a & j - \sigma \\ b & j & c \end{Bmatrix}_q$

$\qquad\qquad$ otherwise.   (32)

Since we have

$$([2e + 1][2f + 1])^{\frac{1}{2}} \begin{Bmatrix} a & b & e \\ d & c & f \end{Bmatrix}_q = (-)^{a+b+c+d} {}_qW^{bdf}_{e-a,c-e,c-a}(c)   \quad (33)$$

where $b \geq |e - a|$, $d \geq |c - e|$, and $f \geq |c - a|$, each Racah coefficient is expressible in this notation.

Now we wish to show that the limit $j \to \infty$ of $_qW^{bdf}_{\alpha\beta\gamma}(j + \gamma)$ is the q-WCG coefficient $_qC^{bdf}_{\alpha\beta\gamma}$. In order to do this we first put $a = j$, $b = b$, $c = j + \gamma$, $d = d$, $e = j + \alpha$ and $f = f$ in Eqs. (27-31) and write $\beta_i = 2j + \delta_i$ where

$$\delta_1 = \min(b + \alpha + d + \beta, \, d - \beta + f + \gamma, \, b + \alpha + f + \gamma)   \quad (34)$$

and δ_2, δ_3 are the pair of integers remaining in the triple

$$(b + \alpha + d + \beta, \, d - \beta + f + \gamma, \, b + \alpha + f + \gamma)   \quad (35)$$

after removing δ_1. Similarly we may write $\alpha_i = 2j + \varepsilon_i$ where $(\varepsilon_1, \varepsilon_2, \varepsilon_3)$ is any permutation of

$$(b + \alpha, \ d + \alpha + \gamma, \ f + \gamma) \tag{36}$$

and $\alpha_4 = b + d + f$. Using these results, we obtain

$$_qW^{bdf}_{\alpha\beta\gamma}(j + \gamma) = \delta_{\alpha+\beta,\gamma}[2f + 1]^{\frac{1}{2}}(-)^{b+d+\gamma+\delta_1}\Delta(bdf)$$

$$\times \left([b - \alpha]! \ [b + \alpha]! \ [d + \beta]! \ [d - \beta]! \ [f + \gamma]! \ [f - \gamma]!\right)^{\frac{1}{2}}$$

$$\times \left(\frac{[2j + 2\alpha + 1]^{\frac{1}{2}}([2j + \delta_1 - b - d - f + 1])_{b+d+f+1}}{\left(([2j + \alpha - b + 1])_{2b+1}([2j + \alpha + \gamma - d + 1])_{2d+1}([2j + \gamma - f + 1])_{2f+1}\right)^{\frac{1}{2}}}\right)$$

$$\times \frac{{}_4\phi_3\left(\begin{array}{cccc} q^{\varepsilon_1-\delta_1} & q^{\varepsilon_2-\delta_1} & q^{\varepsilon_3-\delta_1} & q^{b+d+f-2j-\delta_1} \\ q^{-2j-\delta_1-1} & q^{\delta_2-\delta_1+1} & q^{\delta_3-\delta_1+1} & \end{array} ; q, q\right)}{[\delta_2 - \delta_1]! \ [\delta_3 - \delta_1]! \ [\delta_1 - \varepsilon_1]! \ [\delta_1 - \varepsilon_2]! \ [\delta_1 - \varepsilon_3]!}. \tag{37}$$

In letting $j \to \infty$ we use the limit

$$\frac{([a - 2j])_n}{([b - 2j])_n} \longrightarrow q^{\frac{n}{2}(b-a)} \tag{38}$$

in which we assume $q > 1$; the case $q < 1$ is obtained by replacing $q \leftrightarrow q^{-1}$. The limit of the ${}_4\phi_3$ function in (35) is then

$$_3\phi_2\left(\begin{array}{ccc} q^{\varepsilon_1-\delta_1} & q^{\varepsilon_2-\delta_1} & q^{\varepsilon_3-\delta_1} \\ q^{\delta_2-\delta_1+1} & q^{\delta_3-\delta_1+1} & \end{array} ; q, q\right). \tag{39}$$

In order to find the limit of the remaining j-dependent factors in (37) we use

$$([a + 2j])_n \longrightarrow \frac{q^{\frac{n}{4}(2a+4j+n-1)}}{(q^{\frac{1}{2}} - q^{-\frac{1}{2}})^n} \tag{40}$$

for $q > 1$, which from (37) gives the q-factor q^{μ}, where

$$\mu = \tfrac{1}{4}(b + d + f + 1)(2\delta_1 - b - d - f) - \tfrac{1}{2}\alpha(b + d) - \tfrac{1}{2}\gamma(d + f + 1). \tag{41}$$

Taking the case $\delta_1 = b + \alpha + \beta + d$, we find that the limit of (37) is *precisely* the q-Wigner-Clebsch-Gordan coefficient (20):

$$\lim_{j\to\infty} {}_qW^{bdf}_{\alpha\beta\gamma}(j + \gamma) = \begin{cases} {}_qC^{bdf}_{\alpha\beta\gamma} & \text{if } q > 1 \\ {}_{q^{-1}}C^{bdf}_{\alpha\beta\gamma} & \text{if } q < 1 \end{cases} \tag{42}$$

where in the last equation we used the invariance of the q-Racah coefficient under $q \leftrightarrow q^{-1}$.

Many properties of the q-Clebsch-Gordan coefficients follow from those of the q-Racah coefficients using (42). For example, as noted by Nomura [31] the invariance of the (q-6j) symbol under exchange of the first and third columns leads directly to the symmetry (24).

Concluding Remark:

One of the fundamental problems that Professor Doebner pioneered concerned quantization on topologically non-trivial manifolds [41]. Consider for example N identical point particles in $\mathbb{R}^2$, and denote by $\mathbf{X}_n$ the space $\{(x_1, x_2, \ldots, x_n) | x_i \in \mathbb{R}^2; x_i \neq x_j \text{ if } i \neq j\}$. The *physical space* – because of the identity of particles – is not $\mathbf{X}_N$ but $\mathbf{Y}_N \equiv (\mathbf{X}_N - \Delta)/\mathbf{S}_N$, where Δ is the 'diagonal' set ($x_i = x_j$, for some $i \neq j$) and $\mathbf{S}_N$ is the symmetric group. The fundamental group of the physical space $\mathbf{Y}_N$, denoted $\pi_1(\mathbf{Y}_N)$ is the N-string *braid group*, $\mathbf{B}_N$, whose irreps define fundamental particle symmetry types. For two particles the braid group $\mathbf{B}_2$ provides symmetry types (now called anyons [42]) interpolating continuously between symmetric (bosons) and anti-symmetric (fermions), related to the symmetry structure (33b) we have found above for the q-WCG coefficients (although for anyons q is taken to be complex (a phase factor, with $|q|=1$)). It is because of this connection to Doebner's own researches that we hope our contribution to his *Festschrift* will be considered appropriate.

References

1. E. Sklyanin, L. Takhtajan, L. Faddeev: Theor. Math. Phys. **40** 194 (1979) (in Russian)
2. P. Kulish, N. Reshetikhin: Zap. nauch. seminarov LOMI **101** 101 (1981) (in Russian)
3. V.G. Drinfeld: *Quantum Groups*, Proc. Int. Congr. of Math., MSRI Berkeley, CA, (1986) 798;
    Sov. Math. Dokl. **36** 212 (1988)
4. C. Zachos: ANL-HEP-PR-90-61 to appear in *Symmetries in Science V*, ed. by B. Gruber (Plenum, N.Y.)
5. M. Jimbo: Lett. Math. Phys. **11** 247 (1986)
6. A. Shapere, F. Wilczek: *Geometric Phases in Physics*, Advanced Series in Mathematical Physics, Vol. 5 (World Scientific, Singapore, 1989)
7. R. Askey, J.A. Wilson: Siam J. Math. Anal. **10** 1008 (1985)
8. V.G. Drinfeld: Sov. Math. Dokl. **32** 254 (1985)
9. M. Jimbo: Lett. Math. Phys. **10** 63 (1985)
10. L.C. Biedenharn, J.D. Louck: *Angular Momentum in Quantum Physics*, Encyclopedia of Mathematics and Its Applications, **8**, (Addison Wesley, 1981)
11. A.J. Macfarlane: J. Phys. A. Math. Gen. **22** 4581 (1989)
12. L.C. Biedenharn: J. Phys. A. Math. Gen. **22** L873 (1989)
13. C.-P. Sun, H.-C. Fu: J. Phys. A. Math. Gen. **22** L983 (1989)
14. L.C. Biedenharn, M. Tarlini: Lett. Math. Phys. **20** 271 (1990)
15. I.M. Gel'fand, A.V. Zelevinsky: Société Math. de France, Astérisque, hors series, **117** (1985)
16. G. Lusztig: Adv. in Math. **70** 237 (1988)
17. M. Rosso: Commun. Math. Phys. **117** 581 (1988)

18.   V. Pasquier, H. Saleur: Nucl. Phys. **B330** 523 (1990)
19.   A.N. Kirillov, N.Yu. Reshetikhin: "Representations of the Algebra $U_q(s\ell(2))$, q-Orthogonal Polynomials and Invariants of Links", USSR Academy of Sciences (preprint) 1988
20.   L. Vaksman: Sov. Math. Dokl. **39** 467 (1989)
21.   M. Jimbo: Commun. Math. Phys. **102** 537 (1986)
22.   H. Ruegg: J. Math. Phys. **31** 1085 (1990)
23.   W. N. Bailey: *Generalized Hypergeometric Series* (Cambridge Univ. Press, Cambridge, 1935, reprinted Hafner, New York, 1972)
24.   L.J. Slater: *Generalized Hypergeometric Functions* (Cambridge University Press, Cambridge, 1966)
25.   R. Askey, J.A. Wilson: Memoirs Amer. Math. Soc. **319** (1985)
26.   S.C. Milne: Advances in Math. **72** 59 (1988)
27.   G. Gasper, M. Rahman: *Basic Hypergeometric Series*, Encyclopedia of Mathematics and its Applications, **35** (Cambridge University Press, 1990)
28.   J. Thomae: J. für Math. **87** 26–73 (1879)
29.   L.C. Biedenharn, J.D. Louck: *Racah-Wigner Algebra in Quantum Theory*, Encyclopedia of Mathematics and Its Applications, **9** (Addison Wesley, 1981)
30.   M. Huszár: Acta Phys. Acad. Scient. Hung. **32** 181–185 (1972)
31.   M. Nomura: J. Math. Phys. **30** 2397 (1989);
       J. Phys. Soc. Jap. **58** 2694 (1989); *ibid.* **59** 439 (1990)
32.   D.B. Sears: Proc. London Math. Soc. (2) **53** 158–180 (1951)
33.   S.D. Majumdar: Prog. of Theor. Phys. **20** 798–803 (1958)
34.   M.E. Rose: *Multipole Fields*, p. 92 (Wiley, New York, 1955)
35.   A. Erdélyi: Math. Rev. **14** 642 (1957)
36.   H.A. Jahn, K.M. Howell: Proc. Cambridge Phil. Soc. **55** 338 (1959)
37.   K.S. Rao, T.S. Santhanam, K. Venkatesh: J. Math. Phys. **16** 1528 (1975)
38.   K.S. Rao, K. Venkatesh: "Representation of the Racah coefficient as a generalized hypergeometric function," in Proceedings Fifth International Colloquium on Group Theoretical Methods in Physics, ed. by R.T. Sharp and B. Kolman, pp. 649–656 (Academic Press, New York, 1977)
39.   K. Venkatesh: J. Math. Phys. **19** 1973 (1978); *ibid.*, 2060
40.   L.C. Biedenharn: J. Math. and Phys. (MIT) **XXXI** 287 (1953)
41.   H.D. Doebner, J. Tolar: in *Symposium on Symmetries in Science*, Carbondale, Illinois, 1979, ed. by B. Gruber and R.S. Millman, p. 475 (Plenum, New York, 1980)
42.   L. Biedenharn, E. Lieb, B. Simon, F. Wilczek: Physics Today. p. 90 (Aug. 1990)

Symmetry Groups and Spectrum Generating Groups

Arno Bohm

Department of Physics
The University of Texas
Austin, Texas 78712

Abstract: In this article we want to explain how groups are used in quantum physics. Conventionally, in physics a group is used as the group of motion of an elementary particle. These groups are the symmetry groups. Their irreducible representation spaces describe the space of physical states of an elementary particle. Most elementary particles become, when the energy increases, extended quantum physical objects. The symmetry group then describes the center of mass motion of the extended object. In addition to the center of mass motion, an extended object can perform intrinsic motions. Examples of intrinsic motions are rotations about the center of mass or vibrations of various kinds. Intrinsic motions are described by the spectrum generating group (also called dynamical group). Different intrinsic motions are described by different spectrum generating groups. Simple intrinsic motions, like rigid rotations of a dumbbell about its center of mass, are described by small groups (groups with a small number of parameters). Complicated intrinsic motions, like combinations of rotations and various kind of oscillations, require a large spectrum generating group.

An irreducible representation space of the spectrum generating group describes the space of all physical states of the extended physical object. Subgroup reduction of this irreducible representation space then gives the entire spectrum of the extended object. The various elementary particles are described as different states of the extended object and these different states are different states of the intrinsic motion. The spectrum generating group approach thus analyzes an extended physical object in terms of its intrinsic motions.

In this article we only want to explain the principles of the group theoretical approach in quantum physics. Therefore we use only the simplest example, the rigid rotator. In Sect. 1 we briefly review the quantum mechanical notions. Section 2 presents the symmetry group, Sect. 3 the spectrum generating group in nonrelativistic physics. Section 4 describes how these ideas can be generalized to the nonrelativistic case, when the combination of symmetry group and spectrum generating group is no more trivial.

1 Preliminaries

In quantum mechanics [1] a physical system is described by a linear scalar product space Φ (physicists call Φ the Hilbert space, though it is usually only a dense subspace of the Hilbert space $\mathcal{H} : \Phi \subset \mathcal{H}$; we will mention these distinctions only parenthetically).

Physical states (pure states) are vectors $\varphi \in \Phi$ (with φ and $e^{i\alpha}\varphi$ describing the same state), physical observables are linear operators A in Φ.

The quantities measured in an experiment are a property of the observable *and* of the state and given mathematically by quantities like

$|\langle \varphi \mid \psi \rangle|^2$:   the probability to measure the property of the state φ on a physical system which is in a state ψ

$|\langle \psi \mid A\varphi \rangle|^2$:   the transition probability from a state φ into a state ψ caused by the observable A,

$\langle \varphi \mid A\varphi \rangle$:   the expectation value (average value) of the observable A in the state φ.

$|\langle s, x_i \mid \psi \rangle|^2$:   the probability density for finding the state ψ at the position x_i.

The probability density is the important quantity that we will use to introduce the symmetry group. Therefore we will define it a little better in the following: Let us consider the particular observables $Q = Q_1, Q_2, Q_3$ (self-adjoint, strongly commuting with each other) which represent the components of the position. Then one can show that:

There exists a set of (generalized) eigenvectors[1] $|x_i, s \rangle$ of Q_i such that every $\psi \in \Phi$ can be written as

$$\psi = \sum_s \int_{-\infty < x_i < +\infty} \mathrm{d}^3 x \, |x_i, s ><s, x_i \mid \psi > \; . \tag{1}$$

The set of values x_i (in this case $-\infty < x_i < +\infty$) is called the (continuous) spectrum of Q_i. s is the degeneracy label which we often suppress.

This is of course just an infinite continuous (because the spectrum is continuous) generalization of the basis vector expansion of a vector V in the three-dimensional space $V = \sum_{k=1,2,3} e_k (e_k \cdot V)$ where the spectrum consists only of three values labeling the three directions. The generalized basis vectors are usually "δ- function normalized":

$$\langle x_i' \mid x_i \rangle = \delta(x_1' - x_1)\delta(x_2' - x_2)\delta(x_3' - x_3) \, .$$

---

[1] More precisely there exists a set of antilinear continuous functionals on the space $\Phi, |x_i, s\rangle \in \Phi^\times \supset \mathcal{H} \supset \Phi$, such that $\langle Q_i\varphi \mid x_i \rangle = x_i \langle \varphi \mid x_i \rangle$ for every $\varphi \in \Phi$, where $\langle \varphi \mid x_i \rangle = \overline{\langle x_i \mid \varphi \rangle}$ denotes the value of the antilinear functional $|x_i\rangle$ at the vector $\varphi \in \Phi$ [1]. This statement (1) can be proven and is called the nuclear spectral theorem, but it was used by Dirac long before the underlying mathematics had been created.

As is clear from (1), they are only determined up to an arbitrary phase factor:

$$| x_i \rangle \text{ and } | x_i \rangle' = e^{i\alpha(x_i)} | x_i \rangle, \tag{2}$$

where the real α can depend upon x_i, are equally good basis vectors (with the same δ-function normalization). The component $< s, x_i | \psi >$ of ψ along the basis vector $| x_i, s >$ is in physics usually called the wave function

$$\psi_s(x_i) = < s, x_i | \psi >,$$

it represents the probability density; i.e. the probability to find the state in the interval Δx is

$$\int_{\Delta x} d^3 x < \psi | x_i >< x_i | \psi > = \int_{\Delta x} d^3 x | < x_i | \psi > |^2 \approx \Delta x | < x_i | \psi > |^2 .$$

Because of (2) one can always replace $\psi(x_i)$ by $\psi'(x_i) = e^{i\alpha(x_i)} \psi(x_i)$.

2 Symmetry Groups

2.1 Definition of the Symmetry Group in Quantum Mechanics

Let us consider a state that has been prepared such that it is aligned along a particular direction (e.g. the e_1-axis as in Fig. 1). This state can decay and (one of) its decay products are registered by the detector. The counting rate of the detector (number of clicks) at the position x_i is proportional to $| < x_i | \varphi > |^2$.

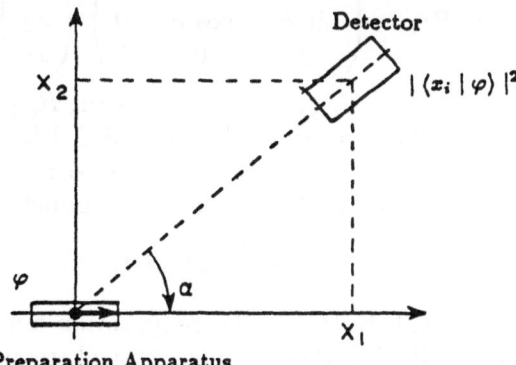

Fig. 1. Preparation apparatus and registration apparatus (detector) in an experiment on microphysical systems. The preparation apparatus prepares the state φ. The detector measures the probability for the state φ at the position x_i: $| \langle x_i | \varphi \rangle |^2 \Delta x$.

We can rotate the detector relative to the state (Fig. 1a) or the state relative to the detector (Fig. 1b).

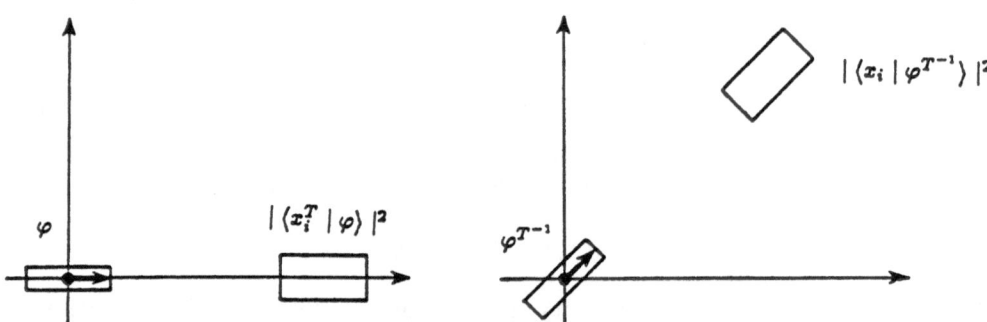

Fig. 1a. The symmetry transformation of the detector changes the observable. Now the detector measures the probability of φ at the transformed position x_i^T : $|\langle x_i^T \mid \varphi \rangle|^2 \, \Delta x_i$.

Fig. 1b. The symmetry transformation of the preparation apparatus changes the state. Now the detector measures the probability of the transformed state $\varphi^{T^{-1}}$ at the position x_i : $|\langle x_i \mid \varphi^{T^{-1}} \rangle|^2 \, \Delta x_i$.

It is intuitively clear that the counting rate for φ at x_i^T is equal to the counting rate for $\varphi^{T^{-1}}$ at x_i: i.e.

$$\left|\langle x_i^T \mid \varphi \rangle\right| = \left|\langle x_i \mid \varphi^{T^{-1}} \rangle\right| . \tag{3}$$

This is the symmetry condition.

We will denote by R the rotation that transforms the detector position from x_i to x_i^T:

$$x_i^T = R x_i = \begin{pmatrix} \cos\alpha & -\sin\alpha & 0 \\ \sin\alpha & \cos\alpha & 0 \\ 0 & 0 & 1 \end{pmatrix} \begin{pmatrix} x_1 \\ x_2 \\ x_3 \end{pmatrix} .$$

We can consider two rotations $R_{(1)}, R_{(2)}$ and perform $R_{(1)}$ after $R_{(2)}$ which we call the product $R_{(1)} R_{(2)}$. The rotations with this definition of a product form a group. Let us denote by $R_i(\alpha)$ the rotation about the i-th axis by an angle α . One can show that any rotation can be obtained as the product

$$R(\alpha, \beta, \gamma) = R_3(\alpha) R_2(\beta) R_3(\gamma)$$

(α, β, γ are the Euler angles). Rotating the preparation apparatus by a rotation R^{-1} (by angle $-\alpha$) is described in the space of physical states Φ (or $\mathcal{H}$) by a linear operator $U(R^{-1})$. Thus the rotation R^{-1} is associated to an operator

$$R^{-1} \;\rightarrow\; U(R^{-1}) \tag{4}$$

and the rotated state $\varphi^{T^{-1}}$ is given by

$$\varphi^{T^{-1}} = U(R^{-1})\varphi .$$

The symmetry condition (3) thus reads

$$|< R\boldsymbol{x} \mid \varphi >| = |< \boldsymbol{x} \mid U(R^{-1}) \mid \varphi >| \,. \tag{3'}$$

U must be a unitary operator if the probabilities of φ^T and φ are to be the same. Let us now consider the rotations R_1, R_2 and $R = R_1 R_2$ $(R^{-1} = R_2^{-1} \cdot R_1^{-1})$. From (3) one obtains

$$
\begin{aligned}
|\langle \boldsymbol{x} \mid U(R^{-1}) \mid \varphi \rangle| &= |\langle R\boldsymbol{x} \mid \varphi \rangle| = |\langle R_1 R_2 \boldsymbol{x} \mid \varphi \rangle| = |\langle R_2 \boldsymbol{x} \mid U(R_1^{-1}) \mid \varphi \rangle| \\
&= |\langle \boldsymbol{x} \mid U(R_2^{-1}) U(R_1^{-1}) \mid \varphi \rangle| \,.
\end{aligned}
\tag{5}
$$

As this must be true for all basis vectors $\mid x_i >$ and all states φ it follows that

$$U(R_2^{-1})U(R_1^{-1}) = \omega(R_2^{-1}, R_1^{-1})\dot{U}(R_2^{-1} \cdot R_1^{-1}) \tag{6}$$

where $\omega(R_1, R_2)$ is a phase factor, $|\omega| = 1$, which appears because (5) is an equality for the absolute values. If in the place of (5) one would have the corresponding expressions for the matrix elements:

$$\langle \boldsymbol{x} \mid U(R^{-1}) \mid \varphi \rangle = \langle \boldsymbol{x} \mid U(R_2^{-1})U(R_1^{-1}) \mid \varphi \rangle \tag{5}$$

then in place of (6) one would have

$$U(R_2^{-1})U(R_1^{-1}) = U(R_2^{-1} \cdot R_1^{-1}), \tag{6'}$$

i.e. $\omega(R_1, R_2) = 1$ would hold.

We compare this with the definition of a representation of a group G [2]: The correspondence

$$g \to T_g$$

which associates with every $g \in G$ a linear operator in the space $\mathcal{H}$ is called a linear projective representation if

$$
\omega(g_1, g_2) \cdot T_{g_1 g_2} = T_{g_1} T_{g_2} \quad \text{with} \quad |\omega(g_1, g_2)| = 1 \,,
$$
$$
T_e = 1 \ (e = \text{unit element of } G) \,.
$$

If also $\omega(g_1, g_2) = 1$ for every $g_1, g_2 \in G$ then $g \to T_g$ is a (proper) representation of G.

From this we see that $R \to U(R)$ of (4) is a projective representation of the rotation group. For $\omega(R_1, R_2) = 1$ the correspondence (4) is a representation of the rotation group. Further all $U(R)$ are unitary operators, $(U\varphi, U\psi) = (\varphi, \psi)$, thus we have a unitary projective representation or a unitary proper representation of the rotation group.

The representation $R \to U(R)$ is also assumed to be a continuous representation. This means that the operators $U_i(\alpha) = U(R_i(\alpha))$ (where $R_i(\alpha)$ is a rotation about the i-th axis by an angle α), called a one parameter group of operators, are continuous function of all parameters, i.e. that

$$U(\alpha) \to U(\alpha_0) \quad \text{if} \quad \alpha \to \alpha_0$$

in a certain sense which in the Hilbert space $\mathcal{H}$ is defined as

$$\| U(\alpha)\varphi - U(\alpha_0)\varphi \| \to 0 \qquad \text{as} \qquad \alpha \to \alpha_0 \qquad \text{for all} \qquad \varphi \in \mathcal{H}.$$

One can show that then for any arbitrary (non-compact) group at least on a dense subspace $D \subset \mathcal{H}$, called space of differentiable vectors [3], the following operator can be defined by differentiation:

$$\frac{1}{i} J_j \varphi \equiv \lim_{h \to 0} \frac{U_j(\alpha_0 + h) - U_j(\alpha_0)}{h} \varphi \equiv \left. \frac{dU(\alpha)}{d\alpha} \right|_{\alpha = \alpha_0} \varphi \qquad (7)$$

for every $\varphi \in \mathcal{H}$ (or at least for every $\varphi \in D$ where D is left invariant by J_j).

$J_i (i = 1, 2, 3)$ are linear operators which are self-adjoint, $J_i^\dagger = J_i$. They are the generators of the group representations. One can differentiate $U_i(\alpha)$ an arbitrary number of times and expand it into a Taylor series about $\alpha = 0$. The result is (for any arbitrary non-compact group at least on a dense subspace $A \subset D \subset \mathcal{H}$ called the space of analytic vectors [3]):

$$U_j(\alpha) = 1 + \alpha \frac{1}{i} J_j + \frac{\alpha^2}{2!} \left(\frac{1}{i} J_j \right)^2 + \ldots \frac{\alpha^n}{n!} \left(\frac{1}{i} J_j \right)^n + \ldots \equiv e^{-i\alpha J_j}. \qquad (8)$$

One says that J_i generates the transformation $U_j(\alpha)$. The properties of the generators, their commutation relations, follow from the property of the group. For the rotation group one derives:

$$J_i J_k - J_k J_i \equiv \left[J_i, J_k \right] = i\varepsilon_{ikj} J_j, \quad i.e. \quad \left[J_1, J_2 \right] = i J_3 \text{ (and cyclic)}. \qquad (9)$$

For a general group G one obtains similarly the commutation relation of the corresponding Lie algebra:

$$\left[X_i, X_j \right] = i C_{ij\ell} X_\ell, \qquad C_{ij\ell} \in \mathbb{R} \text{ are the structure constants.} \qquad (10)$$

Summarizing we conclude: Symmetry transformations (which are in quantum physics transformations of the detector relative to the preparation apparatus) are described by a continuous unitary (projective) representation of the symmetry group. The generators of the representation defined by differentiation of the group representation $G \ni g \to U(g)$ fulfill the commutation relations of the Lie algebra of the symmetry group as in (10).

The operators X_i (generators of the representation) represent physical observables.

The representation space $\mathcal{H}^{(U)}$ in which U acts, is the space of physical states of the physical system. $\mathcal{H}^{(U)}$ and $g \to U(g)$ specify this physical system.

A representation space is irreducible if it has no (nontrivial) subspace invariant under U. *Irreducible* representation (irrep) spaces describe *elementary* physical systems.

For the specific example of the rotation group SO(3) the generators J_i represent angular momentum. Equation (7) and therewith (9) is the most general definition of angular momentum.

The Casimir operators $\mathcal{C}$ are the algebraic functions of the generators X_i which commute with $U(g)$:

$$\left[\mathcal{C}, U(g)\right] = 0 \qquad \text{and} \qquad \left[\mathcal{C}, X_i\right] = 0.$$

The eigenvalues of $\mathcal{C}$ give the quantum numbers that specify the elementary physical system.

For the specific case of SO(3) the Casimir operator is given by $\mathcal{C}(\text{SO}(3)) = \sum_{i=1}^{3} J_i^2 \equiv \boldsymbol{J}^2$. The eigenvalues of $\boldsymbol{J}^2$ are $j(j+1)$.

For irreducible proper representations: $j = 0, 1, 2, 3 \ldots$.

For projective (double valued) irreducible representations: $j = \frac{1}{2}, \frac{3}{2}, \frac{5}{2} \ldots$.
As the symmetry condition of (3) requires only projective representations, we expect integer *and* half-integer values of j to be realized in nature, which is actually observed. The group which has the same set of irreducible representations as the symmetry group has projective representations is called (following Wigner) the quantum mechanical symmetry group [4]. The quantum mechanical rotation group is the group SU(2). SU(2) is the group of operators $U(R)$ representing the rotation R.

There is an infinite set of groups of operators $U^{(j)}(R)$ one for each value $j = 0, \frac{1}{2}, 1, \frac{3}{2}, 2 \ldots$ (representations of SU(2)) each acting in a $(2j + 1)$ dimensional space $\mathcal{R}^{(j)}$. Each $\mathcal{R}^{(j)}$ is the space of physical states of an elementary physical system, which we call elementary rotator, j is the quantum number, called angular momentum, which characterizes the elementary rotator.

The electron – when translations and charges are ignored – is an example of a physical system corresponding to $\mathcal{R}^{(1/2)}$.

The eigenvalues of the operator J_3 are denoted by j_3 :

$$J_3 f = j_3 f, \ f \in \mathcal{R}^{(j)}.$$

These $(2j + 1)$ eigenvalues are $j_3 = -j, -j + 1, -j + 2, \ldots, +j$. They are also called the weights of the irrep $g \to U^{(j)}(g)$, and the eigenvectors $f = f_{j_3}$ (there are $(2j + 1)$ eigenvectors, one for each value of j_3, and they are orthogonal) are called the weight vectors of the irrep. The graphical representation of the weights is called the weight diagram. Examples of weight diagrams are given in Fig. 2.

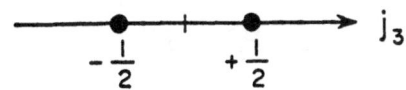

Fig. 2a. Weight diagram of $\mathcal{R}^{(j=\frac{1}{2})}$.

Fig. 2b. Weight diagram of $\mathcal{R}^{(j=2)}$. The number j_3 is a quantum number giving the value of the component of angular momentum (or helicity). The weight vector f_{j_3} describes the state of the elementary rotator whose component is j_3 (in which a measurement of the observable j_3 gives always the value j_3).

2.2 The Quantum Mechanical Symmetry Group of Nonrelativistic Space-Time

The group of symmetry transformations of nonrelativistic space time is larger than SO(3). It is the Galilei group [5] $\mathcal{G}$, which in addition to the rotations consists of translations ($x_i \rightarrow x_i + a_i$), pure Galilei transformations ($x_i \rightarrow x_i + v_i t$), and time translations ($t \rightarrow t + \tau$). The generators and their commutation relations are the following:

J_i generating rotations and fulfilling the commutation relations (9).
P_i generating translations and fulfilling $[P_i, P_j] = 0$.
K_i generating pure Galilei transformations and fulfilling $[K_i, K_j] = 0$.
H generating time translations and fulfilling $[K_i, H] = iP_i$.

The P_i and K_i are J-vector operators (i.e. $[J_i, P_j] = i\varepsilon_{ijk}P_k$) and generate with the J_i two three-dimensional Euclidean groups $E(3)_{P_i\,J_i}$ and $E(3)_{Q_i\,J_i}$. Though translations and pure Galilei transformations commute one *derives* for a *projective* representation

$$[K_i, P_j] = im\mathbb{1}\delta_{ij} \tag{11}$$

where m is a real number (the infinitesimal exponent) which, together with other numbers, characterizes the irreps of the quantum mechanical Galilei group. If one defines the operator

$$Q_i = \frac{1}{m}K_i \tag{12}$$

then (11) takes the form:

$$[Q_i, P_j] = i\mathbb{1}\delta_{ij}. \tag{13}$$

Thus the canonical (Heisenberg) c.r. follow from the symmetry condition (3) for the Galilei group but only because projective representations and not just proper representations are admissible according to (3). The most general definition of the position operator is thus given by (12) with K_i defined by a differential quotient as in (7) with $U_i(\alpha)$ being a one-parameter subgroup of pure Galilei transformations.

The quantum mechanical Galilei group has three invariant (Casimir) operators:

$$M :\quad \text{(from the phase) with an eigenvalue } m \text{ and the} \qquad (14)$$
physical interpretation of mass.

$$S^2 = (\boldsymbol{J} - \boldsymbol{Q} \wedge \boldsymbol{P})^2 :\quad \text{with an eigenvalue } s(s+1) \text{ and the} \qquad (15)$$
physical interpretation of spin.

$$U = H - \frac{1}{2m}\boldsymbol{P}^2 :\quad \text{with an eigenvalue } u \text{ and the}$$
physical interpretation of intrinsic energy.

The irreps of the quantum mechanical Galilei group are, therefore, characterized by the three numbers (m, u, s). The representations (m, u, s) and $(m, 0, s)$ are projectively equivalent and the value of u is irrelevant for elementary physical systems (energy is defined only up to an additive constant). The elementary physical systems described by the Galilei group – the nonrelativistic elementary particles – are therefore characterized by mass m and spin s. The operators Q_i, P_i, J_i (defined by expressions like (7)) are the center of mass position, the total momentum and the total angular momentum observables. The operator

$$S_i = J_i - (Q \wedge P)_i \qquad (16)$$

is the angular momentum relative to the center of mass (spin). One can show that the spin operator S_i commutes with the "orbital" variables Q_i, P_i, H:

$$[S_i, P_j] = 0 \ , \quad [S_i, Q_j] = 0 \ , \quad [S_i, H] = 0 \ , \qquad (17)$$

and the S_i fulfill the commutation relations

$$[S_i, S_j] = i\varepsilon_{ijk} S_k. \qquad (18)$$

Therefore one can choose in the irreducible representation space $\mathcal{H}^u(m, s)$ of the irreducible representation (m, u, s) the basis system

$$\mid \boldsymbol{p}, m, u; s, s_3 \rangle = \mid \boldsymbol{p}, m, u \rangle \otimes \mid s, s_3 \rangle \qquad (19)$$

where the orbital variables P_i, Q_j act only on the first factor and the S_i act only on the second factor which spans an irrep space $\mathcal{R}^{(s)}$ of the quantum mechanical rotation group generated by S_i. The space $\mathcal{H}(m, u, s)$ is thus a direct product

$$\mathcal{H}(m, u, s) = \mathcal{H}_{\text{orb}} \times \mathcal{H}_{\text{int}} \qquad \text{with} \qquad \mathcal{H}_{\text{int}} = \mathcal{R}^{(s)}. \qquad (20)$$

The vectors $\mid p_i, m, u \rangle$ denote eigenvectors of P_i with eigenvalue p_i. In place of momentum eigenvectors one could as well have chosen eigenvectors $\mid x_i, m, u \rangle$ of the position operator Q_i.

3 Spectrum Generating Groups

Whether a physical object is elementary or not does not depend upon the physical object but upon our way of looking at it. Molecules are elementary when looked at with very low (10^{-4} eV) energy. Nuclei are elementary when looked at with energies less than 10^3 eV. Hadrons are elementary when looked at with energies less than 10^8 eV. Above those energies the physical systems are extended objects with intrinsic structure. Intrinsic degrees of freedom can be excited and the extended object can perform intrinsic motion. These intrinsic motions may be the motions of constituents (atomistic models), but they can also be collective motions [6] (collective models). Usually atomistic models are considered more fundamental than collective models but in quantum physics there is no fundamental difference between these two models as they just correspond to two different choices of basis vectors in the same Hilbert space. The important quantities are the observables which are the generators of the motions.

The generators Q_i, P_i, J_i of the Galilei group describe the motion of the object as a whole; the center of mass motion:

Q_i is the center of mass (c.m.) position
P_i is the center of mass (c.m.) momentum or total momentum
J_i is the total angular momentum which is, according to (16), the sum of orbital angular momentum and spin.

The center of mass motion of an extended object, which is the total motion when the extended object is considered as a structureless elementary system, is described by the generators of the symmetry group. Above the threshold of elementariness the extended object can perform intrinsic motion (motion relative to the center of mass) and can change the state of this intrinsic motion. The motion generated by the S_i of (16) (rotations relative to the center of mass) is an example of an intrinsic motion, transitions between states with different spin is an example of a change of the intrinsic state.

Different extended objects have different (groups of) intrinsic motions. Some objects may be rigid and capable only of rigid body rotations about their center of mass. Others may be elastic, performing oscillations. Some may consist of constituents, others may be better visualized as continuous matter distributions. The variables of the collective intrinsic motions together with the S_i generate the Spectrum Generating Group (SGG) [7].

To explain this let us consider an extended object (as shown in Fig. 3a) with a center of positive charge at $\boldsymbol{Q}_{(+)}$, center of negative charge at $\boldsymbol{Q}_{(-)}$, and dipole moment $e\boldsymbol{\xi} = \boldsymbol{d}$. What are the c.r. of ξ_i, or what group do they generate? First it is natural to assume that ξ_i is an intrinsic vector operator, i.e.

$$[S_i, \xi_j] = i\varepsilon_{ijk}\xi_k. \tag{21}$$

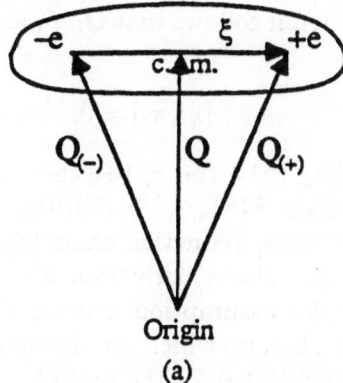

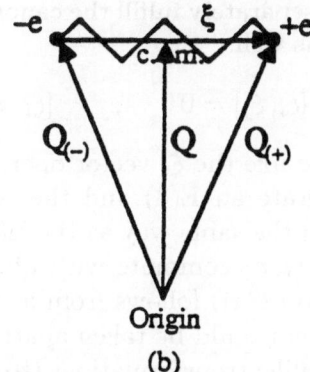

Origin
(a)

Origin
(b)

Fig. 3a. Extended object with center of positive charge $Q_{(+)}$ and center of negative charge $Q_{(-)}$. The dipole moment is $e\xi$. Due to constraints, the relative position ξ and momentum π of the centers of charge need not be independent canonical variables.

Fig. 3b. Vibrating dumbbell consisting of two unconstrained charges.

The commutator of ξ_i and ξ_j may be anything. The three simple alternatives, in which the commutator closes with the S_i into a group, are:[2]

$$[\xi_i, \xi_j] = 0 \qquad \text{leading to a group } E(3)_{\xi,S} \supset SO(3)_S, \tag{22a}$$

$$[\xi_i, \xi_j] = -i\varepsilon_{ijk}S_k \quad \text{leading to a group } SO(3,1)_{\xi,S} \supset SO(3)_S, \tag{22b}$$

$$[\xi_i, \xi_j] = +i\varepsilon_{ijk}S_k \quad \text{leading to a group } SO(4)_{\xi,S} \supset SO(3)_S. \tag{22c}$$

To show that (22a) is *not* the obvious choice for the intrinsic position operators we reproduce here its standard derivation and show that it is based on questionable assumptions:

Let the extended object be a rigid or vibrating dumbbell consisting of two mass-points, as shown in Fig. 3b, with $Q_{(+)}, P_{(+)}$ and $Q_{(-)}, P_{(-)}$ which separately fulfill the canonical c.r.. Then, in addition to the center of mass position and total momentum operators (which are generators of the Galilei group),

$$Q_i = \frac{m_{(-)}Q_{(-)i} + m_{(+)}Q_{(+)i}}{m_{(-)} + m_{(+)}} \quad , \qquad P_i = P_{(-)i} + P_{(+)i}, \tag{23}$$

one introduces the relative position and relative momentum operators,

$$\xi_i = Q_{(+)i} - Q_{(-)i} \quad , \qquad \pi_i = \frac{\mu}{m_{(+)}}P_{(+)i} - \frac{\mu}{m_{(-)}}P_{(-)i} \tag{24}$$

---

[2] We write the generators of the group – representing physical observable – as subscripts in order to distinguish between isomorphic groups which have different physical interpretation.

where $\mu = m_{(+)}m_{(-)}/(m_{(+)} + m_{(-)})$ is the reduced mass. If both $Q_{(+)i}, P_{(+)i}$ and $Q_{(-)i}, P_{(-)i}$ separately fulfill the canonical c.r., then it follows that Q_i, P_i and ξ_i, π_i fulfill them as well:

$$[\xi_i, \xi_j] = 0 \quad , \quad [\xi_i, \pi_j] = i\mathbb{1}\delta_{ij} \quad , \quad [\pi_i, \pi_j] = 0. \tag{25}$$

As the π_i are like the ξ_i vector operators fulfilling (21), the π_i together with the S_i also generate an E(3) and the reduction chain $E(3)_{\pi_i S_i} \supset SO(3)_{S_i}$ can be considered in the same way as the $E(3)_{\xi_i S_i} \supset SO(3)_{S_i}$ reduction chain [8]. It also follows that ξ_i, π_j commute with Q_i, P_j. From the above derivation we see that (25) (i.e. also (22a)) follows from a very particular assumption, namely that the physical system could be taken apart into point objects, which are independently related to Galilei transformations (so that (13) is fulfilled for the pair $Q_{(-)i}, P_{(-)j}$ and for the pair $Q_{(+)i}, P_{(+)j}$). That the Galilei groups has anything to do with the intrinsic motion of an extended object, even if it is as simple as the dumbbell, is of course a very questionable assumption. And for a more complicated extended object (like e.g. most nuclei) one would not expect the relative positions of the centers of charge to have any connection to the Galilei group. Coordinates and momenta of volume elements of a rigid or elastic body need not be canonical variables in the usual sense, due to constraints; in particular the Dirac brackets of the intrinsic coordinates need not be zero. Thus there is no reason to prefer (22a) over (22b) or (22c). Intrinsic positions (and momenta) and, as a consequence, dipole and quadrupole operators do not need to commute. Therefore the groups SO(3,1) and SO(4) must be considered as feasible options. Indeed in molecular and in nuclear physics [8] (the corresponding assumption for the quadrupole operators) the hypothesis (22c) has been used extensively. In hadron physics the (covariant version of the) relation (22b) – together with the corresponding relation for the π_i – has led to good phenomenological results [9]. If the intrinsic motion consists only of rigid rotations about the center of mass of the extended object, then S_i and ξ_i are the only intrinsic observables involved. The groups of equations (22a) - (22c) are thus the possible spectrum generating groups of the rigid rotator. We choose (22b), i.e. we pick $SO(3,1)_{\xi_i S_i}$ as the spectrum generating group, and will now explain the concept of the spectrum generating group using this simple example for the rigid rotator.

As the c.m. motion is independent of the intrinsic motion the ξ_i (and also the π_i) commute with the c.m. observables Q_i and P_i

$$[\xi_i, Q_j] = 0 \quad [\xi_i, P_j] = 0$$

(one can also derive this using (23) and (24)). From (26) and (17) it follows again that the representation space is the direct product of the orbital space $\mathcal{H}_{\text{orb}}$ and the intrinsic space $\mathcal{H}_{\text{int}}$

$$\mathcal{H} = \mathcal{H}_{\text{orb}} \times \mathcal{H}_{\text{int}}$$

and that ξ_i like the S_i acts only in $\mathcal{H}_{\text{int}}$. As the ξ_i are, according to (21) vector operators with respect to the angular momentum S_i they transform from a state with spin s into a state with spin $s + 1$ and $s - 1$ (in general also into a state with spin s). Therefore, the intrinsic space $\mathcal{H}_{\text{int}}$ for a structured extended object is no more an irrep space $\mathcal{R}^{(s)}$ of $SO(3)_{S_i}$ but a direct sum of many of these:

$$\mathcal{H}_{\text{int}} = \Sigma_s \oplus \mathcal{R}^{(s)}. \tag{28}$$

Which values of s occur in this sum depends upon the choice of the irreducible representation of the spectrum generating group.

The irrep spaces $\mathcal{H}$ of $SO(3,1)$ are characterized by two numbers (k_0, c) [11]. As an example we choose the irrep space $\mathcal{H}(k_0 = 0, c)$. A graphical representation of the reduction of an irrep of a non-compact group [here $SO(3,1)$] with respect to its maximal compact subgroup [here $SO(3)$] is given by the K-type [2b] (also called weight diagram). Figure 4(b) shows as an example the K-type of the irrep $(k_0 = 0, c)$ of $SO(3,1)$. It depicts single irrep spaces $\mathcal{R}^{(s)}$ of $K = SO(3)$ by dots and there is one dot for each value $s = 0, 1, 2, \ldots$, because each $\mathcal{R}^{(s)}$ with these values occurs once. The K-type is a shortened version of the collection of weight diagrams for the $SO(3)$-irreps with $s = 0, 1, 2 \ldots$ examples of which were shown in Fig. 2. Each dot in the K-type stands for a series of dots in the corresponding weight diagram as shown in Fig. 4a. Each dot in the weight diagrams of Fig. 4a represents a pure physical state (with definite value of s and s_3), each dot in the K-type represents therefore a mixture of physical states [1] (with definite value of s and with all s_3 values being equally probable). The arrows in Fig. 4a and Fig. 4b show that the operators ξ_i transform between vectors corresponding to these dots (only one arrow is shown). The irrep space $\mathcal{H}(k_0, c)$ reduces, when $SO(3,1)$ is restricted to the subgroup $SO(3)$, into the direct sum of irreducible representation spaces $\mathcal{R}^{(s)}$:

$$\mathcal{H}(k_0, c) = \sum_{s=k_0}^{\infty} \oplus \mathcal{R}^{(s)}. \tag{29}$$

The spectrum generating group $SO(3,1)_{\xi_i S_i}$ can describe many different rigid rotators. These different rigid rotators are characterized (in addition to the system constants like the moment of inertia discussed below) by the two numbers (k_0, c) that characterize the irrep of the SGG. The integer or half integer k_0 determines – according to (29) – the spin spectrum and c determines the values of the dipole matrix elements. The choice of (k_0, c) is thus specific for the particular physical system; if our rigid rotator has spin $s = 0$ states, we have to choose $k_0 = 0$ and some c. (There are many examples of molecular rotators for which $k_0 > 0$). With $(k_0 = 0, c)$ chosen, we have for the space of physical states

$$\mathcal{H} = \mathcal{H}_{\text{orb}} \times \mathcal{H}_{\text{int}} = \mathcal{H}_{\text{orb}} \times \mathcal{H}(k_0 = 0, c)$$

$$= \sum_{s=0,1\ldots}^{\infty} \oplus \mathcal{H}_{\text{orb}} \times \mathcal{R}^{(s)} = \sum_{s=0,1}^{\infty} \oplus \mathcal{H}(m, u, s) \tag{30}$$

where (20) was used. The space of physical states of the extended object is the direct sum of spaces of the elementary systems (20) with different intrinsic angular momentum quantum number $s = 0, 1, \ldots$.

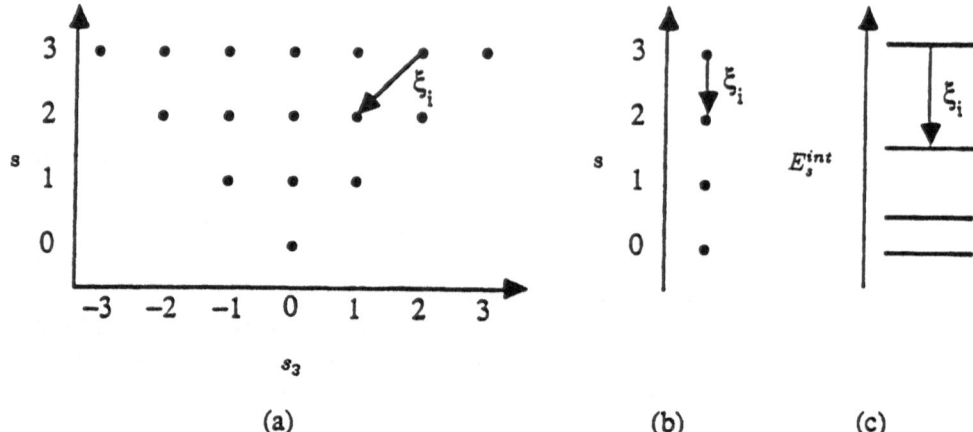

Fig. 4. (a) A collection of weight diagrams of representations of SO(3). (b) K-type of the irreducible representation ($k_0 = 0, c$) of SO(3,1). (c) Corresponding energy diagram of a rigid rotator.

Though for elementary systems the value of the intrinsic energy u is irrelevant, the relative energy of the states with different s is important. Therefore one chooses of all the projectively equivalent $\mathcal{H}(m, u, s)$ for a given set of values m and s that $\mathcal{H}(m, u(m,s), s)$, which has the "correct" value $u(m, s)$ (we will see below that this u is a function of s only and that it is independent of m).

What is the "correct" value of $u(s)$ is determined by a fundamental relation between the generators of the symmetry group and the spectrum generating group. This fundamental relation is not a relation which can be derived from the properties of the group but a new relation that constrains the Casimir operators of the symmetry group and the spectrum generating group. This constraint relation is characteristic for the physical system. The choice of this defining constraint relation is equivalent to the choice of a Hamiltonian, which also defines the physical system. These constraint relations involve, like the choice of the Hamiltonian, phenomenological constants which characterize the system.

For the rigid rotator the system constant is the moment of inertia I (for the simple dumbbell rotator there is only one number for more general rigid bodies there will be three). The constraint relation for the rotator is then given by the relation between the two invariants U and $\boldsymbol{S}^2$:

$$2IU - \boldsymbol{S}^2 = \text{const.} (= 0). \tag{31}$$

With (31) it follows from (15) that the energy operator is

$$H = \frac{1}{2m}\boldsymbol{P}^2 + \frac{1}{2I}\boldsymbol{S}^2. \tag{32}$$

The spectrum of H in the space (30) is then obtained by applying it to the basis vectors (19):

$$E_s = \frac{1}{2m}p^2 + \frac{1}{2I}s(s+1).$$ (33)

Usually, if the rotator is a diatomic molecule, the kinetic energy $\frac{1}{2m}P^2$ is ignored, and so is the center of mass motion and $\mathcal{H}_{\mathrm{orb}}$. The space of physical states is then only the space of intrinsic motion $\mathcal{H}_{\mathrm{int}} = \mathcal{H}(k_0 = 0, c)$. The energy spectrum is the usual rotator spectrum $E_s = \frac{1}{2m}s(s+1)$. Ignoring the center of mass motion is common practice in nonrelativistic quantum mechanics, but will no more be possible for the relativistic spectrum generating group.

Figure 4(c) shows the energy diagram of the diatomic rigid rotator. To each energy level E_s belongs a space of physical states $\mathcal{R}^{(s)}$, with s being the spin of the rotator. The arrows (only one shown) between the energy levels indicate the radiative transitions as they occur in the process

$$M^*(\text{excited molecule}) \rightarrow M(\text{ground state}) + \gamma.$$ (34)

Thus we have the following correspondences between the mathematical and physical quantities: The irrep space $\mathcal{H}(k_0, c)$, corresponds to the space of physical states of the collective motion $\mathcal{H}_{\mathrm{int}}$. The weight diagram ($K$-type) corresponds to the energy diagram. Each dot of the K-type corresponds to an energy level. The matrix elements of the SGG generator ξ_i in the irrep (k_0, c), $\langle s', s_3' \mid \xi_i \mid s, s_3 \rangle$, correspond to the physical transitions, precisely, the group theoretical quantities $\mid \langle s', s_3' \mid \xi_i \mid s, s_3 \rangle \mid^2$ give the probabilities for radiative transitions (34) between energy levels of the molecule.

4 The Relativistic Case

For the relativistic theory the symmetry group is the Poincaré group $\mathcal{P}_{P_\mu, J_{\mu\nu}}$. The irrep spaces of the quantum mechanical Poincaré group are characterized by two invariants m and s, which have similar meaning as for the Galilei group.

The elementary physical systems described by an irrep space $\mathcal{H}(m, s)$ of the Poincaré group [4] is the relativistic elementary particle.

The generators, momentum P_μ and Lorentz generator $J_{\mu\nu}$, are defined by relations like (7). With them one defines

$$\hat{W}_\mu = \frac{1}{2}\varepsilon_{\mu\nu\rho\sigma}\hat{P}^\nu J^{\rho\sigma} \; ; \; \hat{P}_\mu = P_\mu M^{-1} \; ; \; M = (P_\mu P^\mu)^{1/2}$$ (35)

and the spin tensor

$$\Sigma_{\mu\nu} = \varepsilon_{\mu\nu\rho\sigma}\hat{P}^\rho \hat{W}^\sigma.$$ (36)

The Casimir operators and their eigenvalues are (in analogy to (14) (15)):

$$M^2 : \quad \text{with eigenvalue } m^2.$$ (37)

$$-\hat{W}_\mu \hat{W}^\mu = \frac{1}{2}\Sigma_{\mu\nu}\Sigma^{\mu\nu} : \quad \text{with eigenvalue } s(s+1).$$ (38)

The elementary system is described by the irrep space $\mathcal{H}^n(m, s)$ with Wigner basis [4]

$$| \boldsymbol{p}, m, n \; ; \; s, s_3) \,. \tag{39}$$

The quantum numbers here are the same as in (19) (with n (corresponding to u), describing the internal state of excitation when the elementary particle becomes a structured extended object). But the space of states and the basis vectors are no more a direct product as in (20) and (19) [9b]. Except for these complications the theory of the extended relativistic object is constructed in complete analogy to the nonrelativistic case.

The spectrum generating group now has to contain the intrinsic Lorentz group $SO(3,1)_{S_{\mu\nu}}$, $\mu, \nu = 0, 1, 2, 3$ instead of the intrinsic rotation group $SO(3)_{S_{ij}}$ as in (22b). This intrinsic Lorentz group is not to be mistaken for the $SO(3,1)_{\xi_i, S_{ij}}$ of (22b) and it is also not the Lorentz subgroup $SO(3,1)_{J_{\mu\nu}}$ of the physical Poincaré group $\mathcal{P}_{P_\mu, J_{\mu\nu}}$. It is in a certain sense a generalization of the $SO(3,1)_{\frac{1}{2},\sigma_{\mu\nu}}$ where $\sigma^{ij} = \varepsilon^{ijk}\sigma^k, \sigma^{oi} = -i\sigma^i$ and σ^i are the Pauli matrices [9b].

The simplest example of a spectrum generating group for a relativistic extended object is

$$SO(3,2)_{\Gamma_\mu, S_{\mu\nu}} \supset SO(3,1)_{S_{\mu\nu}} \tag{40}$$

where the Γ_μ fulfill here the same function as the transition operators ξ_i or better π_i in (22b). This $SO(3,2)_{\Gamma_\mu S_{\mu\nu}}$ is, in a certain sense, a generalization of Dirac's group $SO(3,2)_{\frac{1}{2}\gamma_\mu, \frac{1}{2}\sigma_{\mu\nu}}$ and describes, instead of elementary electrons, structured hadrons. The variables of the intrinsic motion are defined in terms of the generators of the symmetry group and the spectrum generating group: the spin tensor:

$$\Sigma_{\mu\nu} = \hat{g}_\mu^\rho \hat{g}_\nu^\sigma S_{\rho\sigma} \quad \text{where} \quad \hat{g}_{\mu\nu} = \eta_{\mu\nu} - \hat{P}_\mu \hat{P}_\nu \text{ (projector into the plane } \perp \text{ to } \hat{P}_\mu), \tag{41}$$

the "intrinsic position" (giving the direction of the interquark axis):

$$\xi_\mu = -S_{\mu\nu}\frac{P^\nu}{(cM)^2} \,, \tag{42}$$

and the "intrinsic momentum"

$$\pi_\mu = -\frac{1}{\alpha'}\frac{1}{cM}\hat{g}_\mu^\sigma \Gamma_\sigma \tag{43}$$

where α' is a constant of dimension $(\text{GeV})^{-2}$, which specifies the system, like the constant I in (31). We have also included the light velocity c in (42) and (43) for later demonstrations.

The commutation relations of the intrinsic position and momenta are

$$[\xi_\mu, \xi_\nu] = -\frac{i}{(Mc)^2}\Sigma_{\mu\nu} \; ; \tag{44}$$

$$[\pi_\mu, \pi_\nu] = -\frac{i}{(\alpha'Mc)^2}\Sigma_{\mu\nu} \; ; \tag{45}$$

$$[\xi_\mu, \pi_\nu] = -i\hat{g}_{\mu\nu}\frac{i}{\alpha'(Mc)^2}\hat{P}_\rho\Gamma^\rho \,. \tag{46}$$

These c.r. can be derived in a straightforward way from the definitions (42) and (43) and from the c.r. of SO(3,2) for the $S_{\mu\nu}$ and Γ_μ.

Like the commutators of ξ_i in (22b) the intrinsic position and the intrinsic momenta have non-commuting components. The only reason for calling ξ_μ a position and π_ν a momentum is, that in a special case the limit of (44) ... (46) for $c \to \infty$ are the usual three-dimensional canonical commutation relations. We shall not show this here in detail but the factor $\frac{1}{c^2}$ in the denominator on the r.h.s. is an indication that position and momentum components become commuting in this limit.

The spectrum is determined as in (29) by the choice of the irreducible representation of the spectrum generating group. There are many classes of irreducible (unitary) representations of SO(3,2). As an illustration we choose just one irrep of $SO(3,2) \supset K = SO(2)_{\Gamma_0} \times SO(3)_{S_{ij}}$ which is usually denoted by $D(3/2, 1/2)$. Its K-type is depicted in Fig. 5. The values (μ, s) characterize the irreps of $K = SO(2)_{\Gamma_0} \times SO(3)_{S_{ij}}$ (μ is the eigenvalue of Γ_0).

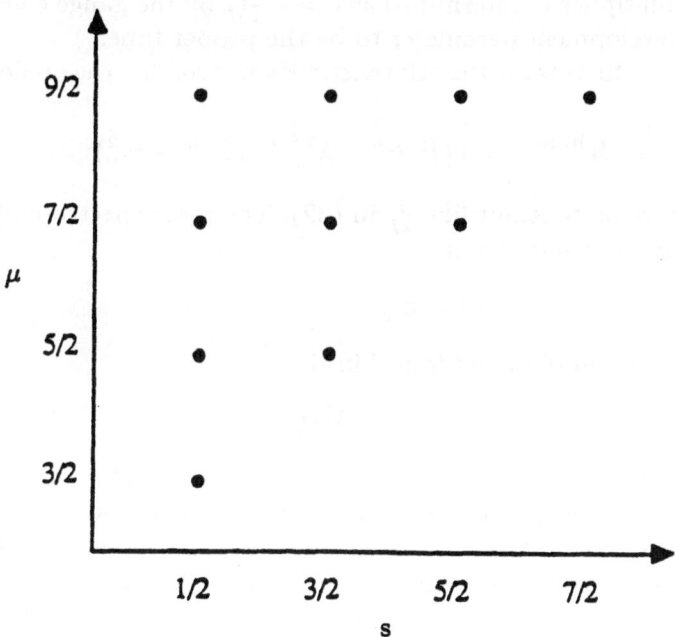

Fig. 5. The K-type of the irreducible representation $D(\mu_{min} = 3/2, s_{min} = 1/2)$ of $SO(3,2) \supset SO(3) \times SO(2)_{\Gamma_0}$. (μ_{min}, s_{min}) which are called the lowest weights, characterize the irreps of $SO(3,2)$. (μ, s) are called the weights, μ is the eigenvalue of Γ_0 and s labels the representations of $SO(3)$.

In the physical interpretation in analogy to Fig. 4b the mathematical representation labels μ and s become the quantum numbers of the physical state represented by the dots of the K-type. Therefore we have for the space of physical states in analogy to (30):

$$\mathcal{H} = \sum_{\mu=\frac{3}{2}}^{\infty} \sum_{s=\frac{1}{2},\frac{3}{2}\cdots\mu-1} \otimes \quad \mathcal{H}^{\mu}(m,s). \tag{47}$$

However (47) is not as easily established as (30), because in the relativistic case $\mathcal{H}$ is not the direct product of an orbital and an intrinsic space.

The mass m in (47) is a function of the discrete quantum numbers μ and s, $m = m(\mu, s)$. Like in the nonrelativistic case (33) for the energy spectrum, the mass spectrum is determined by a constraint relation or, equivalently, by the choice of a relativistic Hamiltonian. The methodology is the same as used in relativistic string theory [11] and is based on Constrained Hamiltonian Mechanics [12]. Different choices for the relativistic Hamiltonian correspond to different intrinsic motions. For the relativistic mass point (spinless relativistic elementary particle) the Hamiltonian [12] is

$$H^{\text{Point}} = v(P_{\mu}P^{\mu} - m_0^2). \tag{48}$$

The Lagrange multiplier is determined as $v = -\frac{1}{2M}$ by the gauge constraint which fixes the time development parameter to be the proper time.

For the relativistic rotator the relativistic Hamiltonian – the analog of (32) – is given by

$$H^{\text{Rotator}} = v(P_{\mu}P^{\mu} - \lambda^2 \frac{1}{2}\Sigma_{\mu\nu}\Sigma^{\mu\nu} - m_0^2) \tag{49}$$

where λ^2 is a systems constant like $\frac{1}{2I}$ in (32). The mass spectrum obtained from (48) via the constraint relation is:

$$m^2 = m_0^2 \tag{48a}$$

and the mass spectrum obtained from (49) is

$$m^2 = m_0^2 + \lambda^2 s(s+1). \tag{49a}$$

The trivial mass spectrum (48a) is the expected result [12b] which is given here only to expose the analogy with the nontrivial relativistic rotator spectrum (49a).

The result (49a) means that the mass values in (47), i.e. the mass of the physical states described by the dot at the position (μ, s) in the K-type of Fig. 5 is a function of the spin quantum number s (but not of the other intrinsic quantum number μ). Therewith we have shown how the spectrum generating group combined with constrained Hamiltonian quantum mechanics (the choice of a Hamiltonian) leads to a nontrivial mass spectrum. This is the relativistic analog of the algebraic method used successfully for nonrelativistic systems.

The relativistic rotator spectrum (49a) is certainly not the full story for any relativistic extended object (hadron) in the same way as (33) gives only part of the molecular spectrum. There will also be radial vibrations described by the vibrational quantum number μ. These can be obtained by a more realistic choice of the relativistic Hamiltonian in place of (49). Transitions can also be described in analogy to the nonrelativistic transitions (34) by introducing electromagnetic coupling into the free Hamiltonian (49). In this way the algebraic method can be extended into the relativistic domain.

The spectrum generating group is the central ingredient of the algebraic method in quantum physics. When using the algebraic method several choices have to be made. As a summary we conclude this article with a table in which we list what these choices are, what they give in general, what has been chosen for them for the nonrelativistic rotator in Sects. 2 and 3 and what has been chosen for them in the relativistic model of this section.

Choices	What They Give	Nonrelativistic Model	Relativistic Model
Motion Symmetry Group	Center of Mass Observables	Galilei Group $\mathcal{G}$	Poincaré Gr. $\mathcal{P}$
Spectrum Generating Group (SGG)	Transition Operators (Intrinsic)	$SO(3,1)_{\xi,S}$	$SO(3,2)_{\Gamma,S}$
Subgroup Chain	Quantum Numbers	s(spin),s_3	μ, s, s_3
Representation of the SGG	Spectrum of Discrete Quantum Numbers	$(k_0 = 0, c)$	$D(3/2, 1/2)$
Constraint Relation	Energies, Masses	Energy Operator H of (32)	Relativistic Hamiltonian of (49)

Acknowledgment

J. Lemke read the manuscript and suggested improvements for which the author wishes to express his gratitude.

References

1.   A. Bohm: *Quantum Mechanics: Foundations and Applications* (Springer Verlag, 2nd edition 1979)

2.   a) A.O. Barut, R. Raczka: *Theory of Group Representations and Applications* (Polish Scientific Publishers, Warsaw, 1977);
      b) David A. Vogan: *Representation of Real Reductive Lie Groups* (Birkhäuser, Boston, 1981)

3.   G. Warner: *Harmonic Analysis on Semi-simple Lie Groups* (Springer Verlag 1972);
      K. Maurin: *General Eigenfunction Expansion and Unitary Representations of Topological Groups* (Polish Scientific Publishers, Warsaw, 1968)

4.   E.P. Wigner: in *Group Theoretical Concepts and Methods in Elementary Particle Physics*, ed. by F. Gürey (Gordon and Breach, N.Y., 1964), p. 37

5.   J.-M. Levy-Leblond: "Galilei Groups and Galilean Invariance", in *Group Theory and its Applications*, ed. by E.M. Loebl (Academic Press, 1971), Vol. II, p. 222

6.   A. Bohr, B. Mottelson: *Nuclear Structure* (Benjamin, 1969), Vol. II;
      G. Herzberg: *Molecular Spectra and Molecular Structure* (D. van Nostrand Comp., 1966)

7.   A. Bohm, Y. Ne'eman, A.O. Barut: *Dynamical Groups and Spectrum Generating Algebras* (World Scientific Publishers, 1988)

8.   F. Iachello, R.D. Levine, O.S. van Roosmalen, et al: J. Chem. Phys. **77** 3047 (1982);
      F. Iachello and A. Arima: *The Interacting Boson Model* (Cambridge University Press, 1987)

9.   a) R.R. Aldinger, A. Bohm, P. Kielanowski, M. Loewe, P. Magnollay, N. Mukunda, W. Drechsler, S. R. Komy: Phys. Rev. D**28** 3020 (1983);
      b) A. Bohm, M. Loewe, L.C. Biedenharn, H. van Dam: ibid. **28** 3032 (1983);
      c) R.R. Aldinger, A. Bohm, P. Kielanowski, M. Loewe, P. Moylan: ibid. **29** 2828 (1984);
      d) A. Bohm, M. Loewe, P. Magnollay, M. Tarlini, R.R. Aldinger, P. Kielanowski: Phys. Rev. Lett. **53** 2292 (1984);
      e) A. Bohm, M. Loewe, P. Magnollay: ibid. **31** 2304 (1985);
      f) ibid. **32** 791 (1985);
      g) A. Bohm, M. Loewe, P. Magnollay, M. Tarlini, R.R. Aldinger, L.C. Biedenharn, H. van Dam: ibid. **32** 2828 (1985)

10.   M.A. Naimark: *Linear Representations of the Lorentz Group* (Pergamon Press, 1964);
      Ref. 1, Append. V.3.

11.   John H. Schwarz: *Superstrings* (World Scientific Publishers, 1985)

12.   a) P.A.M. Dirac: *Lectures on Quantum Mechanics* (Yeshiva University Press, New York 1964);
      b) A.J. Hanson, T. Regge, C. Teitelboim: *Constrained Hamiltonian Systems* (Academia Nazionale dei Lincei, Roma 1976);
      c) N. Mukunda, H. van Dam, L.C. Biedenharn: *Relativistic Models of Extended Hadrons Obeying a Mass-Spin Trajectory Constraint* (Springer-Verlag, New York, 1982)

Spectrum and Character Formulae of so(3,2) Unitary Representations

V.K. Dobrev [1], E. Sezgin [2]

[1]Institute of Nuclear Research and Nuclear Energy, Bulgarian Academy of
Sciences, 72 Boul. Lenin, 1784 Sofia, Bulgaria
[2]Center for Theoretical Physics, Texas A&M University, College Station,
TX 77843, USA

Abstract: We give two derivations of the physically relevant spectrum of unitary representations of the anti de Sitter algebra so(3,2) . One is based on our earlier results on null states of so(3,2) representations. For the other we first obtain the characters of the unitary representations of so(3,2) and then we show their equivalence with the spectrum results

1 Introduction

It is important for physical applications to know the spectrum of the unitary representations of certain Lie algebras. The anti de Sitter algebra so(3,2) has truly remarkable unitary representations known as singletons, which were first discovered by Dirac in 1963 [1]. These representations have been extensively studied by Fronsdal and Flato [2], and Evans [3]. There are two singleton representations, called Di and Rac. In terms of the lowest eigenvalue E_0 of the so(2) which has the interpretation of energy, and the spin s_0 of the state with this energy, the Di has ($E_0 = 1$, $s_0 = 1/2$) while the Rac has ($E_0 = 1/2$, $s_0 = 0$). These representations have remarkably reduced spectrum (weight spaces). Consequently, the singleton field theory has a very large gauge symmetry which enables one to gauge away the singleton fields everywhere except on the boundary of the anti de Sitter space [4]. Moreover, the direct product of two singletons decomposes into infinitely many massless states of the anti de Sitter group [5]. Other interesting unitary representations are the massless ones.

The aim of this paper is to apply some modern tools in the representation theory of Lie algebras, namely, elimination of null states and character formulae in order to give two simple derivations of the spectrum of the unitary representations of so(3,2) . The new results in this paper are the character formulae for these representations.

In Sect. 2 we give preliminaries including facts about the so(3,2) algebra. In Sect. 3 we introduce the lowest weight representations of so(3,2) and recall the list of its unitary representations. In Sect. 4 we recall our earlier results on the null states [6] and show how their elimination leads to the known spectrum (weight spaces) of the unitary representations [1], [2], [3]. In Sect. 5 we derive the character formulae of the unitary representations of so(3,2) and then we show their equivalence with the spectrum results.

2 Preliminaries

The algebra $\mathcal{G} = \mathrm{so}(3,2)$ is 10-dimensional and its generators $M_{AB} = -M_{BA}$, $(A,B = 0,1,2,3,4)$, $\eta_{AB} = \mathrm{diag}\,(+ - - - +)$, obey

$$[M_{AB}, M_{CD}] = i(\eta_{BC}M_{AD} - \eta_{AC}M_{BD} - \eta_{BD}M_{AC} + \eta_{AD}M_{BC}). \tag{1}$$

It is useful to have a finite-dimensional realization of $\mathcal{G}$. For this we introduce four-dimensional γ-matrices as in [7] :

$$\{\gamma_\mu, \gamma_\nu\} = 2\eta_{\mu,\nu}, \quad \eta_{\mu,\nu} = \mathrm{diag}(+ - - -), \quad \mu,\nu = 0,1,2,3. \tag{2}$$

Then we define :

$$\ell_{\mu,4} \equiv -\frac{i}{2}\,\gamma_\mu, \quad \ell_{\mu,\nu} \equiv \frac{1}{4}\,[\gamma_\mu, \gamma_\nu], \tag{3}$$

and $M_{A,B} = i\ell_{A,B}$ satisfy (1). Their hermiticity properties are:

$$M_{AB}^\dagger = M_{AB} \quad \text{for} \quad (A,B) = (0,4),(a,b), \tag{4a}$$

$$M_{AB}^\dagger = -M_{AB} \quad \text{for} \quad (A,B) = (0,a),(a,4), \tag{4b}$$

where $a,b = 1,2,3$. Further we introduce the following notation [7]:

$$M_a^\varepsilon \equiv iM_{0a} - \varepsilon M_{a4}, \quad \varepsilon = \pm. \tag{5}$$

Following [7] we shall use the following basis for $\mathcal{G}$:

$$M^{\varepsilon\varepsilon'} \equiv \frac{1}{2}(M_1{}^\varepsilon + i\varepsilon' M_2{}^\varepsilon), \quad \varepsilon, \varepsilon' = \pm, \tag{6a}$$

$$K^\varepsilon \equiv M_3{}^\varepsilon, \quad \varepsilon = \pm, \quad K^3 \equiv M_{04}, \tag{6b}$$

$$J^{\varepsilon'} \equiv -\varepsilon' M_{23} - iM_{31}, \quad \varepsilon' = \pm, \quad J^3 \equiv M_{12}. \tag{6c}$$

The elements $J^\pm, J^3$ generate an so(3) subalgebra of $\mathcal{G}$ and together with K^3 they generate the maximal compact subalgebra so(3) $\oplus$ so(2) of $\mathcal{G}$, while J^3, K^3 generate the compact Cartan subalgebra $\mathcal{H}$ of $\mathcal{G}$. [A noncompact semisimple Lie algebra may have more than one non-conjugate Cartan subalgebra. The algebra $\mathcal{G} = \mathrm{so}(3,2)$ has three such subalgebras; besides $\mathcal{H}_0 \equiv \mathcal{H}$ we have $\mathcal{H}_1$ generated, say, by M_{01}, M_{23} (or M_{12}, M_{34}) and $\mathcal{H}_2$ generated, say, by

M_{01} , M_{34} . Thus $\mathcal{H}_k$, $k = 0,1,2$, is a Cartan subalgebra with k noncompact generators.]

Using a finite-dimensional representation of $\mathcal{G}$, e.g., as above, and normalizing the Cartan–Killing form $(,)$ on $\mathcal{G}$ as $(X,Y) \equiv (1/2)\mathrm{tr}(X,Y)$, we have (for the nonzero cases):

$$(M^{++}, M^{--}) = (M^{+-}, M^{-+}) = (K^3, K^3) = (J^3, J^3) = 1$$
$$(K^+, K^-) = -(J^+, J^-) = 2 . \tag{7}$$

We note for further reference that the Cartan involution θ is defined by

$$\theta : X \mapsto X , \quad \text{if} \quad X \quad \text{is compact}, \quad (X = K^3, J^3, J^\pm),$$
$$\theta : X \mapsto -X, \quad \text{if} \quad X \quad \text{is noncompact} . \tag{8}$$

We also note the hermiticity properties of the new basis

$$(M^{\varepsilon\varepsilon'})^\dagger = M^{-\varepsilon,-\varepsilon'} ,$$
$$(K^\varepsilon)^\dagger = K^{-\varepsilon} , \quad (K^3)^\dagger = K^3 ,$$
$$(J^{\varepsilon'})^\dagger = -J^{-\varepsilon'} , \quad (J^3)^\dagger = J^3 . \tag{9}$$

By the definition of a Cartan subalgebra $\mathcal{H}$ the elements of the adjoint action ad H can be diagonalized simultaneously in $\mathcal{G}$. For $\alpha \in \mathcal{H}^*$, where $\mathcal{H}^*$ is the dual space of linear functionals over $\mathcal{H}$, let us define:

$$\mathcal{G}_\alpha \equiv \{X \in \mathcal{G} : [H,X] = \alpha(H)X \text{ for all } H \in \mathcal{H}\} . \tag{10}$$

The set $\Delta \equiv \{\alpha \in \mathcal{H}^* : \alpha \neq 0 , \mathcal{G}_\alpha \neq 0\}$ is called the root system of $\mathcal{G}$ relative to $\mathcal{H}$. Every root system has a special basis, the so-called system Δ_S of simple roots, so that every root of Δ can be expressed as a linear combination of the elements of Δ_S with integer coefficients of the same sign. Then the roots for which the coefficients are positive (respectively negative) are called positive (respectively negative) roots; then $\Delta^+(\Delta^-)$ denotes the set of positive (negative) roots and if $\alpha \in \Delta^+$, then $-\alpha \in \Delta^-$.

Further we shall use the decomposition

$$\mathcal{G} = \mathcal{G}_+ \otimes \mathcal{H} \oplus \mathcal{G}_- , \quad \mathcal{G}_\pm \equiv \bigoplus_{\alpha \in \Delta^\pm} \mathcal{G}_\alpha , \tag{11}$$

where $\dim \mathcal{G}_\alpha = 1$. Explicitly for $\mathcal{G} = $ so(3,2) we have:

$$\Delta^\pm = \{\pm\alpha_1 , \pm\alpha_2 , \pm\alpha_3 , \pm\alpha_4 : \alpha_1(K^3, J^3) = (0,1) , \alpha_2(K^3, J^3) = (1,1) ,$$
$$\alpha_3(K^3, J^3) = (1,0) , \alpha_4(K^3, J^3) = (1,1)\} . \tag{12}$$

The simple roots are $\Delta_S = \{\alpha_1, \alpha_2\}$, while

$$\alpha_3 = \alpha_1 + \alpha_2 , \quad \alpha_4 = 2\alpha_1 + \alpha_2 . \tag{13}$$

Let us denote the root space vector of $\mathcal{G}_\alpha$ by X_α . Then, up to a constant we have

$$X_{\pm\alpha_1} = J^\pm, \quad X_{\pm\alpha_2} = M^{\pm\mp}, \quad X_{\pm\alpha_3} = K^\pm, \quad X_{\pm\alpha_4} = M^{\pm\pm}. \tag{14}$$

The simple root system Δ_S is also a basis for $\mathcal{H}^*$. To every root α there corresponds an element $H_\alpha \in h$ which we choose as follows [8]:

$$H_\alpha \equiv [X_\alpha, X_{-\alpha}]/(X_\alpha, X_{-\alpha}). \tag{15}$$

Denoting $H_i \equiv H_{\alpha_i}$, we have

$$H_1 = J^3, \ H_2 = K^3 - J^3, \ H_3 = K^3 = H_1 + H_2, \ H_4 = K^3 + J^3 = 2H_1 + H_2. \tag{16}$$

Finally we can use the duality to define a scalar product in $\mathcal{H}^*$ by setting

$$(\alpha, \beta) \equiv (H_\alpha, H_\beta) \ (= \alpha(H_\beta) = \beta(H_\alpha)). \tag{17}$$

Explicitly for $\mathcal{G} = \text{so}(3,2)$ we have for the nonzero products

$$(\alpha_1, \alpha_1) = 1 = (\alpha_3, \alpha_3), \quad (\alpha_2, \alpha_2) = (\alpha_4, \alpha_4) = 2, \tag{18a}$$

$$(\alpha_1, \alpha_2) = -1, \quad (\alpha_1, \alpha_4) = (\alpha_2, \alpha_3) = (\alpha_3, \alpha_4) = 1. \tag{18b}$$

We shall also need the root system $\Delta^{\mathbb{C}}$ of the complexification $\mathcal{G}^{\mathbb{C}}$ of $\mathcal{G}$. However, Δ and $\Delta^{\mathbb{C}}$ may be identified since $\mathcal{G}$ is maximally split [9] (which means that it has a Cartan subalgebra with all generators noncompact (here $\mathcal{H}_2$)).

3 Lowest Weight Representations of $\mathcal{G}$

Lowest weight representations of a simple Lie algebra $\mathcal{G}$ are characterized by their lowest weight $\Lambda \in \mathcal{H}^*$, and a lowest weight vector $|\Lambda>$ so that

$$H|\Lambda> = \Lambda(H)|\Lambda>, \quad H \in \mathcal{H}, \tag{19a}$$

$$X|\Lambda> = 0, \quad X \in \mathcal{G}_-. \tag{19b}$$

For $\mathcal{G} = \text{so}(3,2)$ we denote $E_0 = \Lambda(K^3)$, $m_0 = \Lambda(J^3)$, $|\Lambda\rangle = |E_0, m_0\rangle$ and we have

$$K^3|E_0, m_0\rangle = E_0|E_0, m_0\rangle, \quad J^3|E_0, m_0\rangle = m_0|E_0, m_0\rangle, \tag{20a}$$

$$X|E_0, m_0\rangle = 0, \quad X = M^{-\pm}, K^-, J^-. \tag{20b}$$

The eigenvalue of K^3 is called the energy since upon contraction of so$(3,2)$ to the Poincaré algebra, K^3 goes to the translation operator P_0. Analogously, J^3 is the third component of the angular momentum. Thus E_0 and m_0 are the lowest values of the energy and third component of angular momentum j, respectively; note that $m_0 \leq 0$. This is why we deal with lowest weight representations, since we would like the energy to be bounded from below. The irreducible subquotients of these representations of so$(3,2)$ will be denoted by $D(E_0, s_0)$, $s_0 = -m_0 \geq 0$.

The representation space $\mathbf{H}$ is given as the direct sum

$$\mathbf{H} = \oplus_{n \in \mathbf{Z}_+} \mathbf{H}_n, \quad \mathbf{H}_n = \oplus_{p \in \mathbf{Z}} \mathbf{H}_{n,p}, \tag{21a}$$

$$\mathbf{H}_{n,p} = \sum_{k_1+k_2+k_3=n} \sum_{k_1-k_2+k_4=p} C_{k_1 k_2 k_3 k_4}(M^{++})^{k_1}(M^{+-})^{k_2}(K^+)^{k_3}(J^+)^{k_4}|E_0,s_0>$$

(21b)

where $C_{k_1 k_2 k_3 k_4}$ are arbitrary constants. One can choose different orders of the factors in (21b) but for each representation this order should be fixed. The scalar product in $\mathbf{H}$ and $D(E_0,s_0)$ is given by a contravariant hermitian form [10] $\phi(u,v)$, $u,v \in \mathbf{H}$ (or $D(E_0,s_0)$) such that

$$\phi(Xu,v) = \phi(u,(\theta X)^+ v), \quad X \in \mathcal{G}.$$

(22)

Unitarity of the representation is equivalent to the positivity of the norm of each nonzero vector $v \in D(E_0,s_0)$, i.e.,

$$\|v\| \equiv \phi(v,v) > 0, \quad v \neq 0.$$

(23)

One assumes that the lowest weight vector has norm equal to 1. The unitary irreducible lowest weight representations of $\mathcal{G}$ are as follows (cf. [1], [2], [3]) :

$$\text{Rac} : D(E_0,s_0) = D(1/2,0), \quad \text{Di} : D(E_0,s_0) = D(1,1/2),$$

(24)

$$D(E_0 > 1/2, s_0 = 0), \quad D(E_0 > 1, s_0 = 1/2), \quad D(E_0 \geq s_0+1, s_0 \geq 1).$$

(25)

The first two are the singleton representations and the last ones for $E_0 = s_0 + 1$ correspond to the spin-s_0 massless representations.

The spectrum (weight spaces) of the unitary representations is well known [1], [2], [3]. Below we shall rederive it from other considerations.

4 Elimination of Null States and Spectrum of Representations

In order to obtain the representations (24), (25) we should eliminate in $\mathbf{H}$ the null states, i.e., nonzero states with zero norms. A systematic way to find all null states is to find the singular vectors (null vectors) of the Verma modules V^Λ [8] with the same lowest weight as $\mathbf{H}$. A Verma module V^Λ is the lowest weight module with lowest weight vector v_0 induced from a 1-dimensional representation $\mathbb{1}^\Lambda$ of a Borel subalgebra $\mathcal{B} = \mathcal{H}^{\mathbb{C}} \oplus \mathcal{G}_-^{\mathbb{C}}$ of $\mathcal{G}^{\mathbb{C}}$, where $\mathcal{G}_-^{\mathbb{C}} = \text{c.l.s.}\{M^{--}, M^{-+}, K^-, J^-\}$, $\mathcal{H}^{\mathbb{C}}$, $\mathcal{G}^{\mathbb{C}}$ are the complexifications of $\mathcal{G}_-$, $\mathcal{H}$, $\mathcal{G}$. The representation $\mathbb{1}^\Lambda$ is defined as $\mathcal{G}_-^{\mathbb{C}} v_0 = 0$, $H v_0 = \Lambda(H)v_0$, $H \in \mathcal{H}^{\mathbb{C}}$. Then V^Λ may be represented as [8]:

$$V^\Lambda \cong U(\mathcal{G}^{\mathbb{C}}) \oplus_{U(\mathcal{B})} v_0 \cong U(\mathcal{G}_+^{\mathbb{C}}) \oplus_{\mathbb{C}} v_0,$$

(26)

where $\mathcal{G}_+^{\mathbb{C}} = \text{c.l.s.}\{M^{++}, M^{+-}, K^+, J^+\}$. If we restrict $\mathcal{G}^{\mathbb{C}}$ to $\mathcal{G}$ and impose $\Lambda(K^3) \in \mathbb{R}$, $\Lambda(J^3) \in (1/2)\mathbb{Z}$ then V^Λ is restricted to $\mathbf{H}$. (This is possible since $\mathcal{G}$ is maximally split.)

A singular vector v_s of a Verma module V^Λ is defined as follows

$$Xv_s = 0, \quad X \in \mathcal{G}_- , \quad Hv_s = \Lambda'(H)v_s , \quad \forall H \in \mathcal{H} , \tag{27}$$

$v_s \neq v_0, \Lambda' \neq \Lambda$. Thus v_s can be represented as (see [8] and (3.43) in [11]):

$$v_s = \mathcal{P}(\mathcal{G}_+)|\Lambda\rangle , \tag{28a}$$

where $\mathcal{P}(\mathcal{G}_+)$ is a polynomial such that

$$[H, \mathcal{P}(\mathcal{G}_+)] = (\Lambda' - \Lambda)(H)\mathcal{P}(\mathcal{G}_+) , \quad \forall H \in \mathcal{H} . \tag{28b}$$

We can extend the scalar product $\phi(u, v)$ by linearity to V^Λ. Then it is clear that singular vectors have zero norm (using property (22)). Moreover $U(\mathcal{G}_+)v_s$ is a submodule of V^Λ isomorphic to the Verma module $V^{\Lambda'}$.

Furthermore a singular vector of a Verma module exists iff the Bernstein–Gel'fand–Gel'fand (BGG) [12] criterion for finite-dimensional $\mathcal{G}$, (or the Kac–Kazhdan [13] criterion for affine Lie algebras), is fulfilled. It is the following condition

$$2(\Lambda - \rho, \alpha) = -m(\alpha, \alpha) , \quad m = 1, 2, \dots , \tag{29}$$

where $\rho \in \mathcal{H}^*$ is defined by the condition that

$$\rho(\alpha_i^\vee) = 1 , \quad \alpha_i^\vee \equiv \frac{2\alpha_i}{(\alpha_i, \alpha_i)} , \quad \forall \alpha_i \in \Delta_S . \tag{30a}$$

For finite-dimensional $\mathcal{G}$, ρ can be defined as the half-sum of all positive roots; for $so(3, 2)$ we have

$$\rho = \frac{1}{2}(\alpha_1 + \alpha_2 + \alpha_3 + \alpha_4) = 2\alpha_1 + \frac{3}{2}\alpha_2 . \tag{30b}$$

Before applying this to our situation we need an explicit expression for Λ. This expression (which follows from (19a), (20), (14)) is:

$$\Lambda = (E_0 - s_0)\alpha_1 + E_0\alpha_2 . \tag{31}$$

Next we substitute in (29) the expressions for Λ and ρ; we also substitute α with the positive roots from (14). Let us denote by m_i the resulting m from (29). Thus we obtain that V^Λ is reducible, if at least one of the following numbers is a positive integer:

$$m_1 = 1 + 2s_0 , \tag{32a}$$

$$m_2 = 1 - E_0 - s_0 , \tag{32b}$$

$$m_3 = 3 - 2E_0 , \tag{32c}$$

$$m_4 = 2 - E_0 + s_0 . \tag{32d}$$

(Note that $m_3 = m_1 + 2m_2$, $m_4 = m_1 + m_2$.) Further we shall need the explicit expression for the weight Λ' in (27) and (28) when (29) is fulfilled. Namely, if in (29) $\alpha = \alpha_i$ and $m = m_i \in \mathbb{N}$ then we have [8], [10], [11]:

$$\Lambda' = \Lambda + m_i\alpha_i . \tag{33}$$

We also note that $\mathcal{P}(\mathcal{G}_+)$ has weight $m_i \alpha_i$ which makes it easier to find it explicitly [14], [11].

Let us recall that if I^Λ is the maximal invariant submodule of V^Λ, then the factor-module $L_\Lambda = V^\Lambda / I^\Lambda$ is irreducible and every lowest (highest) weight representation may be obtained in this way.

Let us analyze the application of (32) and (33) to the irreducible lowest weight representations of $\mathcal{G}$.

First we note that if $m_1, m_2 \in \mathbb{N}$, then the irreducible representations with lowest weight $E_0 = (3-m_1-2m_2)/2$, $s_0 = (m_1-1)/2$ are the finite-dimensional lowest weight representations of $\mathcal{G}$ which are nonunitary; $m_1 = m_2 = 1$ gives the trivial 1-dimensional representation; the fundamental representations are obtained for $m_1 = 1$, $m_2 = 2$ and $m_1 = 2$, $m_2 = 1$.

Further we restrict our attention to the unitary irreducible lowest weight representations of $\mathcal{G}$ given in (24) and (25). We note that (32a) holds always because $s_0 \in \mathbb{Z}_+/2$, and (32b) never holds because $m_2 \leq 1/2$. Next, we note that m_3 is a positive integer only for $E_0 = 1/2, 1$, in which case $m_3 = 2, 1$, respectively. Similarly, m_4 is a positive integer only for $E_0 - s_0 = 1$, and that integer is $m_4 = 1$.

The singular vectors corresponding to these cases are [6]:

$$v_1^+ = (J^+)^{2s_0+1} \oplus v_0 , \quad s_0 \in \mathbb{Z}_+/2 , \tag{34a}$$

$$v_3^+ = [(K^+)^2 + 4M^{+-}M^{++}] \oplus v_0 , \quad m_3 = 2 , \quad E_0 = 1/2 , \ (\Rightarrow s_0 = 0) , \tag{34b}$$

$$v_3'^+ = (s_0 K^+ - M^{+-}J^+) \oplus v_0 , \quad m_3 = 1 , \quad E_0 = 1 , \ (\Rightarrow s_0 \leq 1/2) , \tag{34c}$$

$$v_4^+ = [2s_0(2s_0-1)M^{++} + (1-2s_0)K^+J^+ + M^{+-}(J^+)^2] \oplus v_0 , \quad E_0 - s_0 = 1. \tag{34d}$$

Applying this to our representation space would mean that these null states must vanish :

$$(J^+)^{2s_0+1}|E_0, s_0\rangle = 0 , \tag{35a}$$

$$[(K^+)^2 + 4M^{+-}M^{++}]|E_0 = 1/2, s_0 = 0\rangle = 0 , \tag{35b}$$

$$(s_0 K^+ - M^{+-}J^+)|E_0 = 1 , s_0 \leq 1/2\rangle = 0 , \tag{35c}$$

$$[2s_0(2s_0-1)M^{++} + (1-2s_0)K^+J^+ + M^{+-}(J^+)^2]|E_0 = s_0+1, s_0\rangle = 0 . \tag{35d}$$

Thus for the two singletons we have :

$$\text{Rac}: \quad J^+|1/2, 0\rangle = 0 , \tag{36a}$$

$$[(K^+)^2 + 4M^{+-}M^{-++}|1/2, 0\rangle = 0 ; \tag{36b}$$

$$\text{Di}: \quad (J^+)^2|1, 1/2\rangle = 0 , \tag{37a}$$

$$(K^+ - 2M^{+-}J^+)|1, 1/2\rangle = 0 . \tag{37b}$$

Note that (35d) is zero automatically in these two cases; for the Rac because $s_0 = 0$ and $J^+|1/2, 0 > = 0$ and for the Di because $s_0 = 1/2$ and $(J^+)^2|1, 1/2\rangle = 0$.

Further we shall apply the above considerations in order to rederive the spectrum of the unitary representations of so(3,2) [1], [2], [3].

First let us apply (36) to the basis space $\mathbf{H}$, given in (21). We note that because of $J^+|1/2, 0\rangle = 0$ it follows that $k_4 = 0$ and $|p| \leq n$. Then using (36b) we replace $M^{+-}M^{++}|1/2, 0\rangle = (-1/4)(K^+)^2|1/2, 0\rangle$, thus, we are left only with powers of $M^{+\varepsilon}$ and K^+, where $\varepsilon = \mathrm{sign}(k_1 - k_2)$. So the weight space will consist of states of the type [6]:

$$\mathbf{H}_{n,p} = (M^{+\varepsilon})^{|p|}(K^+)^{n-|p|}|1/2, 0\rangle , \quad \varepsilon = \mathrm{sign}(p), \quad |p| \leq n, \quad (38)$$

i.e., $\dim \mathbf{H}_{n,p} = 1$; that is why this representation is called a singleton representation. Note also that $\dim \mathbf{H}_n = 2n + 1$.

Consider now (37). Because of (37a) we have $k_4 = 0, 1$, and because of (37b) we have $k_3 = 0$. Thus we have [6]:

$$\mathbf{H}_{n,p} = (M^{++})^{(n+p-\varepsilon)/2}(M^{+-})^{(n-p+\varepsilon)/2}(J^+)^{\varepsilon}|1, 1/2\rangle ,$$
$$\varepsilon = (n - p) \bmod 2 \in \{0, 1\} , \quad |p - \varepsilon| \leq n . \quad (39)$$

Thus again $\dim \mathbf{H}_{n,p} = 1$, while $\dim \mathbf{H}_n = 2n + 2$.

Further we analyze the case $s_0 = 0$, $E_0 > 1/2$. In this case (35c) and (35d) are fulfilled automatically. For $\mathbf{H}_{n,p}$ we have (21b) with $k_4 = 0$.

Analogously for $s_0 = 1/2$, $E_0 > 1$, (35d) is fulfilled automatically. For $\mathbf{H}_{n,p}$ we again have (21b) with $k_4 = 0, 1$.

Finally we consider $E_0 = s_0 + 1$, $s_0 \geq 1$. These are the massless representations. Then (35d) is a nontrivial restriction. For $\mathbf{H}_{n,p}$ we have (21b) with $k_1 = 0$, $k_4 \leq 2s_0$.

5 Spectrum of Representations via Character Formulae

Let us recall that each root $\alpha \in \Delta$ defines a reflection s_α in $\mathcal{H}^*$ by the formula

$$s_\alpha(\Lambda) = \Lambda - (\Lambda, \alpha^\vee)\alpha, \quad \alpha^\vee = 2\alpha/(\alpha, \alpha), \quad \Lambda \in \mathcal{H}^*, \quad s_\alpha^2 = 1. \quad (40)$$

These reflections generate a group, called the Weyl group W of $\mathcal{G}$ [8], [10]. [The Weyl groups of a real Lie algebra $\mathcal{G}$ and its complexification differ, unless $\mathcal{G}$ is maximally split as $so(3, 2)$]. W has a finite number of elements if $\mathcal{G}$ is finite dimensional. Actually W is generated by the reflections by simple roots $w_i = s_{\alpha_i}$ which for $so(3, 2)$ obey

$$w_i^2 = 1 , \quad i = 1, 2 , \quad (w_1 w_2)^4 = 1. \quad (41)$$

Since the Weyl group is generated by the simple reflections then every element $w \in W$ may be written as the product of some simple reflections. Every such product which uses a minimal number of simple reflections is called a reduced expression or reduced form for w. The number of simple reflections in a reduced form is called the length of w and is denoted by $\ell(w)$.

For $\mathcal{G} = so(3, 2)$, $|W| = 8$ and all elements of W are given in a reduced form as follows:

$$W = \{1,\ w_1,\ w_2,\ w_1w_2,\ w_2w_1,\ w_1w_2w_1,\ w_2w_1w_2,\ w_1w_2w_1w_2\}\ . \tag{42}$$

Note that $s_{\alpha_3} = w_2w_1w_2$ and $s_{\alpha_4} = w_1w_2w_1$. Further it will be convenient to introduce a shifted action of $w \in W$ by the formula:

$$w \cdot \Lambda \equiv w(\Lambda - \rho) + \rho\ . \tag{43}$$

Using this, (40) and (29) we can rewrite (33) as follows :

$$\Lambda' = \Lambda + m\alpha = \Lambda - (\Lambda - \rho, \alpha^\vee)\alpha = s_\alpha \cdot \Lambda\ . \tag{44}$$

Let $\mathcal{G}$ be any simple Lie algebra. Let Γ , (resp. Γ_+), be the set of all integral, (resp. integral dominant), elements of $\mathcal{H}^*$, i.e., $\Lambda \in \mathcal{H}^*$ such that $(\Lambda, \alpha_i^\vee) \in \mathbb{Z}$, (resp. $\mathbb{Z}_+$), for all simple roots α_i . We recall that for each invariant subspace $V \subset U(\mathcal{G}_+) \otimes v_0 \cong V^\Lambda$ we have the following decomposition

$$V = \bigoplus_{\mu \in \Gamma_+} V_\mu\ ,\quad V_\mu = \{u \in V \mid H_k u = (\Lambda + \mu)(H_k)u,\ \forall H_k\}\ . \tag{45}$$

(Note that $V_0 = \mathbb{C}\, v_0$.) We have $\phi(V_\mu,\ V_\nu) = 0$ if $\mu \neq \nu$. Following [8], [10] let $E(\mathcal{H}^*)$ be the associative abelian algebra consisting of the series $\sum_{\mu \in \mathcal{H}^*} c_\mu e(\mu)$, where $c_\mu \in \mathbb{C}$, $c_\mu = 0$ for μ outside the union of a finite number of sets of the form $D(\Lambda) = \{\mu \in \mathcal{H}^* | \mu \leq \Lambda\}$, using any ordering of $\mathcal{H}^*$; the formal exponents $e(\mu)$ have the properties $e(0) = 1$, $e(\mu)e(\nu) = e(\mu + \nu)$.

The character of V is defined by :

$$\text{ch}\ V = \sum_{\mu \in \Gamma_+} (\dim V_\mu)e(\Lambda + \mu) = e(\Lambda) \sum_{\mu \in \Gamma_+} (\dim V_\mu)e(\mu)\ . \tag{46}$$

We recall [8] that for $V = V^\Lambda$ we have $\dim V_\mu = P(\mu)$, $P(\mu)$ is a generalized partition function, $P(\mu) = \#$ of ways μ can be presented as a sum of positive roots β_j , each root taken with its multiplicity $m_j = \dim \mathcal{G}_{\beta_j}$, (here $m_j = 1$), $P(0) \equiv 1$. Analogously we use [8] to obtain :

$$\text{ch}\ V^\Lambda = e(\Lambda) \sum_{\mu \in \Gamma_+} P(\mu)e(\mu) = e(\Lambda) \prod_{\alpha \in \Delta^+} (1 - e(\alpha))^{-1}\ . \tag{47}$$

The Weyl character formula for the finite-dimensional irreducible lowest weight representations over $\mathcal{G}$ has the form [8]:

$$\text{ch}\ L_\Lambda = \text{ch}\ V^\Lambda \sum_{w \in W} (-1)^{\ell(w)} e(w \cdot \Lambda - \Lambda) = \sum_{w \in W} (-1)^{\ell(w)}\ \text{ch}\ V^{w \cdot \Lambda}\ . \tag{48}$$

Let $\mathcal{G} = so(3,2)$. Denote $t_i \equiv e(\alpha_i)$, $i = 1, 2$, then $e(\alpha_3) = t_1 t_2$, $e(\alpha_4) = (t_1)^2 t_2$. Then (47) and (48) can be rewritten, respectively, as

$$\text{ch}\ V^\Lambda = e(\Lambda)/(1 - t_1)(1 - t_2)(1 - t_1 t_2)(1 - (t_1)^2 t_2)\ , \tag{49}$$

$$\text{ch}\ L_\Lambda = \text{ch}\ V^\Lambda (1 - t_1^{m_1} - t_2^{m_2} + t_1^{m_1} t_2^{m_1+m_2} + t_1^{m_1+2m_2} t_2^{m_2} - \\ - t_1^{2(m_1+m_2)} t_2^{m_1+m_2} - (t_1 t_2)^{m_1+2m_2} + t_1^{2(m_1+m_2)} t_2^{m_1+2m_2})\ . \tag{50}$$

The character formulae for the infinite-dimensional irreducible lowest weight representations over so(3,2) involve less terms than in (50) since the maximal invariant submodules I^Λ of V^Λ are smaller. To find these character formulae we shall use the technique of the reduced Weyl groups [14] which we applied recently to the affine so(3,2) [6]. This means that the character formulae for the infinite-dimensional irreducible lowest weight representations over so(3,2) will look like the Weyl character formula (48), however with W replaced by certain subgroups of W, called reduced Weyl groups.

A reduced Weyl group W_R is defined as follows. It is generated by s_α with $\alpha \in \Delta$ such that the Bernstein–Gel'fand–Gel'fand (BGG) [12] criterion (29) is fulfilled. (Sometimes a single reduced Weyl group is not enough to describe the character formula [14], [6].) We shall give an explicit description for the unitary irreducible lowest weight representations (24), (25).

For the singletons the reduced Weyl group $W_R = W^s$ is generated by $s_{\alpha_1} = w_1$ and $s_{\alpha_3} = w_2 w_1 w_2$. Thus we have:

$$W^s = \{1, w_1, w_2 w_1 w_2, w_1 w_2 w_1 w_2\}. \tag{51}$$

Then the character formula for the singletons is:

$$\operatorname{ch} L^s = \operatorname{ch} V^\Lambda \sum_{w \in W^s} (-1)^{\ell(w)} e(w \cdot \Lambda - \Lambda) = \tag{52a}$$

$$= \operatorname{ch} V^\Lambda (1 - t_1^{m_1} - (t_1 t_2)^{m_1 + 2m_2} + t_1^{2(m_1+m_2)} t_2^{m_1+2m_2}). \tag{52b}$$

More explicitly for the Rac we have $m_1 = 1$, $m_2 = 1/2$ and using (49) we obtain:

$$\operatorname{ch} L_{\text{Rac}} = \operatorname{ch} V^\Lambda (1 - t_1 - (t_1 t_2)^2 + t_1^3 t_2^2) = \tag{53a}$$

$$= e(\Lambda)(1 - (t_1 t_2)^2)/(1 - t_2)(1 - t_1 t_2)(1 - (t_1)^2 t_2) = \tag{53b}$$

$$= e(\Lambda)(1 + t_1 t_2)/(1 - t_2)(1 - (t_1)^2 t_2) = \tag{53c}$$

$$= e(\Lambda) \sum_{n=0}^{\infty} (t_1 t_2)^n \sum_{p=-n}^{n} t_1^p = \tag{53d}$$

$$= e(\Lambda) \sum_{n=0}^{\infty} \sum_{p=-n}^{n} (t_1 t_2)^{n-|p|} t'^{|p|}, \tag{53e}$$

where

$$t' = \begin{cases} t_1^2 t_2 = e(\alpha_4) & \text{for } p \geq 0, \\ t_2 = e(\alpha_2) & \text{for } p < 0. \end{cases}$$

It is clear that the character formula (53) is equivalent to the spectrum description given in (38). Each term in (53d,e) for fixed n, p corresponds exactly to $\mathbf{H}_{n,p}$. Indeed, $t_1 t_2 = e(\alpha_3)$ corresponds to the root vector K^+, $e(\alpha_4)$ corresponds to the root vector M^{++}, $e(\alpha_2)$ corresponds to the root vector M^{+-}, cf. (13).

Analogously the Di we have $m_1 = 2$, $m_2 = -1/2$ and using (49) we obtain:

$$\text{ch } L_{\text{Di}} = \text{ch } V^{\Lambda} (1 - t_1^2 - t_1 t_2 + t_1^3 t_2) = \tag{54a}$$

$$= e(\Lambda) (1 - t_1^2)/(1 - t_1)(1 - t_2)(1 - (t_1)^2 t_2) = \tag{54b}$$

$$= e(\Lambda) (1 + t_1)/(1 - t_2)(1 - (t_1)^2 t_2) = \tag{54c}$$

$$= e(\Lambda) \sum_{n=0}^{\infty} (t_1 t_2)^n \sum_{p=-n}^{n+1} t_1^p . \tag{54d}$$

$$= e(\Lambda) \sum_{n=0}^{\infty} t_2^n \sum_{r=0}^{n} (t_1^{2r} + t_1^{2r+1}) . \tag{54e}$$

It is clear that the character formula (54) is equivalent to the spectrum description given in (39). Each term in (54d) for fixed n, p corresponds exactly to $\mathbf{H}_{n,p}$. Indeed, the terms with $\varepsilon = (n-p) \bmod 2 = 0$, i.e., $(M^{++})^{(n+p)/2}(M^{+-})^{(n-p)/2}$, are represented by $e(\alpha_4)^{(n+p)/2} e(\alpha_2)^{(n-p)/2} = (t_1^2 t_2)^{(n+p)/2} t_2^{(n-p)/2} = t_1^{n+p} t_2^n$ and by the first term in (54e). Analogously, the terms with $\varepsilon = (n - p) \bmod 2 = 1$, i.e., $(M^{++})^{(n+p-1)/2}(M^{+-})^{(n-p+1)/2} J^+$, are represented by $e(\alpha_4)^{(n+p-1)/2}$ $e(\alpha_2)^{(n-p+1)/2} e(\alpha_1) = (t_1^2 t_2)^{(n+p-1)/2} t_2^{(n-p+1)/2} t_1 = t_1^{n+p} t_2^n$ and by the second term in (54e).

Next we consider the case $s_0 = 0$, $E_0 > 1/2$ (cf. (25) and the text after (39)). The reduced Weyl group is generated only by w_1, i.e., $W_R = W^1 = \{1, w_1\}$. This is clear for $E_0 \neq 1$, while for $E_0 = 1$ one should note that the singular vectors $v_3'^+$ in (34c) and v_4^+ in (34d) reduce to v_1^+ in (34a) when $s_0 = 0$. Thus the character formula is:

$$\text{ch } L_{E_0 > 1/2, s_0 = 0} = \text{ch } V^{\Lambda} (1 - t_1) = \tag{55a}$$

$$= e(\Lambda)/(1 - t_2)(1 - t_1 t_2)(1 - t_1^2 t_2) , \tag{55b}$$

which is equivalent to (21b) with $k_4 = 0$.

Analogously for $s_0 = 1/2$, $E_0 > 1$, (cf. (25) and the text after (39)) we have the same reduced Weyl group, i.e., $W_R = W^1 = \{1, w_1\}$. This is clear for $E_0 \neq 3/2$, while for $E_0 = 3/2$ one should note that the singular vector v_4^+ in (34d) reduces to v_1^+ in (34a) when $s_0 = 1/2$. Thus the character formula is:

$$\text{ch } L_{E_0 > 1, s_0 = 1/2} = \text{ch } V^{\Lambda} (1 - t_1^2) = \tag{56a}$$

$$= e(\Lambda) (1 + t_1)/(1 - t_2)(1 - t_1 t_2)(1 - t_1^2 t_2) , \tag{56b}$$

which is equivalent to (21b) with $k_4 = 0, 1$.

Finally we consider the massless representations with $E_0 = s_0 + 1$, $s_0 \geq 1$. Then (34d) is a nontrivial singular vector. Let us define $W' \equiv \{ 1, w_1, w_1 w_2 w_1, w_2 w_1 \}$. We cannot consider the set W' as a subgroup of W, since then the elements of W' will generate the whole W. This means that the approach of the reduced Weyl group is applied only formally here, i.e., in the sense that we conjecturally write the character formula as:

$$\text{ch } L_{\text{massless}} = \text{ch } V^{\Lambda} \sum_{w \in W'} (-1)^{\ell(w)} e(w \cdot \Lambda - \Lambda) = \tag{57a}$$

$$= \text{ch } V^{\Lambda} (1 - t_1^m - t_1^2 t_2 + t_1^m t_2) . \tag{57a}$$

where $m = m_1 = 1 + 2s_0 \geq 3$.

References

1.   P.A.M. Dirac: J. Math. Phys. **4** 901 (1963)
2.   C. Fronsdal: Rev. Mod. Phys. **37** 221 (1965); Phys. Rev. **D10** 589 (1974); Phys. Rev. **D12** 3819 (1975); Phys. Rev. **D26** 1988 (1982); M. Flato and C. Fronsdal: Phys. Lett. **B97** 236 (1980)
3.   N.T. Evans: J. Math. Phys. **8** 170 (1967)
4.   M. Flato and C. Fronsdal: J. Math. Phys. **22** 1100 (1981)
5.   M. Flato and C. Fronsdal: Lett. Math. Phys. **2** 421 (1978)
6.   V.K. Dobrev and E. Sezgin: ICTP, Trieste, preprint IC/90/1 (1990)
7.   H. Nicolai: in *Supersymmetry and Supergravity '84*, ed. by B. de Wit, P. Fayet and P. van Nieuwenhuizen (World Scientific, Singapore, 1984)
8.   J. Dixmier: *Enveloping Algebras* (North Holland, New York, 1977)
9.   S. Helgason: *Differential Geometry, Lie Groups and Symmetric Spaces* (Academic Press, New York, 1978)
10.   V.G. Kac: *Infinite dimensional Lie algebras*, (Birkhäuser, Boston, 1983)
11.   V.K. Dobrev: "Lectures on Lie algebras and their representations: I", ICTP internal report, IC/88/96 (1988)
12.   N.N. Bernstein, I.M. Gel'fand, and S.I. Gel'fand: Funkts. Anal. Prilosh. **5** 1 (1971); English translation, Funct. Anal. Appl. **5** 1 (1971)
13.   V. Kac and D. Kazhdan: Adv. Math. **34** 97 (1979)
14.   V.K. Dobrev: Lett. Math. Phys. **9** 205 (1985);   Talk at the Conference on Algebraic Geometry and Integrable Systems (Oberwolfach, 1984) and ICTP, Trieste, preprint IC/85/9 (1985)

V

General Aspects of Quantum Physics

Quantum Theory of Single Events

A.O. Barut

Physics Department, University of Colorado
Boulder, Colorado, 80309

Abstract: A clear distinction is made between a single individual event denoted by $\psi(x, a)$ depending on the parameters a, and a typical averaged event denoted by $\Psi(x)$. This avoids the notion of the collapse of the wave function, among others. Explicit localized solutions of the wave equations are constructed which move like relativistic particles and obey quantum relations $E = \lambda\omega$ and $P = \lambda k$. Quantum spin correlations are evaluated on the basis of individual events with classical "hidden" parameters.

1 A Single Event Versus a Typical Event

1.1 Experiment Shows Two Different Wave Functions

In an electron interference experiment electrons are sent one after another, are collected as small dots on a screen and the pictures are stored in a memory device [1]. When the experiment begins we see a few isolated dots which get denser and as the number of collected dots becomes very large we see the emergence of interference patterns as a regularity in repeated events.

I shall assume that a single tiny dot represents a single individual electron. This and the regularity in repeated events is shown schematically in Fig. 1. Also shown is the probability of a typical event by the distribution $|\Psi|^2$. I represent the single dot in Fig. 1 by a localized wave lump $\psi(\boldsymbol{x}, t; \boldsymbol{a})$, where $\boldsymbol{a}$ is a set of parameters labeling the position, velocity and the shape of the individual event. The blackening caused by a single event may be proportional to $|\psi|^2$. Clearly ψ and Ψ are entirely different types of entities, hence the different symbols used. One is an objective material individual event; the other is a derived measure of regularity of repeated events and does not exist if the experiment is not repeated. This distinction is fundamental for the present theory, hence we formalize it more precisely:

This is a summary of lectures on a new Foundations of Quantum Theory that I gave over a number of years at Clausthal meetings and dedicated to H.D. Doebner.

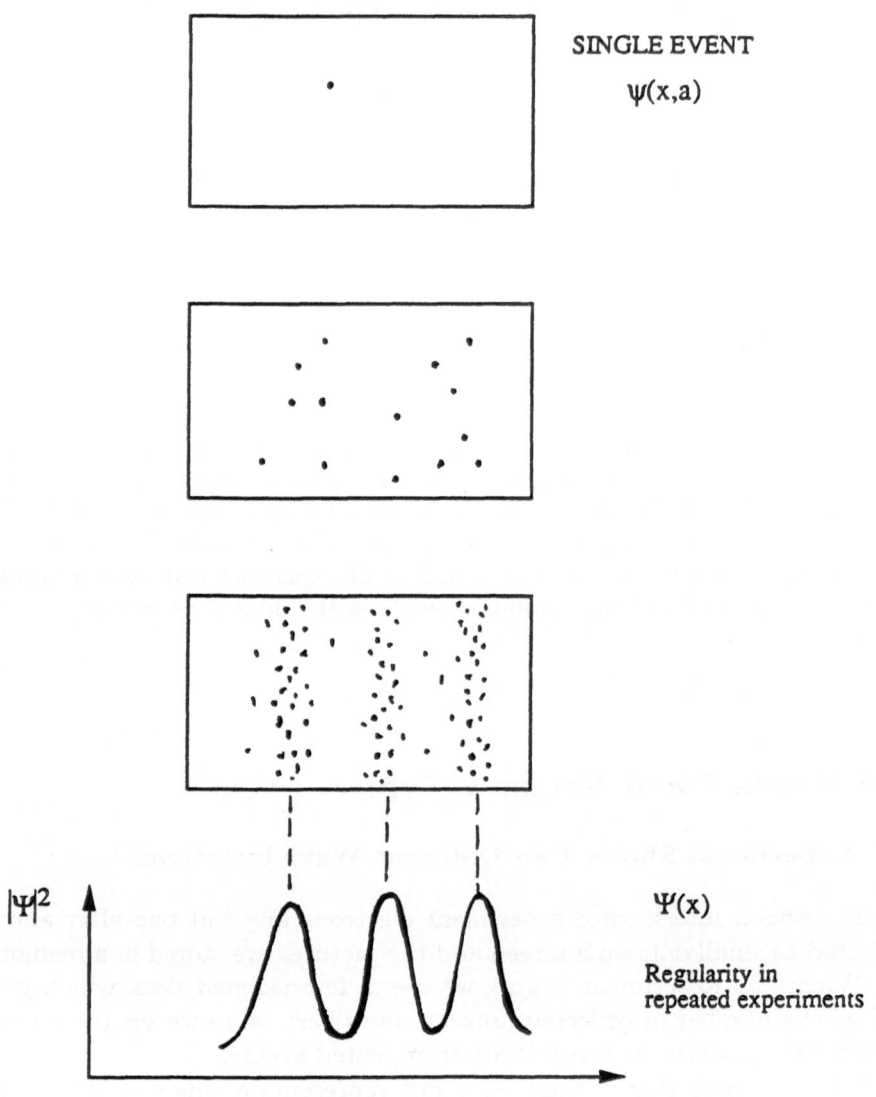

SINGLE EVENT

$\psi(x,a)$

$|\Psi|^2$

$\Psi(x)$

Regularity in
repeated experiments

Fig. 1.

Definition 1 A *single event* is an experimental outcome which can be attributed to an isolated objective individual entity, e.g. a track in a bubble chamber.

Definition 2 A *typical event* is the expected averaged behavior of a single entity in repeated experiments.

Both are observed, though in different ways, and there must be a theory for both. The corresponding wave functions are:

(i) $\psi(\boldsymbol{x}, t; \boldsymbol{a})$ = wave function of the individual with "name" $\boldsymbol{a}$, where $\boldsymbol{a}$ is set of classical "hidden" parameters. The parameters are not really hidden, they are, as we mentioned above, the position, velocity and the shape parameters of a localized wave lump. I prefer, instead of the historical adjective "hidden", the terminology of classical (nonquantized) parameters. This individual wave function is deterministic, objective and highly localized. It is nonstationary, but it is not a wave-packet : we want it to be highly localized and nonspreading as seen in experiments. We shall say more about the spreading a little later. The superposition principle does not hold in the usual sense. The sum of two lumps with different parameters, even if it is a solution of the wave equation, represents two individual particles.

(ii) $\Psi(\boldsymbol{x}, t)$ = the wave function for a typical event without any hidden or classical parameters. The word typical is used here in the same way as we speak for example of a typical American. Thus this wave function is by definition probabilistic. It is not localized, usually represented by a plane wave or by a large wave packet in scattering experiments (of the order of the aperture of the instruments). It obeys the superposition principle : the linear combination of two such wave functions can still represent another state of the same typical event. Finally Ψ can have stationary states, with $|\Psi|^2$ independent of time, unlike ψ .

1.2 These Wave Functions do not Collapse

Let us now see what we gain by clearly distinguishing these two different wave functions.

First of all the introduction of the individual wave function $\psi(\boldsymbol{x}, t; \boldsymbol{a})$ removes completely the need for the notion "collapse of the wave function" and all the attempts to find mechanisms of how the collapse can arise: The localized wave lump ψ is the already collapsed single event, and Ψ never collapses, in fact must not collapse. In the practical use of quantum theory for repeated experiments one needs the uncollapsed form of Ψ and there is never any need to collapse or to reduce the wave function. In the theory of measurement this notion arises, but now we must use ψ. This notion arose historically because one wanted, or one thought, the wave function Ψ to represent *both* the regularity of repeated events and also the individual single event. That is the reason one talks about the discontinuity in the measurement theory : Quantum theory has two kinds of evolutions, we are often told, one the unitary deterministic evolution of Ψ given by the Schrödinger equation; the other the discontinuous probabilistic evolution during the measurement where Ψ is supposed to collapse to an individual event. I think this dilemma towards whose resolution heroic efforts have been spent since the beginning of quantum theory is unnecessary. We cannot describe two entirely different entities by the same mathematical quantity and hope for consistency. What we want to show here is that the evolutions of both ψ and Ψ are continuous and deterministic, but they have different physical meanings. If we are interested in the behavior of an individual particle in a measuring device we must calculate the motion of the localized wave lump everywhere in the instrument, e.g. the motion of the spinning individual atom in the Stern-Gerlach magnet. We cannot represent the measuring device by a black box from which different eigenstates of an operator emerge. The latter is only true for Ψ. However we can pass from the description $\psi(\boldsymbol{x}, t; \boldsymbol{a})$ over to the description Ψ by averaging over the parameters $\boldsymbol{a}$ of ψ as the experimentalist in fact does when he wants to test Ψ. Furthermore in the development of quantum theory several other topics should be formulated in terms of individual wave functions ψ rather than in terms of Ψ in order to avoid inconsistencies. For example, the problem of the classical limit of quantum theory is directly and easily obtained as the limit of the individual wavelets whose crests almost follow the classical trajectories. Classical mechanics is a theory of individual events. On the other hand, as is well known, it is very difficult to obtain classical trajectories from Ψ; it is not just letting $\hbar \to 0$. Similarly, the problems of quantum chaos or quantum statistical mechanics should be formulated in terms of $\psi(\boldsymbol{x}, t; \boldsymbol{a})$ where we can compare these phenomena in a direct one-to-one way with their corresponding classical counterparts.

2 Completeness of Quantum Mechanics and Brief History of the Theory of Single Events

Standard quantum theory deals only with Ψ, i.e. with repeated events, hence with the typical behavior of a single particle. It is of course very successful as such, and we all appreciate it. It cannot make any statements about individual events, but we wish to describe an individual event, hence for this simple reason it is incomplete. The experiment is here ahead of theory: experiments with individual particles abound. It is therefore legitimate, nay a duty, to try to complete the standard quantum theory by adding to it a "Theory of Single Events" in such a way that the standard theory emerges after averaging over the parameters of all the single events observed. Clearly this does not change anything about the standard interpretation, but it adds to it a new dimension so that we can keep quite apart statements made on single events versus on typical events. As we shall see this will be crucial in the discussion of spin correlation experiments. Furthermore, the theory of single events provides us, as mentioned above, a direct approach to classical theory in the limit of vanishing wave properties of the individual wavelets.

It may illuminate to observe that there is a similar situation to the above dichotomy (individual vs. typical) in sociology. Social scientists observe the behavior of an individual. But since one does not know much what to do with this single experiment, one sends questionnaires to a large number of individuals with the hope of deriving the laws for the typical behavior. A typical person does not exists as such materially as an individual does. Moreover, we may get a different typical person if we perform a weighted average. One can well imagine to develop an equation for the behavior of a typical person to predict its behavior as well as for each individual.

Max Born [2], when he first applied the wave mechanics to scattering and introduced the statistical interpretation, was well aware of the distinction between individual and typical particles, which later has often been forgotten. He used a plane incoming wave and an outgoing spherical wave which he expanded into plane waves going into different directions and interpreted (in the second paper) the coefficients as the probability amplitudes of scattering from the initial momentum to the final momentum in the direction θ. He asserted that there were no internal properties of atomic systems which would fix and determine a definite result of any collision in a single event. But he added "but it is of course allowed to someone, who is not satisfied with this, to assume that there are additional parameters that are not in the theory which would determine the single event". The additional parameters a that we envisage are indeed not in the theory but they are in the localized solutions as we shall show explicitly.

L. de Broglie [3], however, from the very beginning "tried to imagine a real physical wave which transported minute and localized objects through space in the course of time", and came back to this vision in the later period of his life [5].

It is not a question of which interpretation is the exclusively correct one, Born's or de Broglie's, but there are, and there should be, two different entities ψ and Ψ, depending on what question we ask. De Broglie's ψ is the more elementary, microscopic entity, and Born's Ψ is the derived, "macroscopic" quantity, much more under control.

Having introduced the wave function for single events and its distinction from the usual wave function, what is now the theory of single events, what are the laws or wave equations for $\psi(\boldsymbol{x}, t, \boldsymbol{a})$? It is generally believed that a theory of single events is *in principle* impossible. The best way to answer this belief is to go ahead and construct an explicit theory of single events. We shall use conservatively the usual wave equations also for ψ, but we should be open minded to modifications in the wave equations since the laws of single events must yet be developed inductively.

3 Localized Solutions of Wave Equations Moving Like Particles

3.1 The Scalar Wave Equation

We begin with the ordinary massless scalar wave equation $\Box \varphi = 0$. We look for a solution which is localized at a point $\boldsymbol{x}_0$ of the form

$$\varphi(\boldsymbol{x}, t) = F(\boldsymbol{x} - \boldsymbol{x}_0) \, e^{-i\Omega(t-t_0)}. \tag{1}$$

Thus a point $\boldsymbol{x}_0$ is singled out – this is an individual solution. This localized wave is not static, but oscillates with an internal frequency Ω; it has the parameters $t_0, \boldsymbol{x}_0$ and Ω. Six more parameters will be introduced when we construct a moving solution by a Lorentz transformation, three velocity components and three angles of rotation. Actually, because the massless wave equation is also conformally invariant we could introduce five more parameters corresponding to dilatations and special conformal transformations making a total of 16 parameters. Note that the equation has no parameters. For simplicity, I consider here a spherically symmetric solution; the general case is given elsewhere [4]. In this case the moving solution has the form

$$\varphi(\boldsymbol{x}, t; \boldsymbol{a}) = F\left(r_\perp(t)\right) e^{i(\boldsymbol{k}\cdot\boldsymbol{x} - \omega t)} \tag{2}$$

where $r_\perp$ is the transformed relativistic radial coordinate

$$r_\perp^2 = (r_\mu u^\mu)^2 - r^2$$

with $r^\mu = (t - t_0, \boldsymbol{x} - \boldsymbol{x}_0)$ and $u^\mu = \gamma(1, \boldsymbol{\beta})$ being the relativistic velocity in the direction of the Lorentz boost; $u^2 = 1$, $\gamma = 1/\sqrt{1 - \beta^2}$, $\boldsymbol{\beta} = \boldsymbol{v}/c$. Again for simplicity we consider a pure boost without rotations and dilatations. For the frequency and the wave vector of the moving solution we find

$$\omega = \gamma\Omega \quad , \quad \boldsymbol{k} = \frac{\Omega}{c}\gamma\boldsymbol{\beta} \tag{3}$$

so that we have the dispersion relation

$$(\omega/c)^2 - \boldsymbol{k}^2 = (\Omega/c)^2 \tag{4}$$

which is that of a massive particle, although we have started with a massless equation. The reason is the localizing factor F in the solution instead of the pure plane wave solution. In order to see this in detail, we insert (1) into $\Box\varphi = 0$, and obtain the Helmholtz equation for F

$$\Delta F + (\Omega/c)^2 F = 0 \tag{5}$$

whose general solution can be expressed in terms of Bessel's functions [4]. The simplest spherically symmetric solution in the rest frame that we decided to consider here is

$$F = C\frac{\sin\left(\frac{\Omega}{c}r\right)}{r} . \tag{6}$$

This is a wave concentrated inside a radius $R \sim c/\Omega$ but with a long decreasing oscillatory tail (Fig. 2a).

The moving solution, say in x-direction is

$$\varphi(\boldsymbol{x}, t; \boldsymbol{a}) = C\frac{1}{r_\perp}\sin\left(\frac{\Omega}{c}r_\perp\right) e^{i(\boldsymbol{k}\cdot\boldsymbol{x} - \omega t)}$$

$$r_\perp = \left[\gamma^2(x - vt)^2 + y^2 + z^2\right]^{1/2} \tag{7}$$

and is shown in Fig. 2b for $v/c = .8$. We can associate a group velocity v to this moving lump; the phase velocity is $u = \omega/k = \frac{c^2}{v}$. Hence $uv = c^2$.

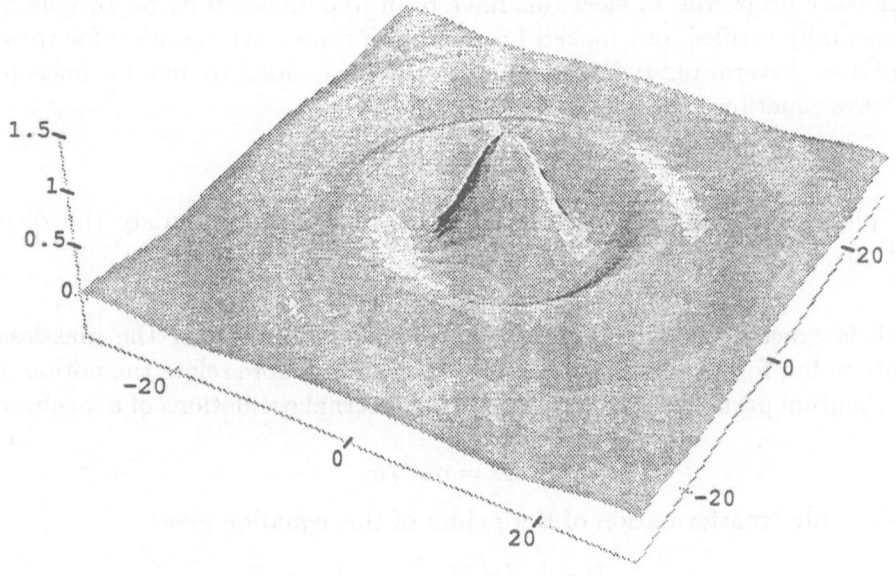

Fig. 2a.

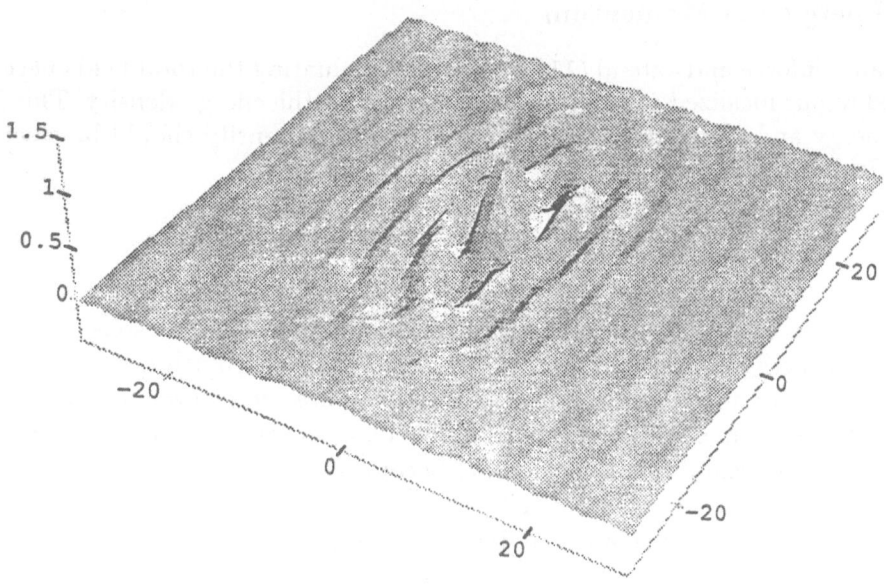

Fig. 2.b.

3.2 Mass and Localization

After wave properties of electrons have been hypothesized by de Broglie and experimentally verified, one looked for an appropriate wave equation for these wave properties. Several physicists independently have added by hand a mass term to the wave equation

$$\left(\Box + \frac{m^2 c^2}{\hbar^2}\right)\phi = 0. \tag{8}$$

The plane wave solutions of this Klein-Gordon equation satisfy the dispersion relation

$$(\omega/c)^2 - \mathbf{k}^2 = m^2 c^2/\hbar^2 \tag{9}$$

which is exactly the same as our dispersion relation (4) for the massless wave equation, but for its localized solutions. We can therefore relate the notion of mass of a quantum particle to the frequency Ω of internal oscillations of a localized wave by

$$\Omega = mc^2/\hbar. \tag{10}$$

The Lorentz transformation of both sides of this equation gives

$$\gamma\Omega = \gamma mc^2/\hbar \quad, \quad \text{or} \quad \hbar\omega = E \tag{11}$$

which is the basic quantum principle which we have now modeled with our localized solutions.

3.3 Energy and Momentum

We can reinforce and extend (11) by actually evaluating the total field energy contained in our localized solution as an integral over the energy density. This is how the energy and momentum of an objective localized entity should be calculated. The field energy is given by

$$E = N_E \int \left\{\left|\frac{1}{c}\frac{\partial\varphi}{\partial t}\right|^2 + |\Delta\varphi|^2\right\} d\mathbf{r} \tag{12}$$

where N_E is a normalization constant, also for dimensional reasons. For a plane wave this energy is threefold infinite ∞^3. For our localized solution (1) or (3) it is still one-fold infinite ∞^1. A method to obtain a finite energy even for a free particle is to consider a superposition of solutions over a narrow range of frequencies around some basic frequency Ω_0. Equation (1) becomes

$$\varphi(\mathbf{x}, t; \mathbf{a}) = \int d\Omega\, F(\mathbf{x} - \mathbf{x}_0, \Omega) e^{-i\Omega t} f(\Omega) \tag{13}$$

where $f(\Omega)$ is a distribution function. When (13) is inserted into (12) and use is made of the orthogonality relations of Bessel's functions

$$\int_0^\infty J_{\ell+1/2}\left(\frac{\Omega}{c}r\right) J_{\ell+1/2}\left(\frac{\Omega'}{c}r\right) r\, dr = \frac{c}{\Omega}\delta(\Omega - \Omega') \tag{14}$$

we obtain the finite result [4]

$$E = \lambda \int \Omega |f(\omega)|^2 \, \mathrm{d}\Omega = \lambda \bar{\Omega} = \lambda \Omega_0 \tag{15}$$

with $\int |f(\omega)|^2 d\Omega = 1$, $\bar{\Omega}$ is the expectation value of the frequency and λ is a proportionality constant. In the moving frame (15) becomes $E = \lambda \omega$ as in (11). The field momentum

$$\boldsymbol{P} \sim \int \frac{\partial \phi^*}{\partial t} \Delta \phi \, \mathrm{d}\boldsymbol{r} \tag{16}$$

evaluated in a similar manner gives

$$\boldsymbol{P} = \lambda \boldsymbol{k} \tag{17}$$

with the same proportionality constant λ so that we get from (4) and (11) the relativistic energy momentum relation

$$(E/c)^2 - \boldsymbol{P}^2 = \lambda^2 \left(\Omega/c\right)^2 = (\lambda/\hbar)^2 m^2 c^2 \,. \tag{18}$$

We cannot determine, of course, within this simple linear wave equation model that the value of λ is the Planck's constant $\hbar$, but perhaps after the interactions are introduced. Note that the standard quantum theory neither calculates $\hbar$, nor the relations (15)–(17); we have at least modeled the latter equations.

According to our discussion in Sect. 1, the localized solution is an individual, and the plane wave solution characterizes the typical behavior, i.e. Ψ, so that we can write our solution in a more suggestive way to be used later as

$$\psi(\boldsymbol{x}, t; \boldsymbol{a}) = F(\boldsymbol{x}, t; \boldsymbol{a}) \Psi(\boldsymbol{x}, t) e^{-i\Omega t} \tag{19}$$

where ψ is a solution of the massless wave equation and Ψ is a plane wave solution of the massive equation. Equation (19) realizes the early idea of de Broglie that a localized particle is guided by the (plane) wave Ψ [3,5].

3.4 Other Wave Equations

Similar solutions have been obtained for Maxwell, Dirac and other wave equations [6]. The general conclusions are the same. For our purpose in this talk it is sufficient to have the simplest case of the scalar wave equation.

3.5 Localized Solutions of the Schrödinger Equation

The Schrödinger equation is obtained as a limit from the Klein-Gordon equation after separating a fast oscillating term $\exp(-i\Omega t)$. The massless wave equation has of course no nonrelativistic limit, but the localized solution (2) has a limit for small velocities. We can proceed in two equivalent ways. Either we take the usual Schrödinger equation and look for localized solutions, or add a mass term to the Schrödinger equation. The difference is just the factor $\exp(-i\Omega t)$. At any rate the result is that the Schrödinger equation

$$i\hbar\frac{\partial\psi}{\partial t} = -\frac{\hbar^2}{2m}\Delta\psi \tag{20}$$

has a localized solution of the form

$$\psi(\boldsymbol{x},t;\boldsymbol{a}) = F(\boldsymbol{x} - \boldsymbol{x}_0 - \boldsymbol{v}t)e^{\frac{i}{\hbar}(m\boldsymbol{v}\cdot\boldsymbol{x} - \frac{1}{2}mv^2 t)}e^{-i\Omega t}, \tag{21}$$

where F satisfies the same equation as (5)

$$\Delta F + \frac{2m}{\hbar}\Omega F = 0 \tag{22}$$

and

$$\boldsymbol{k} = \frac{m}{\hbar}\boldsymbol{v} \ , \ \omega = \Omega + \frac{1}{2\hbar}mv^2 = \Omega + \frac{\hbar k^2}{2m} \ . \tag{23}$$

Equation (21) is the Galilei boosted rest frame solution $Fe^{-i\Omega t}$ as in (1), and can also be written in the form (19) as

$$\psi(\boldsymbol{x},t;\boldsymbol{a}) = F(\boldsymbol{x} - \boldsymbol{x}_0 - \boldsymbol{v}t)\Psi(\boldsymbol{x},t)e^{-i\Omega t} \tag{24}$$

where Ψ is the usual plane wave solution of the Schrödinger equation for free particles. In the calculation of phase differences the phase Ωt drops out.

The field energy and momentum calculated as before with

$$E = \frac{i\hbar}{2}\int d\boldsymbol{r}\left[\psi^*\frac{\partial\psi}{\partial t} - \psi^*\frac{\partial\psi^*}{\partial t}\right]$$
$$\boldsymbol{P} = \frac{\hbar}{2}\int d\boldsymbol{r}\left[\psi^*\frac{\nabla}{i}\psi - \psi\frac{\nabla}{i}\psi^*\right] \tag{25}$$

give, using the normalization condition $\int \psi^*\psi\, d\boldsymbol{r} = 1$,

$$E = \hbar\omega \quad \text{and} \quad \boldsymbol{P} = \hbar\boldsymbol{k}\,. \tag{26}$$

Now, because $\hbar$ is already in the wave equation, we get the correct proportionality constant $\hbar$.

We wish now to generalize the solution (24) to the Schrödinger equation with a potential

$$i\hbar\,\psi_{,t} = -\frac{\hbar^2}{2m}\Delta\psi + V(\boldsymbol{x})\psi\,. \tag{27}$$

We thus look for a solution of the form [7]

$$\psi(\boldsymbol{x},t;\boldsymbol{a}) = F(\boldsymbol{x} - \boldsymbol{g}(t;\boldsymbol{a}))\,\Psi(\boldsymbol{x},t)e^{-i\Omega t} \tag{28}$$

where Ψ is the standard solution of the Schrödinger equation in the potential V, e.g. sum of stationary solutions, and the localization function F again satisfies (22). We obtain the following condition

$$i\dot{\boldsymbol{g}}\cdot\nabla F(\boldsymbol{x} - \boldsymbol{g}) = \frac{\hbar}{m}\nabla F(\boldsymbol{x} - \boldsymbol{g})\cdot\frac{\nabla\Psi}{\Psi}\,. \tag{29}$$

For $V = 0, \Psi \sim e^{i(\mathbf{k} \cdot \mathbf{z} - \omega t)}, \frac{\nabla \Psi}{\Psi} = i\mathbf{k}$ and (29) is solved for $\dot{\mathbf{g}} = \mathbf{v}$ and (28) reduces to (21). In order to have more freedom we may introduce an additional phase in (28) which in the calculation of $|\psi|^2$ does not contribute:

$$\psi(\mathbf{x}, t; a) = F(\mathbf{x} - \mathbf{g})\Psi(\mathbf{x}, t)e^{-i\Omega t}e^{\frac{i}{\hbar}H(\mathbf{x}, t)}, \tag{30}$$

then the conditions are

$$H_{,t} = \frac{\hbar^2}{m}\left(\frac{\nabla F}{F} \cdot \frac{\nabla \Psi}{\Psi} - \frac{1}{2\hbar^2}(\nabla H)^2\right)$$

$$2m\dot{\mathbf{g}} \cdot \frac{\nabla F}{F} = \left(\frac{\nabla F}{F} + \frac{\nabla \Psi}{\Psi}\right) \cdot \nabla H + \Delta H, \tag{31}$$

the phase H may be even non-integrable.

3.6 Relation to Coherent States

It is interesting to compare the nonspreading solution (30) with Schrödinger's nonspreading coherent state for the harmonic oscillator. Due to the special form of the harmonic potential we can write

$$V = \frac{1}{2}m\omega^2 \mathbf{x}^2 = \frac{1}{2}m\omega^2 \left[(\mathbf{x} - \mathbf{g})^2 - \mathbf{g}^2 + 2\mathbf{x} \cdot \mathbf{g}\right] \tag{32}$$

so that the first term $(\mathbf{x} - \mathbf{g})^2$ can be put into the localization function. Writing

$$\psi = G(\mathbf{x} - \mathbf{g})e^{\frac{i}{\hbar}H(\mathbf{z}, t)} \tag{33}$$

where G satisfies

$$-\frac{\hbar^2}{2m}\Delta G(\mathbf{x} - \mathbf{g}) + \frac{1}{2}m\omega^2(\mathbf{x} - \mathbf{g})^2 G(\mathbf{x} - \mathbf{g}) = EG(\mathbf{x} - \mathbf{g}) \tag{34}$$

we obtain with $\nabla H = m\dot{\mathbf{g}}$

$$H_{,t} = \frac{1}{2}m\dot{\mathbf{g}}^2 + E + m\omega^2 \mathbf{x} \cdot \mathbf{g} - \frac{1}{2}m\omega^2 \mathbf{g}^2 \, .$$

A solution is found to this equation if $\mathbf{g}(t)$ is the classical orbit satisfying $\ddot{\mathbf{g}} = -\omega^2 \mathbf{g}$. And if we take E to be the classical energy $E = \frac{1}{2}m\dot{\mathbf{g}}^2 + \frac{1}{2}m\omega^2 \mathbf{g}^2$ the solution ψ of the Schrödinger equation (33) becomes

$$\psi = G(\mathbf{x} - \mathbf{g})e^{\frac{i}{\hbar}(m\dot{\mathbf{g}} \cdot \mathbf{z} - m\int^t \dot{\mathbf{g}}^2 \, dt)} \tag{35}$$

which is the generalization of Schrödinger's coherent state for an arbitrary solution G of the three-dimensional time-independent Schrödinger equation. The coherent states are localized within a size of the order of Bohr radius, whereas our solutions are localized within a size of the order of the Compton wavelength of the electron.

3.7 Limiting Cases

For the interpretation of the localized solutions (28) it is important to discuss two limits:

(i) If $\Omega \to \infty$, $\frac{1}{r} \sin \left(\frac{\Omega}{c} r \right) \to \delta(r)$, hence (28) becomes

$$\psi(\boldsymbol{x}, t; g) \to \delta \left(\boldsymbol{x} - \boldsymbol{g}(t) \right) \Psi e^{-i\Omega t} . \tag{36}$$

This is the limit of localization to a point particle, and we obtain the classical trajectory $\boldsymbol{x} = \boldsymbol{g}(t)$.

(ii) We now average (28) over the parameters $\boldsymbol{a}$. We integrate over the parameters

$$\int d\boldsymbol{a} F \left(\boldsymbol{x} - \boldsymbol{g}(t, \boldsymbol{a}) \right) \Psi e^{-i\Omega t} . \tag{37}$$

The F's at different positions add up to a constant, possible of infinite value, which we renormalize so that the result is

$$\Psi e^{-i\Omega t},$$

up to an undetermined constant, which is then determined by Born's interpretation.

Therefore the single quantum entity ψ is something in between these two limits, it has both particle and wave properties taken from the two limiting cases. It describes now more faithfully the experimental situation, for a single electron is neither a classical point particle, nor an infinitely extended plane wave, but something in between. And that is what we wanted to model.

3.8 Wave packets and nonspreading solutions

It is true that Born's scattering theory can be formulated in terms of wave packets rather than plane waves and by averaging over the parameters of the wave packets. But there is a condition on the size of the wave packets; they have to be large compared to the characteristic size of the scattering potential. They are of the order of the size of the aperture of the beams, $\sim 10^{-3}$ cm in atomic experiments compared to the Bohr radius $\sim 10^{-8}$ cm. The reason is that if we would take a wave packet of the order of the radius of an atom, let alone the size of the electron, it would spread according to the formula

$$\left(\Delta x_t^2 \right) = \frac{t^2 \hbar^2}{4m^2(\Delta x_0^2)} \tag{38}$$

so that after $t \sim 10^{-5}$ sec it would reach a size $\Delta x_t \sim 1$ m! Thus very large wave packets are essentially equivalent to plane waves. But we wish to describe a single electron by a wavelength of very small size which does not spread. For a fixed frequency Ω our solutions do not spread. However, in order to obtain in addition a finite field energy we have taken a superposition of frequencies which may cause a small spreading. On the other hand, we have considered so far linear wave equations. Now charged particles have self-energies which add nonlinear terms to the wave equation. It is possible that these nonlinear terms which must be present would imply further localization without spreading.

4 Single Spin and the Calculation of Quantum Correlations

4.1 Definition of Correlation Functions

The idea of developing a practical quantum theory based on the wave functions $\psi(\boldsymbol{x}, t, \boldsymbol{a})$ for a single event is the following. Calculate the value of an observable A for a single event with parameters $\boldsymbol{a}, A(\boldsymbol{a})$, and then average over $\boldsymbol{a}$. For two observables A and B the correlation function is

$$E(A,B) = \frac{<AB> - <A><B>}{[<A^2><B^2>]^{1/2}} = \frac{\int d\boldsymbol{a} A(\boldsymbol{a}) B(\boldsymbol{a}) - \int d\boldsymbol{a} A(\boldsymbol{a}) \int d\boldsymbol{a} B(\boldsymbol{b})}{[\int d\boldsymbol{a} A^2(\boldsymbol{a}) \int d\boldsymbol{a} B^2(\boldsymbol{a})]^{1/2}}.$$
(39)

This formula is of course valid in classical physics as well as in the standard quantum theory where $< |A| >$ means the quantum mechanical average of the operator A in a state $| >$. In this expression the values of the observables $A(\boldsymbol{a}), B(\boldsymbol{a})$ are deterministic. Probability comes in only in the distribution of the parameters $\boldsymbol{a}$ which is determined by the experimental setup. Usually for similarly repeated events the values of $\boldsymbol{a}$ are equally distributed.

We shall next evaluate these correlations for the much discussed spin variables.

4.2 What is an Individual Spin?

Ever since Pauli introduced his spin formalism with σ-matrices as a "classically not explainable two-valuedness" it has been tacitly assumed that not only on the average but also individually the measurement of any spin component can yield only two sharp values. The belief that a single atom can only have two spin values is so ingrained that it is almost universally never even stated that this is a new assumption about single events. It is of course true in standard quantum theory, that is at the level of repeated events, i.e. at the level of Ψ, that the measurement of spin components yields on the average two values. But as we have seen standard quantum theory cannot make any statements about single events, i.e about the wave function $\psi(\boldsymbol{x}, t; \boldsymbol{a})$ of an individual with parameters $\boldsymbol{a}$, which remains to be explored. The basis for the above assumption is the Stern-Gerlach experiment which however refers to repeated experiments and results in two broad maxima. We will argue now that this assumption of two-valuedness cannot be true for a single event and shall modify it without changing anything in the standard quantum theory for repeated events. The legitimate question has been posed whether we can reproduce observed quantum correlations by averaging over the parameters of individual events. The crucial point is what are the parameters for an individual spin. If we can completely reproduce observed correlations by a model of spin we have solved our problem.

In spite of Pauli's remark a single spin can actually be accurately modeled, like we modeled above the coordinates and velocities; the spin should be a dynamical variable on the same footing. The best relativistic spin model is perhaps that of a point charge performing a helical motion as its natural motion in such a way that the relative orbital angular momentum of the helical motion (called zitterbewegung) gives the spin magnetic moment with correct g-factor. This motion when quantized gives precisely the Dirac theory of the electron [8]. For our purpose here we shall use a nonrelativistic limit and model spin as a direction in space ; a vector S in the direction parametrized by the angles $\theta, \varphi : S = S(\theta, \varphi)$. The angles (θ, φ) are now the parameters a of the individual spin wave function ψ. This is in fact how we intuitively picture the spin.

Let us apply now this model to the EPR-spin correlations of two spins in a total spin singlet state. We measure the spin component of S_1 in some direction $\hat{a}$ and the spin component of S_2 in some direction $\hat{b}$ and evaluate the correlation (39) by integrating over all the angles. With $A = S_1 \cdot \hat{a}$ and $B = S_2 \cdot \hat{b}$, (39) becomes

$$E(A, B) = \frac{\int d\theta \sin\theta \, d\varphi \, S_1(\theta, \varphi) \cdot \hat{a} \, S_2(\theta, \varphi) \cdot \hat{b}}{[\int d\theta \sin\theta \, d\varphi \, (S_1 \cdot \hat{a})^2 \int d\theta \sin\theta \, d\varphi \, (S_2 \cdot \hat{b})^2]^{1/2}} . \tag{40}$$

In the singlet state, for example, the spins are correlated back to back

$$S_2 = -S_1 \tag{41}$$

and assuming all single events to be equally probable, we can evaluate the integrals. The result is [9]

$$E(A, B) = -\hat{a} \cdot \hat{b} \tag{42}$$

which is exactly the quantum mechanical correlation calculated by the same formula (39) as

$$E(A, B) = \frac{\langle \sigma_1 \cdot \hat{a} \sigma_2 \cdot \hat{b} \rangle}{\left[\langle |(\sigma_1 \cdot \hat{a})^2| \rangle \langle |(\sigma_2 \cdot \hat{b})^2| \rangle \right]^{1/2}} = -\hat{a} \cdot \hat{b} \tag{43}$$

The quantities $A(a), B(b)$ can also be interpreted as the expectation values of the operators in "localized" spin coherent states: $S(\theta, \varphi) = {}_c< \theta, \varphi|S|\theta, \varphi >_c$

The result (42) seems to contradict the well known discussions leading to Bell's inequalities. The difference comes from what kind of assumptions we make about individual events. In deriving Bell's inequalities one assumes that the values of the observables $A(a)$ and $B(b)$ for each individual are discrete and equal to $\pm 1/2$ no matter what direction $\hat{a}$ we choose and this simultaneously. In our calculation these observables have continuous values, yet we obtain the correct quantum correlations, and the discrete case gives the wrong answer. So the whole issue really boils down to the question of what a single spin is.

This question can be perhaps answered by carefully monitoring the Stern-Gerlach experiment for individual atoms, by eliminating all effects of broadening of the two peaks to see if a residual broadening remains, and by taking into account the effect of the fringe fields. We have at the present time no complete theory of the Stern-Gerlach apparatus; it is very difficult to solve the nonlinear equations of the motion of spins through the apparatus.

We must be careful, also for spin, to distinguish between the wave function Ψ for repeated events and ψ for single events. The standard theory deals with the former for which the superposition principle holds. Thus superposition states like $\frac{1}{\sqrt{2}}(|\uparrow> +|\downarrow>)$ always mean statistical states, 50% up and 50% down in repeated experiments and it is used as such in practice. A single spin can never be in a superposition state. I think this distinction again eliminates a number of dilemmas in thinking about spin problems.

Similar correlations can be evaluated on the basis of (40) for triplet and other states [10] as well as for the recently discussed correlations between three or four spins [11]. There is always a classical correlation of individual spin vectors $S_i(\theta_i, \varphi_i)$ which when averaged over angles reproduces the quantum correlations.

5 Conclusions

We have seen that for many interpretational questions in quantum theory it is not only necessary to attempt a theory of single events, but it is also possible despite some existing impossibility "theorems". These theorems have all some assumptions whose physical validity is challenged here and counterexamples are given explicitly.

Single events have classical (unquantized) parameters in space-time and in spin space and for each value of these parameters we get a single individual localized event.

The probabilistic wave function Ψ arises from the individual $\psi(a)$ by averaging over the parameters a and answers to questions concerning repeated events. This view overcomes the problem of the collapse of the wave function which becomes unnecessary.

Problems concerning the classical limit of quantum theory, quantum chaos and quantum statistical mechanics are best treated in terms of the wave function $\psi(a)$ rather than Ψ.

In particular we account for spin correlations by evaluating them classically for individual spins and integrating over their parameters, i.e. not assuming that each individual spin is already quantized.

Finally, the Schrödinger's cat being an individual with name Kedi, with a wave function ψ(Kedi) and not Ψ, like an individual atom or an individual DNA molecule, is never in a superposition state!

References

1.   A. Tonomura *et al.*: Amer. J. Phys. **57** 117 (1989)
2.   M. Born: Z. Physik **37** 863 (1926); **38** 803 (1926)
3.   L. de Broglie: C.R. Acad. Sci. **B180** 98 (1925); Ann. Phys. (Paris) **3** 22 (1925)
4.   A.O. Barut: Phys. Lett. **A143** 349 (1990)
5.   L. de Broglie: C.R. Acad. Sci. **B277** 71 (1973);
     L. Mackinnon: Lett. Nuovo Cim. **32** 125 (1985)
6.   A.O. Barut, A. Grant: Found. Phys. Lett. **3** 303 (1990)
7.   A.O. Barut: Found. Phys. **20** (1990), No.10
8.   A.O. Barut, N. Zanghi: Phys. Rev. Lett. **52** 2009 (1984)
9.   A.O. Barut, P. Meystre: Phys. Lett. **A105** 1021 (1984)
10. A.O.Barut: in *Symposium on the Foundations of Modern Physics*, ed. by P. Lahti and P.Mittelstaed (World Scientific, Singapore, 1985, p. 321)
11. D.M. Greenberger, M. Horne, A. Zeilinger: In *Bell's Theorem, Quantum Theory and Conception of the Universe*, ed. by M. Kafatos (Kluwer Academic, Dordrecht, 1989) p. 69

Symmetry, Entropy and Complexity

Yuval Ne'eman[*]

Raymond and Beverley Sackler Faculty of Exact Sciences
Tel Aviv University, Tel Aviv, Israel 69978
and
Center for Particle Theory[‡]
University of Texas, Austin, Texas 78712, USA

Abstract: We review the role of symmetry in Physics and its interrelationship with order and with information, in the light of modern approaches to the concepts of entropy and of complexity versus disorder.

1 Introduction

It is with great pleasure and deep appreciation that I dedicate this essay to the 60th anniversary of Professor H.D. Doebner. Throughout the last fifteen years I have gained much insight from following his work in the intersection of differential geometry and physics. I have also greatly enjoyed the topical conferences he has organized in this field – not the least for the opportunity it provided for discussions with Doebner himself, enhanced by the hospitality of the Doebner household and conversation with Mrs Doebner and their daughter. Through his initiative, Clausthal has become a Mecca for workers in geometry and group theory as applied to physics. In some ways, it revives the memories of nearby Göttingen in the hallowed days of Gauss and Hilbert. Heine's Harz Mountains travelogue missed an important intellectual site by coming 150 years too early.

My debt to Professor Doebner goes even further. He has introduced many students to research in physics; considering his interest in groups and geometry, it is thus not surprising that I should have come across several of them in my work. Beyond that, however, is the fact that I have developed extensive collaborations with two of them. In my work on the extension of the idea of symmetry to spectrum generating groups, I have collaborated since 1968 with Arno Bohm, who independently conceived ideas similar to mine in 1965, when we introduced SGG

[*] Wolfson Chair Extraordinary in Theoretical Physics
[‡] Supported in part by Grant DE-FGO5-85ER40200

as algebraic systems connecting all solutions of a quantum problem, i.e. all energy levels. My work on world-spinors and on the gauge approach to quantum gravity was triggered in 1977 when reading articles by Friedrich W. Hehl on gravity with torsion, which he succeeded in putting on the same footing as other gauge theories. This was the start of an extremely pleasant and fruitful ongoing collaboration. I have thus greatly benefited indirectly as well from Prof. Doebner's efforts as a teacher.

2 Symmetry Implies Abstraction and Loss of Information

Physics is an experimental and observational science and thus deals with the "real world". Its method, however, uses abstraction. The aim is to achieve a unified and coherent presentation of all natural phenomena. To treat different phenomena in a single formulation, physics has to strip away the circumstantial details and identify the essentials and discover the common denominators and their constrained behaviour – the laws of physics. This then implies sweeping generalizations and a loss of information about the individual systems. The more phenomena are encompassed by a law – the more it has to become simple and rely on less specification. The information about the individual systems is left to the boundary conditions, if at all.

Symmetry laws are in that category. They represent negative statements embodying powerful generalizations. They are "Postulates of Impotence" [20], though highly potent ones. IMPOTENCE, because they state that it is impossible to prefer one frame over the rest. If a crystal is hexagonal, it has a symmetry under rotations by $360/6 = 60$ deg., and it is impossible to select one face out of the six as a "preferred" face. To the extent that we wanted to preserve the identity of one of the faces – it is lost in the symmetry. Notice the closeness to the classical concept of entropy in such a symmetry law; as to leaving the information to the boundary conditions – the modern theory of Chaos tells us that this is often also a way of loosing information, since many dynamical systems lead to entirely different evolutions even though the initial conditions may be so close as to be indistinguishable.

The Principle of Covariance in Einstein's General Theory of Relativity states that it is impossible to select a preferred reference frame – i.e. the laws of gravity do not depend on the selection of a particular reference frame, all reference frames are equivalent. The French saying goes "la nuit, tous les chats sont gris" – at night, all cats are grey – i.e. it is impossible to distinguish or specify a preferred cat. There is then a symmetry between cats, they all look the same.

3 Broken Symmetries – Imposed or Spontaneous

Symmetry can sometimes be in the laws of physics and then has a great range of applications: in the example of Einstein's theory of gravity, for instance, whatever the gravitational problem, the laws will still have to be stated "covariantly" (i.e. independently of the selection of a reference frame, of a coordinate system).

Sometimes, however, there is a symmetry that relates to the boundary conditions. In the Kepler problem (sun and planets) for instance, there is an a priori spherical symmetry in the givens themselves: the sun is assumed to be spherical, and therefore there will be no preferred direction for its gravitational pull – in the way that would happen in a description of gravity in this room, where we would be forced to assume a preference for the downwards direction in the action of gravity. It so happens that the laws themselves also contain no preferred direction and are spherically symmetric, even for this room, even though the boundary conditions are less symmetric.

In fact, the symmetry of the laws is generally greater than that of the givens; in the case of Einstein's theory, for instance, the laws are also locally Lorentz-invariant, which includes, aside from insensitivity to rotations of the system in space, an invariance under accelerating boosts ("special Lorentz transformations").

Sometimes, we are surprised by the amount of symmetry sustained by the boundary conditions. In Cosmology, for example, there is no known a priori reason for the boundary conditions to be very symmetric. They could have been as complicated and asymmetric as we wish – and yet in reality, the observations show that the cosmological boundary conditions are highly spherically symmetric.

In modern treatments, there is a delicate interplay between laws and boundary conditions. We shall see that symmetry has to be broken at some stage, when we deal with the real world. In the words of Francis Bacon, "there is no excellent beauty that hath not some strangeness in the proportion". Rather than break the symmetry of the Laws, it is more convenient – and useful – to find formulations in which the Laws are entirely symmetric, and the symmetry breakdown is "blamed" on some boundary conditions. In Quantum Mechanics, the "real world" is given by the Hilbert Space.

We now return to the breaking of symmetry. This can be explicit in the dynamics: the Hamiltonian or Lagrangian will have a contribution breaking the symmetry in a given direction. We use this approach for Unitary Symmetry [17,7] where the breaking of SU(3) (now known as "flavor" SU(3)) is inserted by postulating a higher mass for the "s" quark. The assumption is that this is due to some new and different interaction. I called it the "Fifth Interaction" when I first suggested [18] that the Strong Interactions are really SU(3) invariant and that the symmetry breakdown was due to another perturbative interaction – as against the non-perturbative features of the "true" Strong Interactions. The Fifth Interaction can now be generalized to cover the force responsible for the "generations" structure displayed by quarks and leptons. In my 1964 papers, I had already suggested that it could also generate the mass of the muon, i.e. be responsible for the apparition of a second generation of quarks and leptons. The present "standard model" in which the interquark Strong Interaction is described by Quantum Chromodynamics ("QCD") indeed postulates an SU(3)-flavor invariant Strong Interaction, because the color-SU(3) gauged by QCD commutes with flavor-SU(3).

The alternative way in which a symmetry can be broken is "Spontaneous" symmetry breakdown. It corresponds to cases in which it is possible to "blame" the symmetry breakdown on the boundary conditions. The Laws are assumed to continue to obey the full symmetry, but the basic state in the Hilbert space, the "vacuum state" does have a preferred direction. If, for instance we are dealing with a type of "charge" (that is not explicitly conserved because the symmetry is broken) we already endow the vacuum with a certain amount of that charge, and the particles built on this vacuum will also have that feature. In this manner, we continue to have a preferred direction imposed by the boundary conditions of the problem, in this case the Hilbert space.

"Spontaneous" is sometimes taken to mean more than that. Since the "states" of the system are solutions of the dynamical equations, we search for equations whose solutions will indeed carry quantum numbers breaking the symmetry. The present work on the STRING, for instance, attempts to find such equations (String Field Theory, or other non-perturbative techniques) that will yield solutions with broken supersymmetry etc. and looking like the real world.

This approach was first introduced in the study of superconductivity in the physics of condensed matter. In that discipline, the method was invented [10] to explain phase transitions, such as the transition in a material between a paramagnetic and a ferromagnetic state when it is cooled down to the critical temperature – or the transition to the superconducting state at very low temperatures (since 1985, the temperatures are no more that low). In a more structural theory of superconductivity [2] we can understand the asymmetric behaviour of the vacuum from the dynamics.

In that problem, a "false vacuum" state is created, when the overall interaction between the electrons and the atomic lattice in the metal produces a "pairing" between electrons: two noncontiguous electrons start acting as if they were bound. This then becomes the lowest-energy "ground" state and acts as a vacuum for that particular situation; but this vacuum is not really a "neutral" empty vacuum, and thus contains characteristics that break the symmetry of the equations.

The method was successfully generalized to the physics of particles and fields [15,16,12]. Here, the assumption of a "directed" vacuum requires the existence of massless particles – massless in the approximation in which all other effects are removed. The massless particles are needed to complete the vacuum's multiplet.

To summarize, all symmetry breakdowns are dynamically caused – almost by definition of what physics is all about. However, explicit symmetry breaking is due to an extraneous force, whereas the spontaneous breakdown is caused by that same force that obeys the symmetry, and corresponds to the mathematical feature that a solution can have less symmetry than the equation.

In an unbroken symmetry, the vacuum is invariant, i.e. if we apply to it the symmetry's transformations, it does not change. In other words, the symmetric vacuum is a "scalar", forming a single-state multiplet. This is the algebraic characterization. But when the vacuum has a direction, applying the symmetry operations to that state should rotate it into some other state. What would that state be like in the case of the vacuum in spontaneous symmetry breakdown? It turns out that a particle with zero mass could serve as a "partner" for our non-single vacuum.

The idea was very successfully applied to the understanding of the Yukawa force. This is the force responsible for the attraction between nucleons (protons and neutrons) in any atomic nucleus. It involves the exchange of pions (the "meson" postulated by Yukawa and Stueckelberg in 1934) between nucleons, like volley balls in that game. The force obeys a certain symmetry called SU(3)×SU(3)-"chiral", because the relevant conserved currents are characterized – on top of the "unitary-symmetry" charges they carry – by left or right "handedness". The two SU(3) in the name of the symmetry correspond to two currents, one an SU(3)-left and the other an SU(3)-right. Note that Parity is conserved because both chiralities are present; it is only when the left-chiral current of SU(3)-left comes by itself - in Fermi's Weak Interaction – that Parity is thereby broken.

The doubling of the SU(3) currents and symmetry is quite analogous to what we observe in the case of angular momentum. In very low energy atomic physics we can have a separate conservation of spin and orbital angular momentum, i.e. two SU(2) currents of angular momentum. However, once we increase the energies involved, the spin and orbital angular momenta mix, and only total angular momentum is conserved. The same happens with the unitary symmetry chiral currents. Once the symmetry is broken, only the sum of SU(3)-left + SU(3)-right subsists as a conserved quantity. This sum is plain SU(3), and in a certain approximation it is even locally conserved. Its currents then couple universally to an octet of spin 1 vector-mesons.

Chiral unitary symmetry together with this "SU(3) gauge" provide a good phenomenological working theory for the physics of hadrons – the hundreds of different particles that feel the "strong" nuclear interaction and that we now consider as consisting of bound systems of either three quarks or a quark and an antiquark. The theory is sometimes described as "current algebra". It fuses two theoretical discoveries of 1959–64: unitary symmetry ("SU(3)") and spontaneous symmetry breakdown using techniques [8] inspired by Heisenberg's version of Quantum Mechanics, the "Matrix Mechanics" [19,1].

4 Symmetry, Order and Information

We already noted the negative correlation between symmetry and information. Symmetry represents a lack of information, an impossibility to specify, to provide identification, which is an important type of information.

Lack of information in large ensembles is traditionally connected with entropy, disorder. However, this statement is not precise enough. Missing information may be connected with disorder, in the sense that it becomes too difficult to specify that information because it relates to myriads of turbulent molecules, for example. In computer language, it would involve myriads of information bits. This type of lack of knowledge is described as "subjective" because it is due to our own limitations.

But in Quantum Mechanics, on the other hand, missing information just corresponds to its inexistence – the physical state has not yet been generated, as long as a measurement has not been performed (a "measurement" in the sense of an irreversible interaction with a macroscopic system). At this stage, all there is is just a wave-function, with a probabilistic interpretation. We know from the many experiments that have realized the EPR idea [6] and applied the test provided by Bell's [3] inequalities that there is no physically concrete "underlying reality" other than the wave-function. This lack of knowledge is then an "objective" lack of information, information that does not yet exist.

In the case of the grey cats of the French proverb, the lack of information is due to darkness – not to inexistence – i.e. to a difficulty in the acquisition of the information, resembling the case of disorder. It is subjective.

Very recently, an advance in the study of "chaotic systems" has revealed the existence of objective entropy in non-quantum situations. There are problems in which an infinitesimal difference in the initial conditions will lead to totally different evolutions of the systems. These are then "unstable" initial conditions, generated in collective states by the internal interactions between the constituents. The phenomenon of turbulence in a liquid or in a gas is one such situation.

The entropy of a symmetry is the magnitude of the "Whittaker impotence" it represents. This can be given a quantitative definition by taking, for instance, the volume of the Lie group – or some quantity related to the group dimensionality. SU(3) invariance is related to an 8-dimensional manifold. However, SU(3) is a broken symmetry. It is broken through the "c" quark being about 30 times heavier than the "a" and "b" quarks. This therefore reduces the overall symmetry, leaving a subgroup U(2) as the residual invariance. U(2) has a 4-dimensional group manifold with a smaller volume and is therefore a smaller symmetry and represents less entropy.

The study of entropy in relation with the need to describe complexity has produced in recent years completely different approaches to the objectivisation of entropy. The aim is to have a description that would represent, for instance, the complexity of a living cell or of an organism.

One such measure was "algorithmic complexity" [13,5]. The quantity characterizing the state is the length of the shortest computer program that can describe the state. It will represent the information content of that state, a kind of inverse of the state entropy.

A crystal can be described by a much shorter list of instructions than a living being (who's DNA is probably the relevant program). This means that the crystal embodies less information and has a higher intrinsic entropy than a living system. On the other hand, a gas with quintillions of quintillions of molecules could only be described by a program listing them all with their locations or momenta (or both, classically), i.e. the state and the design program are of the same magnitude. This corresponds to algorithmic incompressibility. This would either imply that the gas contains a very large amount of information – and little entropy in the usual definition in which information is "subjective" – which is not what we would like to understand by entropy, which should be an objective notion. Should we then define entropy as proportional to the program's length? This seems OK in the comparison between a crystal and a gas, but it would assign a large entropy to a living being or DNA – again a paradoxical result. The missing feature in this analysis is the notion of randomness as against complexity. Both notions require longer programs. The characterization should account for the fact that the tremendous amount of information relating to the initial condition of a gas has little meaning because it is random. Any small change has no effect on the physics. In the case of DNA the list is also enormous, but a tiny change will produce a new and different being. Complexity is not disorder.

This issue is resolved in a proposal due to Bennett [4]. He measures order – the opposite of entropy – by the "logical depth" of the system. It represents the logical length of the program for the realization of the state, once the data is fed. To construct a living cell one would require an extremely long set of instructions. For a crystal, a limited number of steps would suffice. For a gas of molecules, the INITIAL DATA would be of an enormous magnitude, but the instructions program would consist in a trivial "copy that data". This definition therefore does fit the concept of objective entropy. It has since been further developed [14].

We can adapt these concepts to symmetry. Instead of the dimensionality or volume of the group, we could measure the information content of the vacuum, i.e. of the multiplet containing the Nambu-Goldstone boson. One way of measuring this quantity could draw from the structure of the Young tableau for that representation of the group, which is similar to a computer program for its construction.

This does not appear interesting in finite-dimensional Lie groups, but something similar might be possible and helpful in infinite cases such as the presently fashionable group of conformal transformations (transformations preserving angles) in two dimensions – a symmetry of the theory of the Quantum Superstring, a "great hope" at present, as a candidate "Theory of Everything". The subject calls for further investigation. In fact, the present search for an equation or a method that would yield the physical vacuum – one out of billions of allowed ways to go from 10 to 4 dimensions – is precisely the type of case that would fit the above discussion and relate directly the notion of volume in phase space with the volume of a symmetry group.

References

1. S.L. Adler, R.F. Dashen: *Current Algebras* (W.A. Benjamin Inc., New York, 1968)
2. J. Bardeen, L.N. Cooper, J.R. Schrieffer: Phys. Rev. **126** 162 (1957)
3. J. Bell: Rev. Mod. Phys. **38** 447 (1966)
4. C.H. Bennett: Found. Phys. **16** 585 (1986)
5. G. Chaitin (1965): see his book *Algorithmic Information Theory* (Cambridge University Press, Cambridge, 1987)
6. A. Einstein. B. Podolsky, N. Rosen: Phys. Rev. **47** 777 (1935)
7. M. Gell-Mann: "The Eightfold Way", Caltech report CTSL 20 (1961), unpub.

8.   M. Gell-Mann: Phys. Rev. **125** 1067 (1962)
9.   M. Gell-Mann: Phys. Lett. **8** 214 (1964)
10.   V.L. Ginzburg, L.D. Landau: JETP **20** 1064 (1950)
11.   H. Goldberg, Y. Ne'eman: Nuov. Cim. **27** 1 (1963)
12.   J. Goldstone: Nuov. Cim. **19** 154 (1961)
13.   A.N. Kolmogoroff: Probl. Peredachi Inf. **1** 1 (1965)
14.   S. Lloyd, H. Pagels: Ann. Phys. (N.Y.) **188** 186 (1988)
15.   Y. Nambu: Phys. Rev. Lett. **4** 380 (1960)
16.   Y. Nambu, G. Jona-Lasinio: Phys. Rev. **122** 345 (1961), Phys. Rev. **124** 246 (1961)
17.   Y. Ne'eman: Nucl. Phys. **26** 222 (1961)
18.   Y. Ne'eman: Phys. Rev. B**134** 1355 (1964)
19.   Y. Ne'eman: *Algebraic Theory of Particle Physics* (W.A. Benjamin Pub., New York, 1967, 334 pp.)
20.   E. Whittaker: *From Euclid to Eddington* (Cambridge University Press, Cambridge, 1949)

Steps in the Philosophy of Quantum Theory

Th. Görnitz and C. F. v. Weizsäcker

D-8130 Starnberg

Abstract: *1. Interpretation.* The Copenhagen Interpretation (CI) is a minimal semantics to quantum theory, expressing what we know at least. It can be extended into a universal Quantum Theory, applied to the observer as well as to the observed object. *2. A Universal Theory as a Philosophical Problem.* A circular epistemology is proposed, consisting of non-hierarchical realism, empirism, apriorism and evolutionism, combined in a description of time: past as discrete facts, future as continuous possibilities. *3. Quantum Logic and the Reconstruction of Quantum Theory.* Non-distributive logic and Bell's theorem are discussed following Doebner and Lücke. Reconstruction is briefly described. *4. Further Philosophical Questions.* Mind-body problem and holism are briefly discussed.

1 Interpretation

1.1 Steps in Quantum Theory

"Can you tell me what you mean by what you say?" This is the philosophical question, asked by Socrates to the Athenians and by his re-incarnation Niels Bohr to the physicists. Philosophy is historically a posteriori. Philosophy of science asks questions about science already invented. Steps in science seem to come first.

Classical physics was a *continuum* dynamics. In point mechanics, observables like energy, momentum, position, angular momentum admitted a continuum of values. Field mechanics even had an infinite number of degrees of freedom.

Statistical mechanics, as Boltzmann realized, needed *rigid atoms* in order to permit thermodynamical equilibrium.

Planck, for the same reason, applied to the Maxwell field, needed the *quantum of action*, i.e. *discrete energy levels* for the oscillator.

Einstein probably clearly saw the *impossibility* of *any continuum dynamics* for fields, and introduced the *photon.*

Bohr, receiving from Rutherford the first credible model of an *atom*, saw that it was possible only with *discrete energy levels*, too.

Heisenberg introduced the *algebra* of observables, Schrödinger the *wave function*, Born interpreted the wave as expressing *probabilities*, and von Neumann united the mathematical concepts in a *Hilbert space*. Within a separable Hilbert space, observables possessing eigenfunctions have only *discrete* eigenvalue spectra.

It is the von Neumann codification of quantum theory to which the later interpretation debate referred.

1.2 The Interpretation Debate

Let us first describe the more than sixty years of interpretation debate by a parable.

Heisenberg [13] said in connection with quantum theory: "In modern physics, nature now has reminded us clearly that we can never hope to open up the complete field of possible knowledge by starting from a fixed basis of operations. Rather for every essentially new insight we will again come into the situation of Columbus who dared to leave all so far known land in the nearly manic hope (in der fast wahnsinnigen Hoffnung) to find land again beyond the seas." Later, in 1948, he described the process of theoretical physics as an open sequence of "closed theories" ("abgeschlossene Theorien") in which the later one always explains and thus limits the success of its predecessors. Thomas Kuhn [15] described the process as a sequence of paradigms which are used but not fundamentally understood, and of scientific revolutions, arising of taking seriously the interpretational contradictions within the earlier paradigm. Which step beyond quantum theory might be foreshadowed by the interpretation debate?

Heisenberg's parable of Columbus is methodologically not precise. Columbus knew, as Greek astronomy had already known, that the earth is a sphere. His mania was not in the absolutely correct idea that going westward one would come to India. His mania was in daring to do the traveling himself with the available nautical means, and his reward was to discover an unexpected continent for which there had been space enough between the oceans. The inventors of quantum theory were in a less satisfactory position. They discovered and conquered an unexpected continent, but they did not know whether the field of physical knowledge is a sphere, and if so, where the new continent is located on it, and whether there are more continents to be discovered.

1.3 The Copenhagen Interpretation

The Copenhagen interpretation (CI) is an attempt to describe consistently the structure of the new continent without making hypotheses about its location with respect to unknown continents or to the complete field of possible knowledge, not to speak of realities which remain unknown to human beings in principle. CI is epistemological, not ontological. It can be described as aiming at a *minimal semantics* for the quantumtheoretical formalism.

In CI, Quantum theory is a theory on available or possible *human knowledge*. In Hilbert space there are vectors and operators. The operators are possible observables, described as Hamiltonians of measurement interaction (see [21, pp. 531-534] and [9,10]). The vectors (Ψ-functions) define probabilities of finding eigenvalues of the observables.

This solves, by the way, the problem of continuum dynamics: The measurable quantities have *discrete* values, the *continuum* defines probabilities, thus not leading to an ultraviolet catastrophe. A remark is relevant on operators with a continuous spectrum like position, momentum, field properties. The claim of eigenstates to such a spectrum cannot be fulfilled in a separable Hilbert space. We propose to consider the definitions of such continuous observables as only approximately valid. This will be discussed in the chapter on reconstruction.

Probabilities which are determined by law of nature are essentially *conditional probabilities*. In the *quantum theory of measurement* this is described by distinguishing between *preparation*, as defining the condition, and *observation*, as defining the outcome. The result of an observation which had a probability different from one will change the condition and hence the probabilities. This is called the reduction (or, more dramatically, the *collapse*) of Ψ. Thus, in CI, Ψ is essentially an *expression of knowledge*.

This makes inevitable the question of the relationship between *observer* and *object*, between knower and known. Measurement is done by physical interaction between observer and instrument. Are we to describe the observer, too, by quantum theory? If yes, how? If no, why not?

Bohr answered No. He spoke of the "detached observer". This can be methodologically justified by considering CI as minimal semantics. We can express our knowledge of the objects without having a theory on how this knowledge as a mental act is to be objectively described. Bohr himself had an additional ontological reason. At least in his younger years he did not believe that living organisms can be fully described by physics; and he never believed that human consciousness can be so described. We shall deviate from these views of Bohr's; but we continue to accept the CI minimal semantics as a meaningful step of interpretation.

Then the question remains, how far the minimal semantics will force us or at least admit to describe the measurement process by quantum theory (compare [22,23]). Bohr insisted that the instrument must be classically described: A measurement means description in intuitive space-time, and it presupposes strict causality between the observed effects and the state of the object which we want to know. (This is a good Kantian argument.) Even in Bohr's view this did not necessarily mean that quantum theory should not be valid in the instrument; it suffices that the conclusions to be drawn from the observation are classical in the necessary approximation.

CI as a minimal semantics is also not troubled by the famous question: "When is Ψ reduced in the act of measurement?" Since Ψ, according to CI, expresses human knowledge, no harm is done by saying: "It is reduced when the observer becomes aware of the result." But Bohr's postulate of a classical description justifies equally the "Golden Copenhagen Rule": "No harm is done in assuming that Ψ is reduced when an irreversible process has taken place in the measuring instrument." This eliminates the impression of "subjectivity", since an irreversible fact is in principle accessible to every observer.

1.4 Universal Quantum Theory

Today it seems to be a possible second step after CI, to fully accept the unchanged von Neumann formalism, but to apply it to all real events without restriction, as a universal theory. Hypothetically to retain the formalism seems justified by its success through 60 years, and by the failure of the only competitor which could be experimentally tested: the theory of local hidden variables. Hypothetically to extend it to universal validity seems justified by the success of physicalism in biology and by progress in cosmology.

The question is whether universal quantum theory admits an interpretation without paradoxes. This question leads into two problem fields:

1. Can and must CI be replaced by a different interpretation?
2. Does "universal" mean to apply the theory also to the mind of the observer?

Problem 1. Several proposals have been made which maintain to retain the mathematical structure and the experimental results of quantum theory but not the Copenhagen Interpretation. We have discussed the proposals made by Kochen [14], by Deutsch [4] who follows Everett [7], and by Cramer [3]; this discussion was given in our papers [9–11]. We shall not repeat it here, but mention our conclusion.

We maintain:

A. If two theories are mathematically isomorphic and predict identical experimental results, then there must be a possible dictionary translating their verbal expressions into each other. It must then be possible to consider them as different formulations of one theory, looking at it from different directions. The question then is how to interpret this theory in the traditional language. Since CI is the minimal semantics of quantum theory, the new proposals must be expressible in the CI language as far as CI can be applied. Hence they will not replace, only contribute to CI if CI can be extended to universal validity. This will lead to Probl. 2.

B. Attempts have been made to change not only the interpretation but also the formalism of quantum theory, mainly in order to avoid apparent paradoxes in the theory of measurement. The main problem arose from considering Ψ not epistemologically but as an "objective reality"; this view then needed an explanation of the "collapse of Ψ" by observation. One proposal – local hidden variables – was experimentally excluded; others still await a test or will not admit an experimental distinction from quantum theory in a foreseeable future (e.g. Ghirardi et al. [8]). In the present paper we shall not consider these proposals since we feel that existing quantum theory permits a consistent interpretation.

Problem 2. Here the problem is not whether quantum theory can be applied to the human brain; if we presuppose physicalism as correct in biology as most biologists do today, then the application to the human brain is a consequence. The task is to eliminate the so-called mind-body problem. This is automatically achieved if we "reconstruct" quantum theory as a general theory for predictions on *any empirically decidable alternatives*: "abstract quantum theory", cf. [21, Chapt. 8],[6]. The only precondition (and hence, limitation) of this view lies in the assumption that mental states can be objectively ("decidably") observed by introspection and communication. We shall return to the philosophical problems of this view in Sect. 4.

2 A Universal Theory as a Philosophical Problem

2.1 The Problem

"Can you tell what you mean by trusting in a universal theory?" Columbus trusted in the earth being a sphere, with greet success. The postulates of abstract quantum theory can be formulated on a half page of print, at least for a mathematically educated reader; the physicists' community trusts today in approximately a billion single empirical facts agreeing with quantum theory and so far in none which would convincingly contradict the theory. How can such a universal theory be possible? What is the "sphere of human knowledge" which admits a universal theory?

Plato, Hume, Kant, and Popper agree that the strict validity of a universal proposition cannot be empirically proved; as Hume pointed out, because at least the future cases of empirical application of the proposition cannot be known at present. This reminds us of the basic role of *time* in empirical science; experience can be defined as learning from the past for the future. Popper's statement that a universal proposition can at least be empirically falsified by one counter-example is correct only as far we can trust the propositions in which we interpret the counter-example.

The problem is hard, and we do not propose a final solution, but steps which might be useful (cf. [21, pp. 622–627]).

2.2 Pragmatism and Hierarchism

The average attitude of scientists in view of our problem is pragmatic. "We are successful; let us continue." In Kuhn's language this is the mentality of "normal science": puzzle-solving under a successful paradigm. It is essential for this attitude, not to ask why the paradigm is so successful. The opposite is true for scientific revolutions. The paradigm of Newton's mechanics was overcome and thereby, justified within its limits, by Mach's critical analysis of its concepts and Einstein's positive answer to Mach's questions.

The opposite attitude to "normal" pragmatism may be called hierarchism. If we formulate the meaning of a paradigm in propositions, these may be used as fixed axioms, from which a scientific discipline might be logically deduced; Newton's mechanics is a good example. The origin of this kind of science is the great Greek discovery of deductive mathematics. But how to find adequate axioms for physics?

2.3 Realism and Empiricism

Physics rests on experience and speaks about reality. Thus the philosophy of physics was tempted to formulate basic axioms either on reality (ontology) or on experience (epistemology). Both ways of theory-building were intermediately successful, none of them, however, had conclusive success.

Classical physics gave us a picture of reality: matter and fields, deterministically interacting in space and time. Many physicists, down to our days, have a nostalgic longing for this picture of reality. CI, the minimal semantics of quantum theory, makes use of this picture only for describing our sensual experience, not for describing basic physical reality. Can we find a classical or semi-classical picture of reality behind CI?

Empiricism or "positivism" denies the legitimacy of this wish; it even questions the positive meaning of concepts like "reality". The Vienna school tried to make axiomatic use of the data of sense-perception. But Popper rightly pointed out, that sensual experience justifies no universal proposition. He proposed a progress of hypothetical pictures of reality which are used as long as they are not falsified by empirical counter-examples. Yet he could not explain why any "picture of reality" should have such horrend success as classical mechanics or even more quantum mechanics. Seeing his lack of explanatory power he called his view a "robust realism". It is a belief, nothing more.

2.4 Kant's Apriorism

Kant offers an answer to the question: We possess cognition a priori, which applies to the experience but does not depend on experience. But how might that be possible?

Kant's first answer: Cognition a priori is a fact in mathematics. I need no special experience for understanding that $2 \times 2 = 4$, that $17 \times 19 = 323$. I have certainly often empirically tested $2 \times 2 = 4$, but never $17 \times 19 = 323$, but I am as sure of one of the equations as of the other. We understand mathematical truths by constructing them ourselves.

Kant maintains that physics can be equally constructed a priori. The construction is done in our originary "forms of intuition", space and time, by means of our originary conceptual categories like substance and causality. Without categories no universal propositions, without universal propositions (like, e.g., the law of causality) and forms of intuition no conceptually expressible experience, hence no physics. The principles of physics apply always *in* experience because they are preconditions *of* experience. There is no contradiction between the universal laws and the behaviour of the reality as described by physics, because this reality is precisely what *we* can know. It is our construct. One might ponder whether perhaps no conceptual experience might be possible at all. Then there would be no conceptually thinking human beings. But *if* experience is possible, then it has to agree to the laws a priori.

Modern physics has made it practically impossible to be a strict Kantian. Relativity and quantum theory deny precisely those universal laws which he considered as a priori true. But it is neither our tendency to accept hierarchic theories as definite nor to deny their value as steps in a proceeding way of understanding. Realism offered the fruitful model of classical physics, positivism offered the fruitful criticism of this model, apriorism offers a hope of understanding why universal laws should hold in experience.

2.5 Evolutionism

How can we possess cognition a priori? Konrad Lorenz said: Because our ancestors have acquired it in the process of evolution as an adaptation to reality. We possess an inborn intuition of three-dimensional space because without this our ancestors, the monkeys and apes, would have fallen from the trees when jumping. Our inborn forms of cognition are adapted to reality because they are a gift of an evolution in the real world. This is, in the present authors' view, a very profound step forward in the philosophy of science. But it is certainly not, as Popper thinks, a vindication of the realism of classical physics, beyond the trivial fact that classical physics is well adapted to the macroscopic bodies with a temperature far from absolute zero, from whose perception its special structure was derived in the history of science. The jumping apes needed no more then being oriented in the same macroscopic world. Beyond this, some of our classical theories may even be products not of evolutionary inborn ideas but of the history of a special – here the occidental – civilization. Thus it is not at all clear that our spatial intuition is strictly Euclidean. In true fact it seems to be unprecise, but adaptable to that wonderful Greek invention, the Euclidean geometry.

2.6 Circular Epistemology and Quantum theory

We would conclude that neither the merely pragmatic attitude nor any strict hierarchism is adequate. There is a circle of mutual explanation through which we must go repeatedly. Nature is older than the human species; the human species is older than natural science. Our concepts are inherited from evolution and from cultural history; our description of evolution and of cultural history is done in our available concepts. Let us keep this in mind when now we return to quantum theory.

The basic idea in using this circular epistemology in interpreting quantum theory is as follows.

We first ask whether we can apply Kant's idea that laws of nature apply always in experience because they are preconditions of experience. This is now not an axiomatic statement but a working hypothesis. Which preconditions of experience would we accept a priori? If experience means to learn from the past for the future, then any empirical science presupposes an understanding of past and future. We consider past events as *facts*, future events as *possibilities*. We hypothetically try to do this by forming *two theses*, also to be called "*six words*":

A. Facts are discrete.
B. Possibilities are continuous.

This corresponds to the historical description of quantum theory given in Sect. 1.

But can the "six words" be considered as presuppositions of all possible experience? In Kant's view, preconditions of experience must be knowledge a priori. We cannot maintain that the two theses have this kind of a priori evidence; else classical continuum dynamics would never have been invented. Hydrodynamics, e.g., considers the continuously distributed field of velocities certainly as a continuous field of facts. Similarly, Schrödinger certainly considered his Ψ as a continuous factual field; this, precisely is the reason why the "collapse of Ψ" seems so paradoxical. If Ψ expresses only possibilities, then the collapse means the transition from possibility into fact. We have not yet tried to describe this in detail (see Sect. 4.4); but certainly it is a well-known everyday event and not a paradox. A new fact implies new possibilities. But all these considerations are not a priori in the sense of Kant. The continuous field of facts seems clearly acceptable to our intuition and our reasoning. So, what do we then mean by the two theses as preconditions of experience?

A first step towards an answer: In a circular epistemology, preconditions of experience do not need to be consciously present to our mind. In evolution, consciousness emerges from a sea of unconscious ways of behaviour. Modern psychology knows of the immense subconscious basis of our conscious perceptions. "Consciousness is an unconscious act". Thus the two theses may describe structures of organic or even inorganic nature which belong to the modes of time and thus are *objective* preconditions of experience without being known a priori.

Indeed: Possibilities can be quantified as probabilities, and these admit all real numbers between 0 and 1 as values; thus they can, and in an indeterministic theory even must be described in a continuum. If facts, on the other hand, are past events, we can only know them, when they have produced an irreversible process, leaving a document in nature or in our memory. Irreversible processes, however, lose their phase relation with neighbouring possible processes and thus become separated, i.e. discrete.

3 Quantum Logic and the Reconstruction of Quantum Theory

3.1 The Role of the Continuum

Traditionally, the structures of logic and of mathematics were considered as prior to and hence as independent of physics. Birkhoff and von Neumann were the first to realize that this must not necessarily be so in the case of logic. Hence we must now consider this step in the philosophy of quantum theory. But we shall see, that the step is connected with a change, perhaps not in the structure, but in the interpretation of the mathematical continuum. Hence we begin with a historical remark on this concept.

Aristotle defined "continuum" as a quantity which can be indefinitely subdivided into smaller quantities of the same structure. Geometry and motion, in modern terms space and time, were his examples; in this application, "continuum", in his philosophy, is a concept of physics. "Indefinitely" does not mean what since Cantor we would call "actually infinite"; it corresponds to the modern term "potentially infinite", to which mathematicians even in the time of Gauß and Cauchy adhered. Aristotle insisted (Physics, Book 8; cf. [19, Sect. IV.4.]) that, e.g., the path on which Achilles reaches the tortoise does not "consist of" infinitely many parts into which it is intellectually divided in the refutation of Zenos's Paradox. Physically, this path can be divided only by moving to the point where we want to divide it, coming to rest there, and then moving again; an activity which takes a finite time. Hence in a finite time-span we can only perform a finite number of divisions. This very consideration shows that Aristotle treats the continuum as a concept of physics.

When Cantor introduced the idea of actual infinity, and introduced the absolutely non-Aristotelian idea that a continuum is a non-countable set of points, he was criticized by philosophers and traditional mathematicians. They said: Number (called natural number in modern mathematics) expresses our ability of counting. Counting, as Kant, and in our century Brouwer pointed out, is done in time. We can count beyond any number reached in actual counting – in the language of the present paper a number reached by counting is a fact, the next numbers describe possibilities. Thus counting is indefinite but not actually infinite. Cantor cleverly replied: By your argument you admit that, while counted facts are always a finite set, the possible numbers form an infinite set, and mathematics is a science of concepts, i.e. of possibilities.

In quantum theory we describe possibilities as continuous. We shall have to investigate, whether Cantor's idea of the continuum as a point set then stays adequate.

3.2 Non-distributive Logic and Quantum Theory

The classical logic of propositions can be mathematically expressed in a Boolean algebra. This algebra is connected with set theory. For simplicity's sake we consider a finite set, e.g. of three elements a, b, c. They should express three mutually exclusive statements (in physics: time-dependent propositions like "the z-component of the angular momentum 1 is $+1$, or 0, or -1"; or they might, equivalently, mean the states whose presence is expressed in the corresponding statement). Any statements x,y,... permit the logical functions

$x \wedge y$:   "x and y",

  in set theory: the intersection of the subsets x and y,

$x \vee y$:   "x or y",

  in set theory: the join of the subsets x and y,

$\neg x$:   "non x",

  in set theory: the complement of the subset x.

The elements (in our example a, b, c) of the considered set are called the "atoms" of the lattice. For brevity we shall sometimes write xy for $x \vee y$. Then the corresponding Hasse diagram is given by Fig. 1.

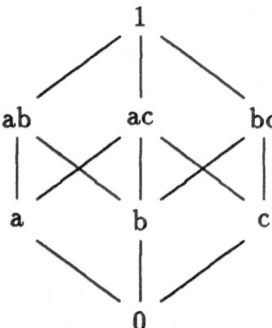

Fig. 1.. The lines in the diagram describe implications, e.g. $a < ab$, a is a subset of ab, or a implies ab. $1 = abc$ is the complete set, or "the true proposition", 0 is the null set, or "the false proposition".

In the logical interpretation referring to events in time, we shall also call the elements of a set containing n elements an "n-fold alternative": one of them must be true at any moment, and then none of the others can be true. We shall have to consider two theorems that hold in a Boolean lattice:
The first distributive law

$$xy \wedge z = (x \wedge z) \vee (z \wedge y), \tag{1}$$

and the law of double negation

$$\neg(\neg x) = x. \tag{2}$$

Instead of such a Boolean lattice, Birkhoff and von Neumann considered the lattice of the subspaces of a Hilbert space. In our example, let the Hilbert space be 3-dimensional, corresponding to the three values of spin-1 z-component. Here

$x \wedge y$ is again the intersection, now of the subspaces x and y,

$x \vee y$ is the subspace, linearly composed from x and y,

$\neg x$ is the subspace of all vectors orthogonal on x.

Now the atoms of the lattice are a continuously infinite set: all vectors of the given space. In this lattice, the law (2) still holds: every subset x is orthogonal on the subset consisting of all subsets which are orthogonal on x. But (1) does no longer hold in general. Let, e.g., x and y be two different vectors, and z a vector in their plane xy, but different from both x and y. Then:

$$xy \wedge z = z, \tag{3}$$

$$\text{but} \quad x \wedge z = y \wedge z = (x \wedge z) \vee (y \wedge z) = 0 \,. \tag{4}$$

The "logic" formulated in this kind of lattices is now generally called "quantum logic". We shall call it here, more specifically, "non-distributive logic".

There have been debates whether such a mathematical formalism deserves the name of "logic". We shall not enter into the philosophical backgrounds of this debate in the present paper. We only mention a few steps. Lorenzen has tried to deduce a constructive logic from rules of operational thinking [16] or of dialogue [17], thus excluding formal attempts like "quantum logic". Mittelstaedt [18] applied Lorenzen's methods to the description of physical experiments, thus justifying "quantum logic" as a logic. Weizsäcker [21, Chapt. 2] follows this path, describing it as a logic of temporal statements.

A different path was opened by Doebner and Lücke [5]. In the so-called "orthodox" tradition, as codified by von Neumann, the inevitability of probabilities in quantum mechanics has been interpreted as an inevitable indeterminism. Quantum logic was understood as a consequence and hence as corroboration of this indeterminism. Attempts at an experimental direct finding of deterministic hidden variables behind quantum mechanics have so far not been successful. But it seems still impossible, theoretically to exclude such a "hidden determinism" [2,1,22]. Our interpretation of CI as "minimal semantics" leaves this discussion open. Now Doebner and Lücke have shown that a deterministic hypothesis behind quantum mechanics can easily produce a non-distributive lattice, hence logic.

They have proved that a non-distributive logic can be directly embedded into a Boolean logic of "hidden variables". "Direct embedding" here means retaining the new relations $\wedge$ and $\vee$ from the full Boolean lattice. The method of "direct embedding", however, does not yet produce the full quantum theory. It does not produce the quantum-mechanical *violation of Bell's inequality*. Yet this can be achieved by adequate additional assumptions.

Doebner and Lücke have shown that their non-distributive quantum logic with direct embedding satisfies Bell's inequality, if we assume that the probabilities of combined, but independent measurements can be factorized into products of the probabilities of the separate measurements. In the simplest example this is easily shown. Consider two independent binary alternatives. Call the probability for the outcome a and b in the first alternative p_a, p_b, in the second alternative q_a, q_b, and the probability of an outcome x in the first *and* y in the second alternative r_{xy}. If the alternatives are independent (e.g., with a local interaction law, simultaneous and at a spatial distance), we would classically expect

$$r_{xy} = p_x q_y \,. \tag{11}$$

Now define "Bell's quantity" B_{xy} by

$$B_{xy} \overset{\text{def}}{=} p_x + q_y - r_{xx} + r_{yx} - r_{xy} - r_{yy} \overset{\text{def}}{=} p_x + q_y - C_{xy} \,. \tag{12}$$

From (11) we conclude

$$C_{xy} = +p_x q_x - p_y q_x + p_x q_y + p_y q_y = p_y(q_y - q_x) + p_x(q_y + q_x) = p_y(q_y - q_x) + p_x \,, \tag{13}$$

since

$$q_y + q_x = 1.$$ (14)

Further, due to (14)

$$q_y - q_x = 2q_y - 1,$$ (15)

and

$$B_{xy} = q_y + p_y - 2p_yq_y = p_y(1 - q_y) + q_y(1 - p_y) \geq 0.$$ (16)

This is Bell's inequality, valid since $1 \geq p_y \geq 0$, $1 \geq q_y \geq 0$.

In this derivation, no use has been made of the assumption that the independence of the two results is produced by spatial distance; this is only one example of such independence. Since Bell's inequality seems now to be definitely violated in the special case of local distance, the preconditions of independence, where we introduce it, might equally lead to a wrong result. We tentatively conclude the *holism* of quantum theory: There are no strictly independent events.

We end this section by mentioning some unresolved problems. We have not investigated how far quantum logic implies full quantum theory. Non-distributive logic is certainly not sufficient. The basic additional point seems to be the symmetry which establishes pure states. Further, we have not found out how far a completed quantum theory admits a testable determinism in the assumed hidden variables. This would depend on an interpretation of these variables and their time-dependence. Accepting holism, they might just express the influence of the outer world on the object. This, again, would presuppose a theory of space and time in the quantum context.

In the following section we shall give a very brief outline of our own attempt at reconstructing quantum theory, including the theory of the space-time continuum, from simple postulates.

3.3 Reconstruction

The ensuing postulates try to formulate simple preconditions of quantum theory as a *theory of human knowledge*. This expresses the same tendency as our interpretation of CI as a *minimal semantics*. What, *at least*, is to be assumed for such a theory? (See [21, Chapt. 8] and [6,23]).

Postulates:

0. Holism.
   The reality is a whole (eine Ganzheit), not strictly separable into parts.
1. Alternatives.
   In a good approximation there are separable finite (n-fold) empirically decidable alternatives.
2. Indeterminism.
   Let x and y be two mutually exclusive states, then there are "intermediate" states z with conditional probabilities $p(z, x)$ and $p(z, y)$ of finding x or y if z is present, such that both probabilities are neither zero nor one.

3. Kinematics.

States belonging to an alternative according to Post. 2 change continuously in time such that the conditional probabilities are not altered: $p[(x,t),(z,t)] = p[(x,0),(z,0)]$.

Consequences:

Assuming that any kinematical law fulfilling Post. 3 is permissible, we can conclude that Post. 2 implies a symmetry between the states z which permits the representation in a complex space with probabilities defined by a Hermitian metric. Thus "abstract quantum theory" is derived.

Any alternative then possesses a continuum of vectors, called "pure states" in traditional quantum theory. By the conditional probabilities $p(z,x)$ etc. these vectors are connected such that if z is present there is a probability to find x. In this sense the continuum of states expresses possibilities. In this direct experimental sense, z and x are not disjoint elements of a set but have a probability of being formal "identical". If "continuum" is considered as a concept of physics, expressing possibilities, it is an inadequate way of speech to call the continuum a set of "points". Of course, this expression is used by defining the "set of all states z" which belongs to a given discrete alternative. But then the question arises, which measurement would permit to distinguish two states z_1 and z_2, connected by $p(z_1,z_2) \neq 0,1$. It might be done by statistical measurements, with always a limited precision.

Our reconstruction does not exclude the possibility that behind the postulated indeterminism there is a deterministic lattice as studied by Doebner and Lücke. The present paper, as said before, cannot yet study the structures of the proposed hidden variables.

Another consequence is independent of this question. Every finite or countably infinite discrete alternative can be decided by successive decision of binary (2-fold) alternatives ("yes-no decisions"). The quantum theory of the binary alternative has the symmetry group $SU(2) \times U(1)$. $SU(2)$ has a natural representation in a 3-dimensional real space, where $U(1)$ may be used to describe time-dependence. We suppose that thus the three-dimensionality of position space is a necessary consequence of abstract quantum theory [21, Chapts. 9–10]. This, again, lies beyond the present paper; but it might be mentioned as indicating the probable universality of quantum theory.

4 Further Philosophical Questions

4.1 Minimal Semantics and Beyond

Describing CI as a minimal semantics of quantum theory, we try to keep free from "ideologies". Interpretations beyond CI are permissible if they can be formulated with sufficient clarity to admit a discussion, if possible even an experimental decision. It is, however, important to see which consequences can be drawn from the theory *without* adding new interpretations.

If the consideration on three-dimensional real space as a consequence of the possible reduction of all alternatives to successive binary alternatives should turn out to be successful, then the space-time continuum would not be "behind" quantum theory as a field in which hidden variables might be located, but would already be a consequence of the minimal semantics of the theory. Therefore we feel that this question ought to be intensively studied. It can easily be shown that the space-time continuum so deduced admits an approximate description by special relativity (in a tangential Minkowski space). Hence it will have as a consequence the existence of particles as irreducible representations of the Poincaré group. We would gladly invite able theoretical physicists to study these consequences. We suppose that they might give a frame in which the Doebner-Lücke background can be interpreted.

4.2 The Mind-Body Problem

If we interpret quantum theory correctly, the idea of two *substances*, viz. the thinking and the extended substance, as proposed by Descartes, is probably a misunderstanding. In a theory of human knowledge, there are two *roles* for parts of reality, the role of the knower and of the known, of subject and of object. The abstract quantum theory as reconstructed by postulates on decidable alternatives is essentially a theory on the time-dependence of *information*. Information can be defined as a measure for the quantity of *form*. Particles as representations of a symmetry group derived from abstract quantum theory are nothing but "agglomerations of form". But whose form is this? Plato considered form as the ultimate reality. Descartes' starting point was the self-awareness of the ego: I can doubt everything but not the fact that I am doubting. Modern psychology knows that this conscious psyche is embedded into an immensely larger sub- or unconscious psyche. Quantum theory does not imply but certainly also not exclude the idea that psyche or spirit is the basic reality. The "thinking substance", if it can know itself and thus decide alternatives on its own state, will, according to the considerations, necessarily also appear as "extended", at least to the approximation in which abstract quantum theory might be applied to these alternatives.

But why, then, are we finding ourselves as conscious beings isolated in an extended world where we so far see no other entities that might be regarded as possessing self-consciousness? Here it is to be remembered that our consciousness is a late step in evolution. It seems to presuppose a complicated organ, the brain and nervous system. The parts of the brain, the nervous cells, contribute to consciousness but seem not to have consciousness of their own. A nervous cell seems to have no private ego; neither has a special emotion, a sense of pain, a perception of colour, which is part of our self-awareness, an ego in its own. As we said before, consciousness rests on the subconscious psyche. Hence ego-consciousness, when it appears for the first time in evolution, must be seen to be unique, solitary in the extended world. What if we wait for another half billion of years of evolution? What if we were able to perceive the psychic aspects of larger parts of the universe?

4.3 Holism

In our postulates in Sect. 3.4, we started with a "zeroth" postulate of holism. As a consequence of quantum theory, this holism is fully recognized by many authors today. The state space of a composite object contains only a set of measure zero of product states in which its parts are in well-defined states of their own. Separate alternatives are thus no more than a useful manipulation of human conceptual thought. How would we have to describe their embedding in the real larger world of which we are only parts?

4.4 Events

We have so far left unconsidered the question how possibilities are transformed into facts. Haag has recently written a paper [12] in which he argues that a special postulate must be added to quantum theory which craves that this transition takes place. This problem was treated in [20] and [21, Chapt. 13.3]. We propose to treat this problem in a separate, later paper.

4.5 Final Remark

Philosophy is done today – not in the past, not in the future, not in eternity. It is reflection on what we seem to know *now*. Hence it may be justified to end with open questions.

Acknowledgement

One of us (Th. G.) is grateful to the Stifterverband der Wissenschaft for financial support for part of this work.

References

1.   J.S. Bell: *Speakable and Unspeakable in Quantum Mechanics* (University Press, Cambridge, UK, 1987)
2.   D. Bohm: "A Suggested Interpretation of the Quantum Theory in Terms of Hidden Variables", I and II, Phys. Rev. **85** 166–179, 180–193 (1952)
3.   J.G. Cramer: Rev. Mod. Phys. **58** 647 (1986)
4.   D. Deutsch: Int. J. Theor. Phys. **24** 1 (1985)
5.   H.D. Doebner, W. Lücke: "Quantum Logic as a Consequence of Realistic Measurements on Deterministic Systems", to app. in J. Math. Phys. (1990/91)
6.   M. Drieschner, Th. Görnitz, C.F. v. Weizsäcker: "Reconstruction of Abstract Quantum Theory", Int. J. Theor. Phys. **27** 289–306 (1987)
7.   H. Everett: Rev. Mod. Phys. **29** 454 (1957)
8.   G.C. Ghirardi, A. Rimini, T. Weber: "Unified Dynamics for Microscopic and Macroscopic Systems", Phys. Rev. D**34**, 470–491 (1986)

9.   Th. Görnitz, C.F. v. Weizsäcker: "Remarks on S. Kochen's Interpretation of Quantum Mechanics"; in *Symposium on the Foundations of Modern Physics*, ed. by P. Lahti and P. Mittelstaedt (World Scientific, Singapore, 1987)

10.   Th. Görnitz, C.F. v. Weizsäcker: "Quantum Interpretations", Int. J. Theor. Phys. **26** 921 (1987)

11.   Th. Görnitz, C.F. v. Weizsäcker: "Copenhagen and Transactional Interpretations" Int. J. Theor. Phys. **27** 237–250 (1988)

12.   R. Haag: "Fundamental Irreversibility and the Concept of Events", DESY 90-049 (1990)

13.   Heisenberg: private communication, (1934)

14.   S. Kochen: "A new interpretation of quantum physics", in *Symposium on the Foundations of Modern Physics*, ed. by P. Lahti and P. Mittelstaedt (World Scientific, Singapore, 1985)

15.   Thomas S. Kuhn: *The Structure of Scientific Revolutions* (University of Chicago Press, Chicago, 1962)

16.   P. Lorenzen: *Einführung in die operative Logik und Mathematik* (Springer-Verlag, Berlin etc., 1955)

17.   P. Lorenzen, K. Lorenz: *Dialogische Logik* (Wiss. Buchgesellschaft, Darmstadt, 1978)

18.   P. Mittelstaedt: *Quantum Logic* (Reidel, Dortrecht, 1978)

19.   C.F. v. Weizsäcker: *Die Einheit der Natur* (Hanser, München, 1971); Engl. ed.: *The Unity of Nature* (Farrar, Straus, Giroux, New York, 1980)

20.   C.F. v. Weizsäcker: "Classical and Quantum Descriptions", in *The Physicist's Conception of Nature*, ed. by J. Mehra (Reidel, Dordrecht, 1973)

21.   C.F. v. Weizsäcker: *Aufbau der Physik* (Hanser, München, 1985)

22.   C.F. v. Weizsäcker, Th. Görnitz: "Quantum Realistic Interpretation", to app. in Found. Phys. (1990)

23.   C.F. v. Weizsäcker, Th. Görnitz: "Quantum Theory as a Theory of Human Knowledge", to app. in *Symposium on the Foundations of Modern Physics*, ed. by P. Lahti and P. Mittelstaedt (World Scientific, Singapore, 1990)